Genetics of Salt Tolerance in Plants

A Central Dogma Perspective and Strategies for Enhancement

Genetics of Salt Tolerance in Plants

A Central Dogma Perspective and Strategies for Enhancement

Edited by

Showkat Ahmad Ganie

School of Life Sciences, University of Essex, Colchester, UK

and

Shabir Hussain Wani

Sher-e-Kashmir University of Agricultural Sciences and Technology of Kashmir, India

CABI

CABI is a trading name of CAB International

CABI
Nosworthy Way
Wallingford
Oxfordshire OX10 8DE
UK

Tel: +44 (0)1491 832111
E-mail: info@cabi.org
Website: www.cabi.org

CABI
200 Portland Street
Boston
MA 02114
USA

Tel: +1 (617)682-9015
E-mail: cabi-nao@cabi.org

A catalogue record for this book is available from the British Library, London, UK.

Library of Congress Cataloging-in-Publication Data

Names: Ganie, Showkat Ahmad, 1981- editor. | Wani, Shabir Hussain, editor.
Title: Genetics of salt tolerance in plants : a central dogma perspective and strategies for enhancement / edited by Showkat Ahmad Ganie and Shabir Hussain Wani
Description: Boston, MA : CAB International, [2024] | Includes bibliographical references and index | Summary: "This book presents the current state of knowledge on plant salt tolerance acquired through meticulous analysis of the processes operating across the Central Dogma with particular focus on the modulation of gene expression, and its implications for improving salt tolerance"-- Provided by publisher
Identifiers: LCCN 2024020375 (print) | LCCN 2024020376 (ebook) | ISBN 9781800623019 (hardcover) | ISBN 9781800623026 (ebk) | ISBN 9781800623033 (epub)
Subjects: LCSH: Plant genetics. | Plants--Effect of salt on. | Plant physiology.
Classification: LCC QK981.5 .G495 2024 (print) | LCC QK981.5 (ebook) | DDC 572/.5238224--dc23/eng/20240510
LC record available at https://lccn.loc.gov/2024020375
LC ebook record available at https://lccn.loc.gov/2024020376

ISBN-13: 9781800623019 (hardback)
9781800623026 (ePDF)
9781800623033 (ePub)

DOI: 10.1079/9781800623033.0000

Commissioning Editor: David Hemming
Editorial Assistant: Emma McCann
Production Editor: Shankari Wilford

Typeset by Straive, Pondicherry, India
Printed and bound in the UK by CPI Group (UK) Ltd, Croydon, CR0 4YY

Contents

Contributors

Adeboye, Kehinde A., Department of Agricultural Technology, Ekiti State Polytechnic, Isan, Nigeria. Email: adeboye@daad-alumni.de

Bamisaye, Victor J., Department of Agricultural Technology, Ekiti State Polytechnic, Isan, Nigeria. Email: bamisayevictor03@gmail.com

Basit, Farwa, Institute of Crop Sciences, College of Agriculture and Biotechnology, Zhejiang University, Hangzhou 310058, China. Email: 11916138@zju.edu.cn

Bello, Usman Rabi'u, Department of Biology, Khadija University, Majia, Jigawa, Nigeria. Email: urbello97@gmail.com

Bhat, Javaid Akhter, Zhejiang Lab, Hangzhou 310012, China. Email: javid.akhter69@gmail.com

Buddiga, Jamindar, Department of Agriculture, Koneru Lakshmaiah Education Foundation, Andhra Pradesh, India. Email: jamindarbuddiga@kluniversity.in

Das, Utpal, VIT School of Agricultural Innovations and Advanced Learning (VAIAL), VIT, Vellore, Tamil Nadu, India. Email: utpaldashorts14@gmail.com

Dharbaranyam, Balasundari, Plant Molecular Biology Laboratory, M.S. Swaminathan Research Foundation, III Cross Street, Taramani Institutional Area, Chennai 600113, India. Email: baladharbaran@gmail.com

Feng, Xianzhong, Zhejiang Lab, Hangzhou 310012, China. Email: fengxianzhong@iga.ac.cn

Gayathri, M., Department of Agriculture, Koneru Lakshmaiah Education Foundation, Andhra Pradesh, India. Email: gayathrimungara@kluniversity.in

Gumi, Abubakar Mohammad, Department of Plant Science, Faculty of Chemical and Life Sciences, Usmanu Danfodiyo University, PMB 2346, Sokoto, Nigeria. Email: gumi.abubakar@udusok.edu.ng; muhammadag@yahoo.co.uk

Hossain, Md Sazzad, Institute of Plant Nutrition and Soil Science, Kiel University, 24118 Kiel, Germany; Department of Agronomy and Haor Agriculture, Sylhet Agricultural University, Sylhet 3100, Bangladesh. Email: shossain@plantnutrition.uni-kiel.de; sazzadmh.aha@sau.ac.bd

Idehen, Emmanuel O., Department of Plant Breeding and Seed Technology, Federal University of Agriculture, Abeokuta, Nigeria. Email: emmaidehen@yahoo.com

Ingle, Krishnananda P., Centro Internacional de la Papa, International Potato Center, CGIAR Research Center, Pune, India. Email: krisona369@gmail.com

McMulkin, Nancy, Plant Molecular Science and Centre of Systems and Synthetic Biology, Department of Biological Sciences, Royal Holloway University of London, Egham, Surrey, TW20 0EX, UK. Email: Nancy.McMulkin.2019@live.rhul.ac.uk

Najar, Mushtaq Ahmad, Department of Biological Sciences, Unified Academic Campus, EN 80, Sector V, Bidhan Nagar, Kolkata 700091, West Bengal, India. Email: mushtaqnajar149@gmail.com

Oyetunde, Oyeboade A., Department of Horticulture, Lagos State University of Science and Technology, Ikorodu, Nigeria. Email: biyi.oyetunde@gmail.com

Patil, Kapil, Department of Horticulture, University of Agricultural Sciences, Raichur, Karnataka, India. Email: kapil0382@gmail.com

Pitann, Britta, Institute of Plant Nutrition and Soil Science, Kiel University, 24118 Kiel, Germany. Email: bpitann@plantnutrition.uni-kiel.de

Prasad, Shiva Sai, Department of Agriculture, Koneru Lakshmaiah Education Foundation, Andhra Pradesh, India. Email: 5257shiv@gmail.com

Quilloy, Fergie Ann, Rice Breeding Innovations Platform, International Rice Research Institute, Metro Manila, Philippines. Current address: School of Agriculture and Food Sustainability, University of Queensland, Brisbane, Australia. Email: f.quilloy@uq.edu.au

Rahman, Md Zillur, Centre for Carbon, Water, and Food, The University of Sydney, Camden, NSW 2570, Australia; Department of Agronomy and Haor Agriculture, Sylhet Agricultural University, Sylhet 3100, Bangladesh. Email: zillur.agronomy@sau.ac.bd; zillur.agronomy@sau.ac.bd

Rajeshwar, J.R., Department of Agriculture, Koneru Lakshmaiah Education Foundation, Andhra Pradesh, India. Email: jrrajeshwar@kluniversity.in

Ramteke, P.W., Department of Molecular Biology and Genetic Engineering, RTM Nagpur University, Nagpur, India. Email: gayathrimungara@kluniversity.in

Ramzan, Hamza, Max Planck Institute for Plant Breeding Research, Cologne, Germany. Current address: Università degli Studi di Milano, Milan, Italy. Email: hamzaramzan953@gmail.com

Reyes, Vincent Pamugas, Graduate School of Bioagricultural Sciences, Nagoya University, Nagoya 464-8601, Japan. Email: reyesvp@nagoya-u.jp

Sakthivel, Kausalya, Plant Molecular Biology Laboratory, M.S. Swaminathan Research Foundation, III Cross Street, Taramani Institutional Area, Chennai 600113, India. Email: kausalyasakthivel01@gmail.com

Senthilmurugan, Palanisamy, Department of Agriculture, Koneru Lakshmaiah Education Foundation, Andhra Pradesh, India. Emai: agilanagri2008@kluniversity.in

Sheraz, Usama, Department of Plant Breeding Genetics, University of Agriculture, Faisalabad 38040, Pakistan. Current address: Wheat Breeding Group, Nuclear Institute for Agriculture and Biology, Faisalabad 38000, Pakistan. Email: usamasheraz.pbg@gmail.com

Shokat, Sajid, Wheat Breeding Group, Nuclear Institute for Agriculture and Biology, Faisalabad 38000, Pakistan. Current address: Plant Breeding and Genetics Laboratory, IAEA Laboratories Seibersdorf, Friedensstrasse 1, 2444 Seibersdorf, Austria. Email: sajid_agrarian@yahoo.com

Shukla, Pradeep Kumar, Sam Higginbottom University of Agriculture, Technology Sciences, Prayagraj, Uttar Pradesh, India. Email: pradeep.shukla@shiats.edu

Singh, Atul, Department of Agriculture, Koneru Lakshmaiah Education Foundation, Andhra Pradesh, India. Email: singhatul9415@gmail.com

Venkataraman, Gayatri, Plant Molecular Biology Laboratory, M.S. Swaminathan Research Foundation, III Cross Street, Taramani Institutional Area, Chennai 600113, India. Email: gayatri@mssrf.res.in

Zahan, Mst Ishrat, Breeding Division, Bangladesh Sugarcrop Research Institute, Ishurdi 6620, Pabna, Bangladesh. Email: ishrat.zahan@bsri.gov.bd

1 Involvement of Genetic Mutations in Plant Salt Tolerance

Usama Sheraz[1†], Hamza Ramzan[2†] and Sajid Shokat[3,4*]

[1]Department of Plant Breeding Genetics, University of Agriculture, Faisalabad, Pakistan; [2]Max Planck Institute for Plant Breeding Research, Cologne, Germany; [3]Wheat Breeding Group, Nuclear Institute for Agriculture and Biology, Faisalabad, Pakistan; [4]Plant Breeding and Genetics Laboratory, IAEA Laboratories, Seibersdorf, Austria

Abstract

The chapter provides valuable insights into the crucial role of DNA mutations, shedding light on how they contribute to the adaptation and resilience of plants to salinity stress. The chapter offers an in-depth analysis of mutation breeding as a fundamental strategy for creating genetic diversity. It emphasizes the major steps involved in generating beneficial mutations. The utility of TILLING (targeting induced local lesions in genomes) in enhancing the accuracy of mutation breeding is examined, highlighting its effectiveness in functional genomics and crop improvement. Furthermore, the chapter comprehensively examines the importance of DNA mutations and identifies significant genetic changes, like SNPs (single-nucleotide polymorphisms) and indels (insertions/deletions), which can contribute to salt tolerance. The chapter gives substantial focus to advanced gene editing techniques, emphasizing the role of CRISPR/Cas9 (clustered regularly interspaced short palindromic repeats/CRISPR-associated protein 9) genome editing in achieving precise genetic modifications. In conclusion, the chapter provides a holistic overview of current and prospective genetic methodologies aimed at augmenting salt tolerance in plants.

1.1 Introduction

Plants often come across a myriad of biotic and abiotic challenges throughout their life cycle, which significantly impedes their food productivity. Salinity stress is one of the major stresses affecting crop productivity throughout the world (Melino and Tester, 2023).

According to the Global Map of Salt-affected Soils (GSASmap), a project initiated by the Global Soil Partnership under the Food and Agriculture Organization of the United Nations, salt-affected soils cover over 833 million ha globally, which constitutes approximately 8.7% of the Earth's surface (FAO, 2024). These soils are predominantly located in naturally arid or semi-arid regions across Africa, Asia, and Latin America. Additionally, it is observed that a notable portion of irrigated soils on all continents, ranging from 20 to 50%, suffer from high salinity levels. Moreover, as per the GSASmap, over 1.5 billion individuals worldwide encounter significant impediments in agricultural and food production due to heightened salinity levels (FAO, 2024).

Based on the plant growth and properties of salt-affected soils, two types of soils have been

*Email: sajid_agrarian@yahoo.com

†These authors contributed equally.

 The Genetics of Salt Tolerance in Plants: A Central Dogma Perspective and Strategies for Enhancement (S. Ganie and S.H. Wani eds)
DOI: 10.1079/9781800623033.0001

classified: (i) saline soils, characterized by the presence of NaCl and Na_2SO_4; and (ii) sodic soils, which predominantly contain sodium salts in the form of Na_2CO_3. Additionally, salinity can be further categorized based on its source: (i) primary salinity, which arises from natural calamities such as prolonged periods of rain and storms leading to the accumulation of salts on water and land; and (ii) secondary salinity resulting from human activities such as deforestation, depletion of fertile land, and excessive irrigation.

In field conditions, plants experience significant stress caused by salinity throughout various stages of development, including seed germination, vegetative growth, and reproductive stages (Ganie, 2020; Hussain *et al.*, 2021). The excessive accumulation of sodium ions in the rhizosphere can modify the texture of the soil. This alteration disturbs the soil's capacity to retain water and its osmotic pressure and leads to decreases in the soil's osmotic potential, porosity, water conductance, and aeration (Zhu, 2001). Consequently, sodium ions restrict the uptake of water and vital nutrients such as nitrogen, sulfur, calcium, and potassium, thereby causing an imbalance in nutrient levels (Paul and Lade, 2014).

Salinity stress induces both osmotic stress and ion toxicity, thereby affecting critical plant functions such as photosynthesis (Feng *et al.*, 2014), cellular metabolism, and nutrient uptake (Van Zelm *et al.*, 2020). Salinity stress in plants disturbs ion homeostasis across diverse tissues, which is considered one of the main hurdles for plants to survive in salt stress (Munns and Tester, 2008; Ganie, 2021). For example, the presence of an excessive amount of sodium and chloride ions can result in an imbalance of ions within a system and subsequently hinder the selective functioning of the root membrane, thus leading to a deficiency in potassium. This disruption in ion balance can in turn result in a deprivation of essential nutrients, impede the efficiency of numerous enzymes, and give rise to the toxic effects of ions and oxidative stress (Guo *et al.*, 2020). Exposure to salinity stress results in a high accumulation level of reactive oxygen species (ROS) and subsequent electrolyte leakage (Shokat and Großkinsky, 2019). This phenomenon is accompanied by the manifestation of ultra-structural irregularities in chlorophyll and mitochondria, as well as the deterioration of cellular membranes. Furthermore, the incursion of salinity stress disrupts essential cellular processes by inflicting damage upon nucleic acids, proteins, and lipids (Arif *et al.*, 2020).

Plants have different kinds of mechanisms to cope with salt stress including the role of different genes and pathways (Shokat and Großkinsky, 2019). It is also obvious that nature plays its role in the form of evolution to screen the plants adapted better to changing environments. These changes ultimately impact the DNA/genetic material of the plant and create mutations that can be used for trait improvement in plants through breeding approaches. Salt tolerance is a complex, quantitatively controlled trait and hard to improve. Breeding for complex traits requires consideration of multiple factors, including genetic, physiological, and molecular understanding of the traits, as well as their interactions with the environment (Tester and Langridge, 2010). Molecular breeding approaches have hastened the process of breeding and are more predictable as compared to traditional approaches (Xu *et al.*, 2012).

In this chapter, we focus on advanced breeding approaches to utilize novel mutations in plant salinity tolerance where mutation breeding, quantitative trait locus (QTL) mapping, genome-wide association studies (GWAS), and precise genome editing – especially the novel emerging technique of clustered regularly interspaced short palindromic repeats/CRISPR-associated protein 9 (CRISPR/Cas9) – will be emphasized. In the recent past these advancements have been used for trait improvement and performed well; therefore, these techniques have the potential to be utilized in breeding programmes to improve plant salt tolerance. Identification of useful mutations within the susceptible genes can play a vital role in achieving salt tolerance and can be a vital resource for crop improvement. These improvements in the salt tolerance of plants will lead to sustainable food production and to achieving the Zero Hunger and Climate Action missions under the umbrella of the Sustainable Development Goals (SDGs) set by the United Nations in the perspective of an ever-increasing population.

1.2 DNA Mutations and Their Role in Salt Tolerance

Mutations are random changes in the DNA sequence that can lead to genetic diversity within a population. These changes can happen naturally or can be induced through various methods like radiation, chemicals, and precise genome editing tools. Mutations can affect a single gene or larger segments of DNA, leading to alterations in different traits of the plant or sometimes the overall phenotype of the plant.

Mutations can occur at any stage of the life cycle of plants and their effects differ accordingly. Based on the stage of occurrence of a mutation in the life cycle of a plant, they are categorized into two types, i.e. haplontic and diplontic phase. During *haplontic* phase the mutation must pass to gametes, while during *diplontic* phase the mutation must pass to next cells during mitosis (Gregory, 1972). Further, the mutation must pass through these two phases to be heritable (Scott and Rescan, 2017). It is important to note that the impact of a mutation depends on various factors, including its location within the genome, and can be point mutations, insertions, deletions, and duplications. Mutations can be neutral, harmful, or beneficial, and their effects may be subject to natural selection, which determines whether they persist in a population over time.

Single-nucleotide polymorphisms (SNPs) and insertions/deletions (indels) play an important role in salt tolerance in plants. Identification of valuable genetic mutations in the form of SNPs and indels can be a source of tackling complex traits like salt stress at genetic level. SNPs can be created artificially by the exposure of DNA to gamma rays and other ionizing radiations, but in this way only random SNPs will be created (Gregory, 1972). If one wants to create trait-specific SNPs, then the CRISPR/Cas9 technique is more reliable. Basically, SNPs are the change in a single nucleotide whereas insertions or deletions (indels) are the addition or removal of nucleotides sequences, respectively, contributing towards the genetic diversity in plants. Researchers can identify specific SNPs and indels in relation to salt tolerance across different plant species by using GWAS and other genetic analyses. These variations may occur in genes related to ion transport, osmotic regulation, and other pathways critical for responding to salinity. For example, 72 loci linked with natural ionomic variation were identified by performing the GWAS analysis on a diverse germplasm of 529 accessions of rice each genotyped at ~6.4 million SNP loci (Yang *et al.*, 2018). In another study on rice, ~3.6 million SNPs were identified by using the sequence data of 517 landraces of rice for 14 agronomic traits including salt tolerance (Huang *et al.*, 2010). Moreover, a GWAS in *Arabidopsis thaliana* identified SNPs associated with natural variation in sodium transporters *AtHKT1;1*. These variations are directly linked with the salt-tolerance mechanisms (Baxter *et al.*, 2010). Another study explored genomic diversity in soybean genomes and identified almost 10 million high-quality SNPs associated with various traits including salinity. While it may not specifically focus on salt tolerance, the methodology is applicable for trait discovery, including salt-related traits (Valliyodan *et al.*, 2016). Additionally, the study of SNPs and indels enhances our understanding towards the evolutionary patterns of salt-tolerance mechanisms in plants which can be further used for sustainable agriculture solutions in saline environments. Overall, SNPs and indels serve as indispensable tools in elucidating the genetic architecture underlying salt tolerance, contributing to advancements in crop improvement strategies. In a recent study, Xu *et al.* identified 54 significant SNPs out of more than 25,000 SNPs in a diverse germplasm of barley (*Hordeum vulgare* L.). Most of the identified SNPs were associated with sodium:potassium ratio and ion homeostasis traits. By combining GWAS and transcriptomics, they found five promising salt-responsive genes, i.e. *PGK2*, *BASS3*, *SI-NAT2*, *AQP*, and *SYT3*, in the hotspot region (Xu *et al.*, 2023). For example, 11 significant SNPs were identified related to salinity. Among the five non-synonymous SNPs that were significantly associated with salt-stress traits, a mutation called T67K was found. This particular mutation has the potential to induce the destabilization of one transmembrane domain in the protein OsHKT1;5. Additionally, an alteration known as P140A was identified, which has been found to greatly enhance the tendency of phosphorylation of the protein OsHKT1;5. Finally, the mutation K24E has the potential to impact the function of the protein by potentially affecting its interaction with other proteins involved in salt-related processes (Negrao *et al.*, 2013). This study found the allelic variants that play an important role in salt tolerance which can be used in future breeding programmes. Based on this type of real-time data, the identification of genetic markers facilitates marker-assisted selection in breeding programmes, allowing for the targeted development of salt-tolerant crop varieties.

A recent comprehensive study identified 23 SNPs in *Gossypium hirsutum* L. (cotton) across seven regions which are associated with two key salt-tolerance traits: relative survival rate (RSR) and salt tolerance level (STL) (Sun *et al.*, 2018). This investigation further screened 280 candidate genes that exhibited varied expression levels under salt-stress conditions. A significant proportion of these genes are implicated in vital biological roles, such as transcription factors, transporters, and enzymes. Notably, many of these genes, including those in the NAC (named after its NAM, ATAF, and CUC domains), MYB (myeloblastosis), NXH (Na^+/H^+ antiporter), WD40 (beta-transducin repeat), CDPK (calcium-dependent protein kinase), LEA (late embryogenesis abundant), and CIPK (calcineurin B-like protein-interacting protein kinase) families, have been previously identified in the context of plant salt tolerance. The study also highlights that mutations in the DNA-binding domains of transcription factors can alter their capacity to bind to specific DNA sites. Such alterations can disrupt the regulatory functions of these genes (Sun *et al.*, 2018). The diversity of transcription factors in plants, encompassing a range of functions including salt tolerance, underscores the complexity of regulatory mechanisms in plant stress responses (Strader *et al.*, 2022). These include MYB, bHLH (basic helix-loop-helix), WRKY (named after its WRKY domain), NAC, and bZIP (basic leucine zipper) transcription factors, all of which are known to play pivotal roles in plants' response to abiotic stresses, particularly in salt tolerance (Joshi *et al.*, 2016). For instance, the homeodomain leucine zipper (HD ZIP) gene family is key in plant adaptation and growth regulation, responding to environmental changes and hormonal signals (Wang *et al.*, 2013). A member of the HD-ZIP II subfamily, *SlABIG1*, responds to various stresses such as salt, drought, high temperature, and cold. Previous research shows that *SlABIG1* is regulated by salt stress. *SlABIG1* is a salt-responsive gene and has been used to enhance the salt tolerance by creating a mutation in the form of loss of function (Ding *et al.*, 2022). Similarly, the response to salt stress has been reported in other members of the HD-ZIP II subfamily, namely *StHOX17*, *StHOX20*, and *StHOX27*. The family of MYB transcription factors is also one of the largest in plants and presents a role in abiotic stresses (Jiang and Rao, 2020). The MYB transcription factors in plants play a crucial role in the regulation of various biological processes. These include the control of primary and secondary metabolisms, the regulation of cell development, synthesis, responses to biotic and abiotic stresses, and a role in signalling (Zhang *et al.*, 2012). Moreover, Fang *et al.* (2021) identified ZmWRKY86 as a transcription factor that regulates salt stress-related genes in maize. The above-discussed transcription factors have a significant function in the salt-stress response; thus, we can leverage their potential by identifying valuable mutations and employing them with the aim of enhancing salt tolerance in crops. For example, it has been discovered that ABA-induced transcription repressors (AITRs) are a revolutionary group of transcription factors that serve as feedback regulators of abscisic acid (ABA) signalling. Traditionally, ABA is reported to play a vital role in the response to different abiotic stresses (Shokat *et al.*, 2020, 2021a,b).

In conclusion, researchers can harness mutations for salt tolerance in plants through a multifaceted approach. Utilizing mutagenesis techniques, advanced genomic tools, and gene editing technologies in the above-mentioned genes can induce genetic variations to enhance salt tolerance in crops. Stacking mutations and employing marker-assisted breeding techniques can further accelerate the development of salt-tolerant crop varieties.

1.3 Role of Mutation Breeding

Extensive research across the globe has been conducted to improve the salt tolerance of crop plants (Shokat and Großkinsky, 2019). Mutation breeding, as a powerful tool to generate genetic diversity, proves to be highly beneficial to improve the output of crops. A pivotal aspect of mutation breeding revolves around the detection and recognition of individuals harbouring a desired mutation, which entails two vital stages:

(i) the screening process; and (ii) the confirmation of mutants. In the process, the seeds undergo a treatment involving mutagenic agents such as physical and chemical mutagens. These treated seeds are then cultivated further to allow for segregation, after which the plants exhibiting advantageous traits are carefully selected to be nurtured in the subsequent generation. To ascertain the efficacy of these plants, multi-location trials are carried out to evaluate their performance. Upon successful evaluation, they are officially introduced as a novel variety, ready for utilization. For example, Lethin *et al.* (2020) developed a wheat mutant line tolerant to salt using the above-mentioned procedure of mutation breeding. Similarly, Sen and Sarsu (2019) developed salt-tolerant wheat mutant progenies using sodium azide mutagenesis and observed higher activities of antioxidant enzymes and lower lipid peroxidation in the mutants under salt-stress conditions. They combined chemical mutagenesis with somatic embryogenesis to obtain salt-tolerant mutant wheat lines, demonstrating the potential of mutation breeding to improve the salt tolerance (Sen and Sarsu, 2019).

In light of the above, it is obvious that mutation breeding has great potential to improve the salt tolerance of crops. Further, a few recent techniques like targeting induced local lesions in genomes (TILLING), a reverse genetics approach, has broadened the scope of mutation breeding as it enables the precise identification of desired mutations (Taheri *et al.*, 2017). Here, we precisely discuss the scope of TILLING.

1.3.1 TILLING

TILLING allows the precise identification of mutations for the genes of interest. Unlike traditional mutation breeding, which involves random mutagenesis across the entire genome, TILLING focuses on a specific gene or group of genes. A chemical mutagen (often ethyl methane sulfonate) is used to induce mutations in plants. The mutagenized plants are grown to extract DNA and then PCR is used to amplify the genes of interest. The PCR products are then screened for mutations using various techniques, such as enzymatic mismatch cleavage or high-resolution melting analysis. Mutations found in the target genes can lead to changes in the gene's function, which can result in altered traits (Triques *et al.*, 2007).

TILLING is a valuable method for functional genomics and crop improvement programmes. It allows scientists/breeders to identify the function of genes and to create novel germplasm by precisely targeting the genes of a specific trait. Navarro-León *et al.* (2020) generated three new CAX1 variants in *Brassica rapa* using TILLING strategy and found that one mutant provided higher tolerance to salinity stress. Another study investigated the expression profile of genes involved in ion homeostasis in *H. vulgare* and observed differences in expression patterns between the mutant and its wild type, suggesting the potential of mutation breeding to alter ion homeostasis (Yousefirad *et al.*, 2018) and to accelerate the breeding process that ultimately contributes towards the development of salt-tolerant plants.

In conclusion, mutation breeding complemented by advanced techniques like TILLING offers a targeted and efficient approach to enhance salt tolerance in crops. Further, precision in TILLING allows for the identification and validation of key genes, facilitating the creation of crops with enhanced resilience to salt stress. This approach not only contributes to sustainable agriculture by addressing environmental challenges but also aligns with regulatory and public acceptance due to its relative simplicity and reduced environmental impact compared to some other genetic engineering methods, and thus can stand as a promising strategy for securing food production in salt-affected regions in the future.

1.4 Precise Genetic Modifications for Salinity Tolerance in Plants

In the era of genomics, nuclease systems attached to specific DNA sequence-binding domains have been emerged to precisely edit the gene(s) of interest. Previously, first-generation genome editing technologies such as zinc finger nucleases (ZFNs), transcription activator-like

Table 1.1. Enhanced salt tolerance in various crop species achieved through the CRISPR/Cas genome editing system.

Crop	Gene targeted	Trait improved	References
Rice (*Oryza sativa* L.)	*OsRR22*	Improved cultivar stability and enhanced yield under saline conditions	Zhang *et al.* (2019); Sheng *et al.* (2023)
Tomato (*Solanum lycopersicum* L.)	*SlHyPRP1*	Enhanced salt tolerance during both germination and vegetative growth phases	Tran *et al.* (2021)
Soybean (*Glycine max* L.)	*GmAITR*	Improved salinity tolerance	Wang *et al.* (2021)

effector nucleases (TALENs), and meganucleases contributed a lot in modern research (Gaj *et al.*, 2013). TALENs have achieved greater success in the recent past due to higher bonding specificity and applied in various plants including *Arabidopsis*, rice, tobacco, and *Brachypodium* (Gaj *et al.*, 2013).

1.4.1 CRISPR-mediated genome editing

Second-generation genome editing technologies, particularly CRISPR/Cas9, present a more efficient, straightforward, and cost-effective approach. Genome editing with CRISPR/Cas9, in particular, has the potential to reduce both development time and costs compared to traditional breeding methods (Malzahn *et al.*, 2017). This technology has been exploited in several model plant species, e.g. *Arabidopsis*, rice, and tobacco. A notable application is the modification of the rice *OsR22* gene, which enhanced salt tolerance in the edited rice lines, exemplifying the potential of CRISPR/Cas9 (Zhang *et al.*, 2019). Using the vectors pC-22-KT12 and pC-22-T3 Cas9, the gene editing efficiency in the first generation (T0) of transgenic plants was found to be 84% and 58.82%, respectively. An important result of this process was the creation of a new third-generation hybrid rice line, named ST1, developed from these genetically modified lines (Sheng *et al.*, 2023). The application of this technology not only has the potential to revolutionize crop improvement strategies but also provides an avenue to bypass regulatory challenges associated with genetically modified organisms (GMOs), by facilitating specific mutations without the generation of transgenic plants (Sheng *et al.*, 2023). Similarly, another study utilized the CRISPR/Cas9 system to target the tomato gene, *SlHyPRP1*, which regulates salt-stress responses (Li *et al.*, 2016). Precise editing of specific domains resulted in enhanced salt tolerance during germination and vegetative growth phases, underscoring the precision as compared to traditional breeding methods and indicating its potential to tackle agricultural challenges (Tran *et al.*, 2021). Another study, aided by CRISPR/Cas9 genome editing, targeted *GmAITR* genes in soybeans, which are associated with abiotic stress responses (Wang *et al.*, 2021). Mutations in these genes resulted in enhanced salinity tolerance, highlighting the potential of CRISPR/Cas9 in addressing challenges posed by environmental stresses in vital crops like soybean (Wang *et al.*, 2021). Table 1.1 shows how the CRISPR/Cas system has improved salt tolerance in different crops.

1.5 Conclusion

In conclusion, the exploration of salinity stress and its impact on crop productivity has underscored the significance of advanced breeding techniques to improve the salt tolerance of plants. Genetic changes, i.e. SNPs and indels, can play a significant role to improve the salinity tolerance of crops. Further, techniques like TILLING can cause mutations in the known genes and germplasm can be created with a wide range of salinity tolerance. Similarly, the application of genome editing, i.e. CRISPR/Cas9, has been particularly promising, offering precise, efficient, and cost-effective solutions to improve salt tolerance in crop plants. A combination of these approaches along with phenotypic evaluation can be adopted to improve the salt tolerance of crops.

References

Arif, Y., Singh, P., Siddiqui, H., Bajguz, A. and Hayat, S. (2020) Salinity induced physiological and biochemical changes in plants: an omic approach towards salt stress tolerance. *Plant Physiology and Biochemistry* 156, 64–77.

Baxter, I., Brazelton, J.N., Yu, D., Huang, Y.S., Lahner, B. *et al.* (2010) A coastal cline in sodium accumulation in *Arabidopsis thaliana* is driven by natural variation of the sodium transporter AtHKT1;1. *PLoS Genetics* 6, e1001193.

Ding, F., Qiang, X., Jia, Z., Li, L., Hu, J. *et al.* (2022) Knockout of a novel salt responsive gene SlABIG1 enhance salinity tolerance in tomato. *Environmental and Experimental Botany* 200, 104903.

Fang, X., Li, W., Yuan, H., Chen, H., Bo, C. *et al.* (2021) Mutation of ZmWRKY86 confers enhanced salt stress tolerance in maize. *Plant Physiology and Biochemistry* 167, 840–850.

FAO (2024) Food and Agriculture Organization of the United Nations I Global Soil Partnership I Global Map of Salt-affected Soils (GSASmap). Available at: https://www.fao.org/global-soil-partnership/gsasmap/en (accessed 2 March 2024).

Feng, Z., Deng, Y., Fan, H., Sun, Q., Sui, N. and Wang, B. (2014) Effects of NaCl stress on the growth and photosynthetic characteristics of *Ulmus pumila* L. seedlings in sand culture. *Photosynthetica* 52, 313–320.

Gaj, T., Gersbach, C.A. and Barbas, C.F. (2013) ZFN, TALEN, and CRISPR/Cas-based methods for genome engineering. *Trends in Biotechnology* 31, 397–405.

Ganie, S.A. (2020) RNA chaperones: potential candidates for engineering salt tolerance in rice. *Crop Science* 60(2), 530–540.

Ganie, S.A. (2021) Amino acids other than proline and their participation in abiotic stress tolerance. In: Wani, S.H., Gangola M.P. and Ramadoss, B.R. (eds) *Compatible Solutes Engineering for Crop Plants Facing Climate Change*. Springer, Cham, Switzerland, pp. 47–96.

Gregory, W. (1972) Manual on mutation breeding International Atomic Energy Agency, Vienna 1970, 237 pp., paper bound $6.00. *Journal of Nuclear Energy* 26, 443–444.

Guo, Q., Meng, L., Han, J., Mao, P., Tian, X. *et al.* (2020) SOS1 is a key systemic regulator of salt secretion and K^+/Na^+ homeostasis in the recretohalophyte *Karelinia caspia*. *Environmental and Experimental Botany* 177, 104098.

Huang, X., Wei, X., Sang, T., Zhao, Q., Feng Q. *et al.* (2010) Genome-wide association studies of 14 agronomic traits in rice landraces. *Nature Genetics* 42, 961–967.

Hussain, S., Hussain, S., Ali, B., Ren, X., Chen, X. *et al.* (2021) Recent progress in understanding salinity tolerance in plants: story of Na^+/K^+ balance and beyond. *Plant Physiology and Biochemistry* 160, 239–256.

Jiang, C.-K. and Rao, G.-Y. (2020) Insights into the diversification and evolution of R2R3-MYB transcription factors in plants. *Plant Physiology* 183, 637–655.

Joshi, R., Wani, S.H., Singh, B., Bohra, A., Dar, Z.A. *et al.* (2016) Transcription factors and plants response to drought stress: current understanding and future directions. *Frontiers in Plant Science* 7, 1029.

Lethin, J., Shakil, S.S., Hassan, S., Sirijovski, N., Töpel, M. *et al.* (2020) Development and characterization of an EMS-mutagenized wheat population and identification of salt-tolerant wheat lines. *BMC Plant Biology* 20, 18.

Li, J., Ouyang, B., Wang, T., Luo, Z., Yang, C. *et al.* (2016) HyPRP1 gene suppressed by multiple stresses plays a negative role in abiotic stress tolerance in tomato. *Frontiers in Plant Science* 7, 967.

Malzahn, A., Lowder, L. and Qi, Y. (2017) Plant genome editing with TALEN and CRISPR. *Cell & Bioscience* 7, 21.

Melino, V. and Tester, M. (2023) Salt-tolerant crops: time to deliver. *Annual Review of Plant Biology* 74, 671–696.

Munns, R. and Tester, M. (2008) Mechanisms of salinity tolerance. *Annual Review of Plant Biology* 59, 651–681.

Navarro-León, E., López-Moreno, F.J., De La Torre-Gonzalez, A., Ruiz, J.M., Esposito, S. and Blasco, B. (2020) Study of salt-stress tolerance and defensive mechanisms in *Brassica rapa* CAX1a TILLING mutants. *Environmental and Experimental Botany* 175, 104061.

Negrao, S., Cecília Almadanim, M., Pires, I.S.S., Abreu, I.A., Maroco, J. *et al.* (2013) New allelic variants found in key rice salt-tolerance genes: an association study. *Plant Biotechnology Journal* 11, 87–100.

Paul, D. and Lade, H. (2014) Plant-growth-promoting rhizobacteria to improve crop growth in saline soils: a review. *Agronomy for Sustainable Development* 34, 737–752.

Scott, M.F. and Rescan, M. (2017) Evolution of haploid–diploid life cycles when haploid and diploid fitnesses are not equal. *Evolution* 71, 215–226.

Sen, A. and Sarsu, F. (2019) Obtaining candidate salt tolerant wheat mutant lines derived from combination of sodium azide mutagenesis and somatic embryogenesis. *Trakya University Journal of Natural Sciences* 20, 129–134.

Sheng, X., Ai, Z., Tan, Y., Hu, Y., Guo, X. *et al.* (2023) Novel salinity-tolerant third-generation hybrid rice developed via CRISPR/Cas9-mediated gene editing. *International Journal of Molecular Sciences* 24, 8025.

Shokat, S. and Großkinsky, D.K. (2019) Tackling salinity in sustainable agriculture – what developing countries may learn from approaches of the developed world. *Sustainability* 11, 4558.

Shokat, S., Großkinsky, D.K., Roitsch, T. and Liu, F. (2020) Activities of leaf and spike carbohydrate-metabolic and antioxidant enzymes are linked with yield performance in three spring wheat genotypes grown under well-watered and drought conditions. *BMC Plant Biology* 20, 400.

Shokat, S., Großkinsky, D.K. and Liu, F. (2021a) Impact of elevated CO_2 on two contrasting wheat genotypes exposed to intermediate drought stress at anthesis. *Journal of Agronomy and Crop Science* 207, 20–33.

Shokat, S., Novák, O., Široká, J., Singh, S., Gill, K.S. *et al.* (2021b) Elevated CO_2 modulates the effect of heat stress responses in *Triticum aestivum* by differential expression of an isoflavone reductase-like gene. *Journal of Experimental Botany* 72, 7594–7609.

Strader, L., Weijers, D. and Wagner, D. (2022) Plant transcription factors – being in the right place with the right company. *Current Opinion in Plant Biology* 65, 102–136.

Sun, Z., Li, H., Zhang, Y., Li, Z., Ke, H. *et al.* (2018) Identification of SNPs and candidate genes associated with salt tolerance at the seedling stage in cotton (*Gossypium hirsutum* L.). *Frontiers in Plant Science* 9, 1011.

Taheri, S., Abdullah, T.L., Jain, S.M., Sahebi, M. and Azizi, P. (2017) TILLING, high-resolution melting (HRM), and next-generation sequencing (NGS) techniques in plant mutation breeding. *Molecular Breeding* 37, 1–23.

Tester, M. and Langridge, P. (2010) Breeding technologies to increase crop production in a changing world. *Science* 327, 818–822.

Tran, M.T., Doan, D.T.H., Kim, J., Song, Y.J., Sung, Y.W. *et al.* (2021) CRISPR/Cas9-based precise excision of SlHyPRP1 domain(s) to obtain salt stress-tolerant tomato. *Plant Cell Reports* 40, 999–1011.

Triques, K., Sturbois, B., Gallais, S., Dalmais, M., Chauvin, S. *et al.* (2007) Characterization of *Arabidopsis thaliana* mismatch specific endonucleases: application to mutation discovery by TILLING in pea. *The Plant Journal* 51, 1116–1125.

Valliyodan, B., Qiu, D., Patil, G., Zeng, P., Huang, J. *et al.* (2016) Landscape of genomic diversity and trait discovery in soybean. *Scientific Reports* 6, 23598.

Van Zelm, E., Zhang, Y. and Testerink, C. (2020) Salt tolerance mechanisms of plants. *Annual Review of Plant Biology* 71, 403–433.

Wang, H., Li, G.-B., Zhang, D.-Y., Lin, J., Sheng, B.-L. *et al.* (2013) Biological functions of HD-Zip transcription factors. *Yi Chuan = Hereditas* 35, 1179–1188.

Wang, T., Xun, H., Wang, W., Ding, X., Tian, H. *et al.* (2021) Mutation of GmAITR genes by CRISPR/Cas9 genome editing results in enhanced salinity stress tolerance in soybean. *Frontiers in Plant Science* 12, 779598.

Xu, C., Song, L.-Y., Zhou, Y., Ma, D.-N., Ding, Q.-S. *et al.* (2023) Integration of eQTL and GWAS analysis uncovers a genetic regulation of natural ionomic variation in *Arabidopsis*. *Plant Cell Reports* 42, 1473–1485.

Xu, Y., Lu, Y., Xie, C., Gao, S., Wan, J. and Prasanna, B.M. (2012) Whole-genome strategies for marker-assisted plant breeding. *Molecular Breeding* 29, 833–854.

Yang, M., Lu, K., Zhao, F.-J., Xie, W., Ramakrishna, P. *et al.* (2018) Genome-wide association studies reveal the genetic basis of ionomic variation in rice. *The Plant Cell* 30, 2720–2740.

Yousefirad, S., Soltanloo, H., Ramezanpour, S., Zaynalinezhad, K. and Shariati, V. (2018) Salt oversensitivity derived from mutation breeding improves salinity tolerance in barley via ion homeostasis. *Biologia Plantarum* 62, 775–785.

Zhang, A., Liu, Y., Wang, F., Li, T., Chen, Z. *et al.* (2019) Enhanced rice salinity tolerance via CRISPR/Cas9-targeted mutagenesis of the OsRR22 gene. *Molecular Breeding* 39, 1–10.

Zhang, L., Zhao, G., Xia, C., Jia, J., Liu, X. and Kong, X. (2012) Overexpression of a wheat MYB transcription factor gene, TaMYB56-B, enhances tolerances to freezing and salt stresses in transgenic *Arabidopsis*. *Gene* 505, 100–107.

Zhu, J.-K. (2001) Plant salt tolerance. *Trends in Plant Science* 6, 66–71.

2 Regulation of Nucleotide Metabolism in Response to Salt Stress

Atul Singh[1]*, Shiva Sai Prasad[1]*, Utpal Das[2], Krishnananda P. Ingle[3], Kapil Patil[4], Pradeep Kumar Shukla[5], P.W. Ramteke[6], J.R. Rajeshwar[1], M. Gayathri[1], Palanisamy Senthilmurugan[1] and Jamindar Buddiga[1]

[1]Department of Agriculture, Koneru Lakshmaiah Education Foundation, Andhra Pradesh, India; [2]VIT School of Agricultural Innovations and Advanced Learning (VAIAL), VIT, Vellore, Tamil Nadu, India; [3]Centro Internacional de la Papa, International Potato Center, CGIAR Research Center, Pune, India; [4]Department of Horticulture, University of Agricultural Sciences, Raichur, Karnataka, India; [5]Sam Higginbottom University of Agriculture, Technology Sciences, Prayagraj, Uttar Pradesh, India; [6]Department of Molecular Biology and Genetic Engineering, RTM Nagpur University, Nagpur, India

Abstract

Salt stress is a significant environmental element that affects the growth and development of plants. To maintain cellular homeostasis and survival under salt stress, plants regulate nucleotide (purines and pyrimidines) metabolism. The nucleotide regulation fluctuates due to the salt stress and can disrupt the equilibrium of purines and pyrimidines production. This modification will alter the availability of precursors for the production of nucleic acids, impacting cell functions. Purines and pyrimidines regulate, break down, and manipulate enzymes responsible for metabolism. Also, it has been found that salt-tolerant plants have faster rates of purines production and lower rates of degradation than salt-sensitive plants, which enables them to maintain significant quantities of nucleotides and nucleic acids under salt stress. These modifications help plants to adapt and become more tolerant to salt stress while also playing a critical role in reducing the harmful effects of salt stress on growth and development. This chapter covers how crucial nucleotide *de novo* and salvage mechanisms enable plants to cope with salt stress.

2.1 Introduction

The development and survival of plants are frequently impacted by salt stress. It is a significant abiotic factor that impacts plant development and output, particularly in dry and coastal areas. A high salt concentration in the soil or water can damage the delicate balance of ions in a plant's cells, leading to physiological and biochemical changes. These changes can cause stunted growth, reduced photosynthesis, and reduced yield (Elsayed *et al.*, 2022). Meanwhile, certain plants have developed adaptations to withstand high salt concentrations, such as avoiding salts with numerous mechanisms and modifications. It is very important to study and understand the salt

*Email: singhatul9415@gmail.com; 5257shiv@gmail.com

 The Genetics of Salt Tolerance in Plants: A Central Dogma Perspective and Strategies for Enhancement (S. Ganie and S.H. Wani eds)
DOI: 10.1079/9781800623033.0002

tolerance of plants in order to achieve food and nutritional security.

An integration of agronomic and genetic approaches can mitigate salt stress. Through the agronomic approach, various cultural practices such as leaching, irrigation management, and gypsum application can alleviate stress (Anand *et al.*, 2021). Additionally, using organic amendments like compost and mulching helps to improve soil structure and water-holding capacity by reducing the salts. On the other hand, the genetic approach is to select or develop salt-tolerant varieties that can cope under saline conditions; also, to use transgenic technology to introduce salt-tolerance genes into crops. Using an integrated approach combining different mitigation strategies is often the most effective against stress. Further, plants have been found to synthesize solutes to minimize the harmful effects and preserve cellular osmotic equilibrium.

Similarly, modification in morphology, such as excluding salt from roots, leaves, and shoots, reduces salts reaching the cells and tissues, which maintains the osmoregulation involving adjusting of cell water content and osmotic potential to maintain cellular turgor (Ganie, 2020, 2021). Also, manipulating ion transport pathways enables the plant to control the absorption and distribution of necessary ions. Additionally, the stimulation of gene expression and modifications improves the plant's capacity to respond to salt stress and activates the protective antioxidant systems, preventing damage.

Nucleotides are essential in building deoxyribonucleic acid (DNA) and ribonucleic acid (RNA), which carry genetic information. They comprise a nitrogenous base, a phosphate group, and a sugar molecule (ribose or deoxyribose). Nucleotides are essential for several plant biological activities, such as protein synthesis, transcription, and DNA replication (Ohta *et al.*, 2003). The modifications at the cellular level, such as in nucleotides (purines and pyrimidines) reported by Witte and Herde (2019), are effective in alleviating stress. As per the report of Chen *et al.* (2011), the level of 3′-phosphoadenosine-5′-phosphate (PAP) was majorly controlled in plants and did not accumulate significantly under stress conditions in *Arabidopsis*; while in mutant plants, high levels of PAP were detected, indicating PAP is critical for stress alleviation. Through the regulation of gene expression, nucleotides also have a significant influence on the management of plant growth and development. Additionally, they function as precursors for synthesizing adenosine triphosphate (ATP), utilized as an energy source by plant cells, and are involved in controlling cellular metabolism and energy transmission (Mostek *et al.*, 2015). Nucleotides transmit energy from light-excited electrons to the Calvin cycle, which is utilized to create sugars during photosynthesis. They are also crucial for maintaining cellular integrity and are involved in plant defence mechanisms, such as the reaction to biotic and abiotic stress (Ashihara *et al.*, 2020).

The purines and pyrimidines are nitrogen-containing substances that significantly function in cellular metabolism. While they are not directly involved in the regulation of metabolism, they play a key role in the phenylpropanoid pathway, a metabolic pathway to synthesize secondary plant metabolites, i.e. lignins, flavonoids, and phenolic acids, from the amino acid phenylalanine. The phenylpropanoid pathway is primarily regulated by environmental conditions and the availability of precursors and regulatory enzymes (Pietrowska-Borek *et al.*, 2020). Allantoin is a purine metabolite involved in various physiological processes, including osmoregulation, stress response, and cellular signalling. It has also been found to regulate transporter proteins and mitigates oxidative stress under stress conditions in *Arabidopsis* (Lescano *et al.*, 2016). It also helps tolerance to saline–alkaline conditions in sugarbeet by activating a self-amplifying feedback loop involving jasmonic acid (JA) and nitric oxide (NO) (Zhang *et al.*, 2023). Under saline–alkaline stress, allantoin can increase levels of JA, a plant hormone involved in stress response and defence, and NO, a signalling molecule that modulates cellular processes such as cell growth, differentiation, and stress response. The increased levels of JA and NO activate a set of genes that are involved in salt tolerance and defence, leading to enhanced tolerance to saline–alkaline conditions (Mukarram *et al.*, 2022; Zhang *et al.*, 2023).

Similarly, cyclic nucleotides are crucial secondary messengers in cellular communication and include cyclic adenosine monophosphate (cAMP) and cyclic guanosine monophosphate (cGMP) which activate cyclic nucleotide-gated (CNG) ion channels (Jha and Pandey, 2021). The concentrations of cyclic nucleotides alter in response to various stress signals, which can activate CNG channels and help calcium and sodium to enter the cells. When calcium enters cells via CNG channels and activates calcium-sensitive signalling pathways, this in turn leads to stress-responsive gene expression, protein synthesis, and the activation of cellular defence systems against stress (Yang *et al.*, 2019). It also modulates several metabolic processes, including ion homeostasis and signal transduction. The cAMP signalling cascade in tomato plants altered several metabolites, including amino acids, carbohydrates, and secondary metabolites, via modulating transcription factors and enzymes engaged in metabolic processes (Duszyn *et al.*, 2019). The transcriptome and metabolome analysis of salt-stressed quinoa seed germination confirmed the significant strategy of seed germination against salt stress by regulating carbohydrate metabolism and antioxidant defence functions (Yan *et al.*, 2022). As per the research, the cAMP signalling cascade can trigger the expression of genes that synthesize protectants like proline and control stress-responsive pathways like the antioxidant defence system. As Kamran *et al.* (2019) report, proline was found to be a significant compatible solute in salt-stressed canola root and shoot tissues and during germination. In plant tissue culture, salt stress leads to the synthesis of purine alkaloids in *Coffea arabica*. Purines metabolites found in coffee beans play a role in tolerance of stress (Frischknecht and Baumann, 1985). This chapter discusses the plant regulation of nucleotides (purines and pyrimidines) in alleviating salt stress.

2.2 Nucleotide Structure

Nucleotides are an essential component of DNA and RNA and play a novel role in cellular processes such as DNA replication, protein synthesis, and energy storage. Three elements make up a nucleotide: a sugar molecule, a nitrogenous base, and a phosphate group (Moffatt and Ashihara, 2002). Similarly, a nucleoside contains only a sugar molecule with a nitrogenous base. The sugar molecule in a nucleotide is usually ribose in RNA and deoxyribose in DNA (Werner and Witte, 2011). The nitrogenous bases are responsible for the genetic information stored in DNA and RNA. The sugar molecule's 5′ carbon has a bonded phosphate group. When nucleotides are linked, the phosphate group of one nucleotide is linked to the sugar molecule of the next nucleotide through a phosphodiester bond, forming a long chain of nucleotides that make up the nucleic acid (Ashihara *et al.*, 2018).

2.2.1 *De novo* synthesis of purines

Plants have cellular metabolism for the *de novo* production of pyrimidine nucleotides, which produce uridine monophosphate (UMP), and purine nucleotides, for the production of adenosine monophosphate (AMP). The purine nucleotides, glycine, and formyltetrahydrofolate are freshly produced during *de novo* biosynthesis from active ribose phosphoribosyl pyrophosphate (PRPP), aspartate, glutamine, and bicarbonate (Moffatt and Ashihara, 2002; Stasolla *et al.*, 2003; Suzuki-Yamamoto *et al.*, 2006; Zrenner *et al.*, 2006; Ashihara *et al.*, 2018). The PRPP amidotransferase, commonly known as PRPP synthase, is essential for nucleotide biosynthesis. The regulation of PRPP amidotransferase activity is indeed influenced by both guanosine monophosphate (GMP) and AMP (Chen *et al.*, 1997; Naik, 2010). The presence of both AMP and GMP can reduce the activity of PRPP amidotransferase, as they compete for the same binding site on the enzyme. GMP can enhance PRPP amidotransferase activity, while AMP has the opposite effect and can inhibit its activity. The inhibition of PRPP amidotransferase by AMP and GMP helps to regulate purine and pyrimidine nucleotide biosynthesis and maintain cellular homeostasis (Bhagavan and Ha, 2015; Hove-Jensen *et al.*, 2017). Further, they are synthesized from simple precursors. The formation of the first ring in the purine ring structure follows the addition of glycine, formaldehyde, and ammonia. Adding a carboxyl group from CO_2 and aspartate forms a complex intermediate, followed by the loss of fumarate, which results in a net gain of an amine. These reactions produce a purine intermediate that will undergo further modifications. This intermediate can then undergo more modifications to produce the final purine nucleotide (Ashihara *et al.*, 2018; Tan *et al.*, 2022). Inosine monophosphate (IMP) is created when a second carbon is added from

N^{10}-formyltetrahydrofolate and the second ring is closed, which can then be further phosphorylated to form other purine nucleotides such as AMP and GMP (Stanbury *et al.*, 2017).

IMP serves as a branch point in the biosynthesis of purine nucleotides. From IMP, adenine nucleotides can be synthesized by adding an amine group, while guanine nucleotides can be synthesized by adding an amine and a carbonyl group. These nucleotides are important skeletons in DNA and RNA and play critical roles in cellular processes such as energy metabolism, gene expression, and cell signalling (Carlucci *et al.*, 1999). Adding an amine group from aspartate and the need for energy from guanosine triphosphate (GTP) are components of the route that converts IMP to AMP (Chou *et al.*, 2005; Suzuki-Yamamoto *et al.*, 2006; Duszyn *et al.*, 2019). This is catalysed by the enzyme IMP dehydrogenase, which utilizes the energy from GTP to transfer the amine group from aspartate to IMP, forming AMP (Sullivan *et al.*, 2015; Meek, 2022). In order to convert IMP into GMP, a glutamine-derived amine group will be added, followed by oxidation. The enzyme xanthine oxidase catalyses the oxidation of IMP to form xanthine monophosphate, while the addition of the amine group from glutamine is catalysed by the enzyme GMP synthase. The GMP is crucial for cellular functions including gene expression, protein synthesis, and energy consumption (Delgado-García *et al.*, 2021). IMP to GMP and the pathway to AMP also require energy from ATP, which is the primary source of energy in cells. Additionally, both the route leading to GMP and the pathway resulting in AMP are blocked by their respective end products. This is a feedback inhibition, where the final product of a metabolism controls the enzymes necessary for its own synthesis. This mechanism helps to maintain a proper balance of these molecules within the cell (Moffatt and Ashihara, 2002).

From the IMP branch point onwards, the balance of purine nucleotides is preserved via controlling PRPP amidotransferase. PRPP amidotransferase is an enzyme essential for the biosynthesis of purine nucleotides because it catalyses the initial committed step in producing purine and pyrimidine nucleotides (Hove-Jensen *et al.*, 2017). Amidotransferase will be inhibited in the presence of an imbalance in AMP or GMP, enabling the processes leading to IMP synthesizing slowly (Duszyn *et al.*, 2019; Pietrowska-Borek *et al.*, 2020). This mechanism helps to adjust the rate of purine nucleotide synthesis based on the cell's requirements, ensuring that a proper balance of these molecules is maintained within the cell to act as building blocks for DNA and RNA synthesis (Hove-Jensen *et al.*, 2017; Ashihara *et al.*, 2020; Pietrowska-Borek *et al.*, 2020). Feedback inhibition is a common regulatory mechanism in metabolic pathways and helps to control the flow of metabolic reactions based on the cell's needs (Jiménez *et al.*, 2005). When purine nucleotides are abundant, the activity of PRPP amidotransferase is inhibited, thus limiting the production of purines and preventing their over-accumulation. This helps to avoid the potentially toxic effects of purines over-accumulation (Hove-Jensen *et al.*, 2017). Also, nucleoside diphosphate kinase and guanine nucleotide-binding protein were found to be involved in nucleotide metabolism and enoyl-acyl carrier protein (ACP) reductase in salt-stressed rice (Dooki *et al.*, 2006) and PRPP in peas (Qureshi *et al.*, 2007).

2.2.2 Salvage of purines

The purines salvage pathway is a metabolic pathway that allows the reuse of purine bases degraded from RNA and DNA rather than synthesizing them *de novo*, also an energy-saving mechanism. The first step in the purines salvage pathway is the conversion of purine bases (such as adenine and guanine) to nucleosides (such as adenosine and guanosine) (Delgado-García *et al.*, 2021) (Fig. 2.1A). Purine nucleoside phosphorylases catalyse this conversion. The nucleosides are then phosphorylated to nucleotides by the action of nucleoside kinases. The nucleotides are then converted to nucleoside triphosphates (ATP and GTP) by the action of nucleoside diphosphokinases (Ashihara *et al.*, 2018). The nucleoside triphosphates will be used directly in cellular processes, such as DNA and RNA synthesis, or further metabolized to form other compounds, such as ATP and GTP (Peterson *et al.*, 1988). With a salvage pathway, the cell will conserve energy and resources that would otherwise be required for the *de novo* synthesis of purines. The regulation of the purines salvage pathway helps to maintain a proper balance of purine nucleotides in the cell.

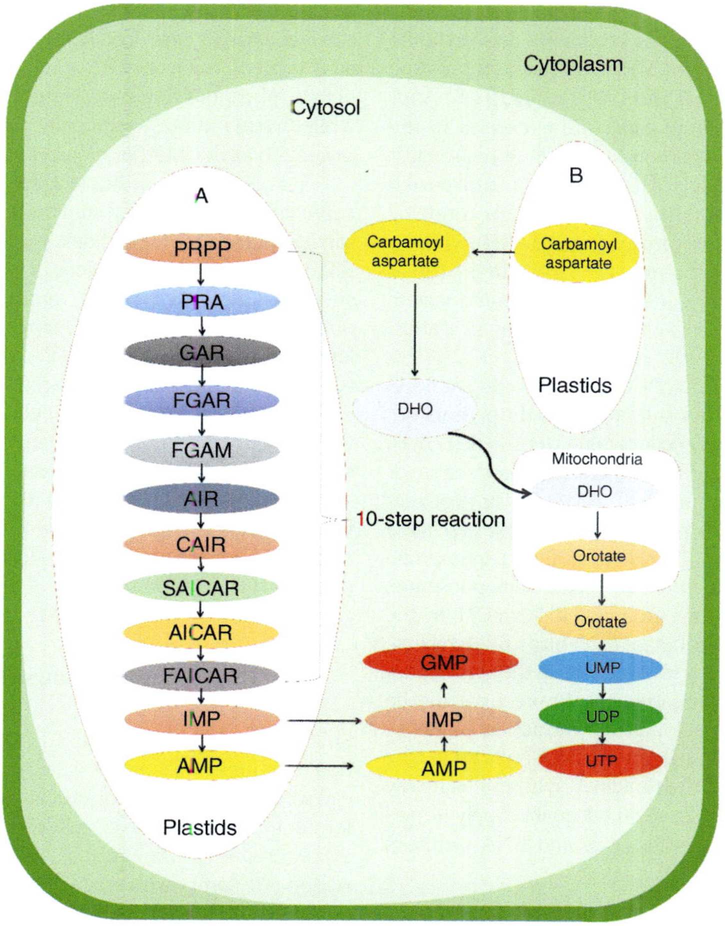

Fig. 2.1. (A) Steps involved in the salvage of purines in plant cells. PRPP (phosphoribosyl pyrophosphate) is used for the synthesis of AMP (adenosine monophosphate) by a ten-step reaction process, released from plastids and converted to GMP (guanosine monophosphate), used for osmotic regulation and ion homeostasis (PRPP = 5-phosphoribosyl-α-1-pyrophosphate; PRA = 5-phospho-β-D-ribosylamine; GAR = 5-phosphoribosylglycinamide; FGAR = 5′-phosphoribosyl-*N*-formylglycinamide; FGAM = 5′-phosphoribosyl-*N*-formylglycinamidine; AIR = 5′-phosphoribosyl-5-aminoimidazole; CAIR = 5′-phosphoribosyl-5-aminoimidazole-4-carboxylate; SAICAR = 5-amino-4-imidazole-*N*-succinocarboxamide ribonucleotide; AICAR = 5-amino-4-imidazolecarboxamide ribonucleotide; FAICAR = 5′-phosphoribosyl-4-carboxamide-5-formamidoimidazole; IMP = inosine monophosphate). (B) Steps involved in the salvage of pyrimidines in plant cells. Carbamoyl aspartate is converted into DHO (dihydroorotate) in a series of enzymatic reactions, and DHO is further converted into orotate by DHOase (dihydroorotase). Orotate is converted into UMP (uridine monophosphate) that is phosphorylated to form UDP (uridine diphosphate) and UTP (uridine triphosphate), important in maintaining cellular homeostasis. (Authors' own figure.)

2.2.3 *De novo* synthesis of pyrimidines

The *de novo* synthesis of pyrimidines begins with condensing of carbamoyl phosphate and aspartate to form orotic acid. The enzyme aspartate transcarbamoylase catalyses the reaction in plastids. It is the first step in the pathway which continues with the reduction of orotic acid to UMP and the further modification of UMP to form the pyrimidine nucleotides such as cytidine monophosphate (CMP) and thymidine monophosphate (TMP), eventually producing deoxycytidine monophosphate (dCMP) and deoxythymidine monophosphate (dTMP) (Witz *et al.*, 2012). After forming, carbamoyl aspartate is carried to the cytosol, where dihydroorotase (DHOase) converts it to dihydroorotate (DHO), which is transported into the mitochondria and oxidized to orotate by dihydroorotate dehydrogenase (DHODH). This reaction provides the necessary reducing equivalents for producing orotic acid, a key intermediate in the pyrimidines synthesis pathway (Katahira and Ashihara, 2002; Hove-Jensen *et al.*, 2017).

The location of DHODH in the mitochondria helps to ensure the oxidative step in the pyrimidines synthesis pathway occurs in an oxygen-rich environment, allowing for the efficient production of orotic acid (Schröder *et al.*, 2005). In the synthesis of cytidine triphosphate (CTP), the first step is the phosphorylation of UMP to uridine triphosphate (UTP) by the action of nucleoside monophosphate kinases. UTP is then aminated to form CTP by the action of CTP synthetase (Katahira and Ashihara, 2002). CTP is a key intermediate in the pyrimidines synthesis pathway, serving as the precursor for the production of RNA and the formation of RNA–DNA hybrid molecules during DNA replication. The regulation of the pyrimidines synthesis pathway ensures that cells have an adequate supply of pyrimidine nucleotides for DNA and RNA synthesis (Stasolla *et al.*, 2003).

2.2.4 Salvage of pyrimidines

The pyrimidines salvage pathway is a metabolic pathway that allows the reuse of bases degraded from RNA and DNA rather than synthesizing them *de novo* (Zrenner *et al.*, 2006). The first step in the pyrimidines salvage pathway is the conversion of pyrimidine bases (such as cytosine and uracil) to nucleosides (such as cytidine and uridine) (Fig. 2.1B). This conversion is catalysed by pyrimidine nucleoside phosphorylases (Witz *et al.*, 2012). The phosphorylation of nucleosides is a key step in the pyrimidines salvage pathway. Nucleosides are phosphorylated to nucleotides by the action of nucleoside kinases. This step is important because it increases the energy content of the nucleotides and makes them suitable for use in cellular processes. The nucleoside kinases involved in the pyrimidines salvage pathway are specific to the phosphorylated type of nucleoside. For example, cytidine kinase phosphorylates cytidine to form CMP, while uridine kinase specifically phosphorylates uridine to form UMP (Peterson *et al.*, 1988).

The conversion of nucleotides to nucleoside triphosphates is a crucial step in the pyrimidines salvage pathway. Nucleotides, such as CMP and UMP, are converted to nucleoside triphosphates, such as CTP and UTP, by nucleoside diphosphokinases. Nucleoside diphosphokinases transfer a phosphate group from ATP to the nucleotide to form a nucleoside triphosphate (He *et al.*, 2021). This step increases the energy content of the nucleoside triphosphates and makes them suitable for use in cellular processes. The nucleoside triphosphates will be used directly in cellular processes, such as DNA and RNA synthesis, or can be further metabolized to form other compounds, such as CTP and UTP.

2.2.5 Subcellular localization

Subcellular localization refers to the accumulation of a nucleotide in a specific location within a cell where a protein, an organelle, or other cellular components are accumulated (Duszyn *et al.*, 2019; Li *et al.*, 2020; Cui *et al.*, 2023). The localization of a component is crucial because it determines its function and interactions with other cellular structures. Nucleotides can be found in different subcellular compartments depending on their role and the stage of cellular metabolism. Nucleotides participate in DNA synthesis, replication, and repair in the nucleus. They are also involved in the regulation of gene expression. In the cytoplasm, nucleotides are used for energy production through ATP synthesis and as a precursor for the synthesis of other cellular compounds (Rubio and Berne, 1980). The purine nucleoside phosphorylases are primarily located in the mitochondria and cytoplasm, where they play a role in regulating energy metabolism.

2.3 Purines

Purines play an active role in the cytoplasm and nucleoplasm by participating in the metabolic processes that generate energy and maintain cellular homeostasis. During cellular respiration, glucose and other fuel molecules are metabolized to produce energy as ATP. In the form of adenine, purines are one of the building blocks of ATP (Bhagavan and Ha, 2015). Purine derivatives, such as adenosine, are essential in regulating cellular signalling in plants. Adenosine acts as a signalling molecule that modulates various physiological processes in plants, including stress response and osmotic adjustment (Akatsu *et al*., 1996; Suzuki-Yamamoto *et al*., 2006; Werner and Witte, 2011). It has been shown to regulate plant stress responses by influencing the activity of various signalling pathways (Chou *et al*., 2005). By controlling the expression of crucial genes involved in seed germination, root development, and floral development, adenosine is reported to function in regulating these processes.

Also, under salt stress, maize root tips showed a decrease in the level of ATP and adenosine diphosphate (ADP), accompanied by an increase in the activity of enzymes involved in ATP synthesis (Peterson *et al*., 1987). The level of inorganic phosphate (Pi) also increased, indicating a possible role of Pi in maintaining energy balance. UTP and uridine diphosphate (UDP) levels also decreased significantly, indicating a disruption in RNA synthesis and turnover (Mitchell *et al*., 2017). This decrease may be due to the increased demand for ATP for maintaining ion homeostasis and osmotic balance. Furthermore, adenosine has been implicated in regulating plant immunity by triggering defence responses in the plant in response to pathogen infection (Duszyn *et al*., 2019). Further, purine derivatives, such as hypoxanthine and xanthine, are crucial in maintaining plant cellular homeostasis. These compounds are involved in various metabolic processes that help to regulate cellular energy balance, prevent oxidative damage, and maintain cellular redox homeostasis (Peterson *et al*., 1988). In a study, xanthine regulated photosynthesis by modulating the activity of the photosynthetic electron transport chain (Miras-Moreno *et al*., 2021).

2.4 Pyrimidines

Pyrimidines are a class of nitrogenous bases that are essential components of nucleic acids. Cytosolic pyrimidines, such as uracil and thymine, are primarily located in the cytoplasm of cells and are involved in various metabolic processes, including DNA and RNA synthesis and degradation (Zrenner *et al*., 2006). In the nucleus, pyrimidines, such as cytosine and thymine, are integrated into the nucleotides of DNA, forming the basis for coding genetic information. The cytosolic pyrimidines, uracil and thymine, play an active role in various metabolic processes in plants. These compounds are involved in DNA and RNA synthesis and degradation (Suzuki-Yamamoto *et al*., 2006). A series of enzymatic reactions regulates the synthesis of uracil and thymine, and the balance of their levels in the cytoplasm is critical for maintaining cellular homeostasis (Wang *et al*., 2020). By modifying the activity of several stress-response signalling pathways, uracil has also been found to function in controlling plant stress responses (Witz *et al*., 2012). In the nucleus, cytosine and thymine are incorporated into the nucleotides of DNA, where they form the basis for coding genetic information. The pairing of cytosine with guanine in DNA provides stability and integrity of the double-stranded structure of DNA and allows for the storage and expression of genetic information. They are also found in the repair and replication of DNA. In a study, cytosine was reported to play a role in regulating DNA repair by modulating the activity of DNA glycosylases, enzymes that remove damaged bases from DNA (Nygaard and Saxild, 2009; Engelking, 2015).

2.5 Nucleotide Regulation

Nucleotide regulation refers to the control of gene expression by regulating the levels of specific nucleotides. It includes various mechanisms such as changes in the rate of synthesis, degradation, or modification of nucleotides, or changes in the interactions between nucleotides and other cellular components.

The enzymes ribose 5-phosphate pyrophosphokinase (PRPP synthetase) and glutamine PRPP amidotransferase perform the first two stages of purines biosynthesis. They are the rate-limiting processes (glutamine PRPP amidotransferase) (Nassar *et al*., 2020). They are responsible for synthesizing the important precursor molecule

PRPP, which is required to synthesize all purine nucleotides in the cell. PRPP synthetase catalyses the initial stage of purines biosynthesis, which is feedback suppressed by purine nucleotides. Similar to the first step, the second phase is similarly feedback suppressed by purine nucleotides and is catalysed by glutamine PRPP amidotransferase (Senecoff *et al.*, 1996; Decros *et al.*, 2019). There are two allosteric sites on glutamine PRPP amidotransferase: one for binding AMP, ADP, and ATP; and the other for GMP, guanosine diphosphate (GDP), and GTP. Additionally, PRPP activates this enzyme, which aids in the cell's ability to synthesize purine nucleotides. The glutamine synthase was found to modulate proline biosynthesis and glycine dehydrogenase in salt-stressed wheat (Caruso *et al.*, 2008). The relative concentrations of AMP and GMP also control the two branches of the pathway that produce AMP and GMP from IMP (Zhang *et al.*, 2023). GMP inhibits the GMP branch (IMP dehydrogenase), whereas AMP inhibits the AMP branch (Gupta and Gupta, 2021). This regulation helps to coordinate the metabolic pathways involved in purines biosynthesis, ensuring that the cell has the appropriate levels of purine nucleotides to meet its metabolic needs. The regulation in pyrimidines biosynthesis is the feedback inhibition of aspartate carbamoyltransferase, which is the first rate-limiting enzyme (Stasolla *et al.*, 2003; Kafer *et al.*, 2004). Additionally, it will be inhibited by feedback inhibition. The expression of genes involved in pyrimidines biosynthesis can be regulated by various environmental and developmental signals, such as light and nutrients (Bein and Evans, 1995; Zrenner *et al.*, 2006).

2.6 Role in Salt Stress

Nucleotides significantly influence a plant's ability to tolerate salt stress. High salt concentrations can cause many physiological and metabolic changes in plants, including alterations in cellular ion balance, oxidative stress, and reductions in growth and productivity (Ganie *et al.*, 2017). Nucleotides alleviate salt tolerance by acting as precursors for synthesizing essential signalling molecules (cyclic nucleotides) which help regulate the plant's response to stress (Peterson *et al.*, 1988). Among them, cAMP has been shown to regulate ion transport, protect against oxidative damage, and modulate gene expression. Additionally, the nucleotides contribute to salt tolerance through their involvement in energy metabolism (Pietrowska-Borek *et al.*, 2020). They synthesize high levels of ATP, which can help to maintain cellular ion balance and energy homeostasis under salt stress, while high levels of GTP have been shown to play an important role in regulating the transport of salts across cellular membranes.

2.6.1 Purines

Under salt stress, changes in the balance of purines biosynthesis and degradation can impact the availability of purines for nucleic acid synthesis and energy production, playing a pivotal role in plant stress signalling and defence mechanisms (Fig. 2.2).

Salt stress in the root tips of maize (*Zea mays*) was found to expedite purines catabolism and prevent pyrimidines salvage (Peterson *et al.*, 1988), which might be the balance between the purines and pyrimidines biosynthesis pathways and the impact of high salt concentrations on plant metabolism. The production of suitable solutes may be connected to promoting pyridine nucleotide catabolism and the last stage of pyrimidines breakdown during salt stress in maize root tips (Suzuki-Yamamoto *et al.*, 2006). The plant may be producing precursors for the synthesis of suitable solutes by promoting the catabolism of pyridine nucleotides and the degradation of pyrimidines, assisting in reducing the consequences of salt stress. Compatible solutes are small, organic compounds that help plants cope with osmotic stress by stabilizing cellular water content. They are produced with exposure to stress, can lessen cellular structure damage, and can safeguard crucial functions, including enzyme functioning and metabolic pathways. As per the report by Peterson *et al.* (1988), under salt stress, a swift modification is seen in maize root tips in the capacity for purines catabolism and a reduction in the capacity for pyrimidines salvage. This shift in metabolic pathways helps the plant to maintain a balance between nitrogen and carbon metabolism, which is critical for growth and survival under stressful conditions (Sun *et al.*, 2021). By increasing the

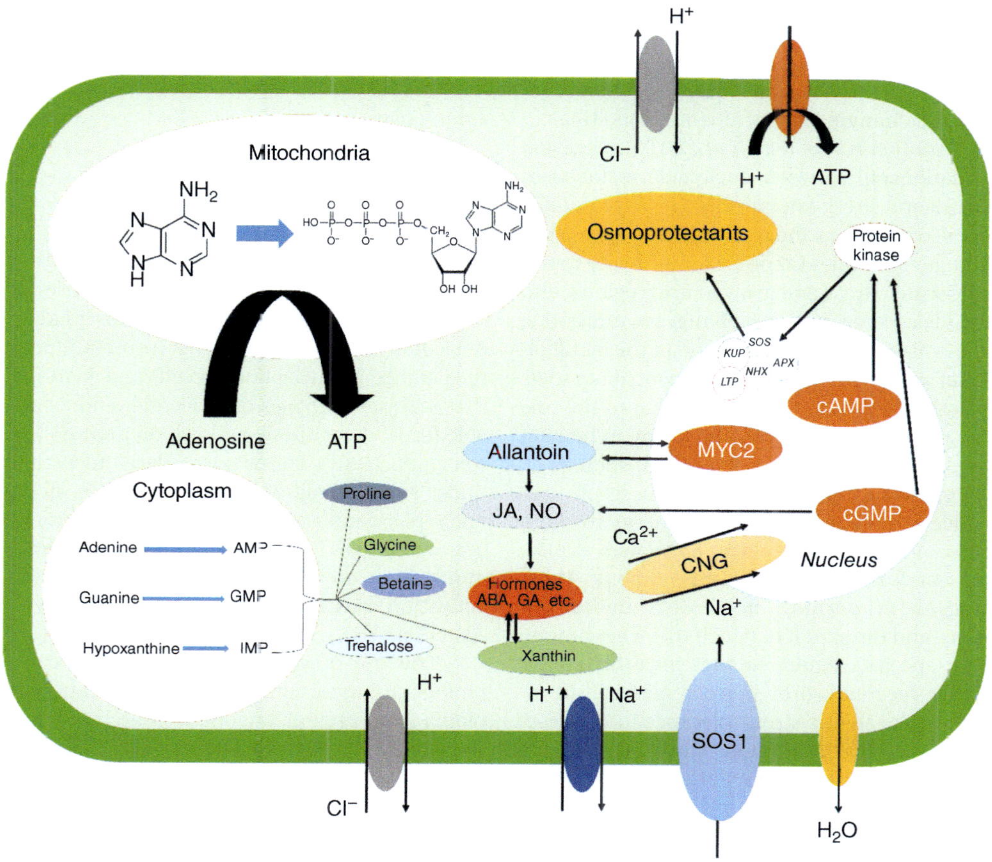

Fig. 2.2. Role of *de novo* synthesis of purines in alleviating the salt stress in plants. Adenosine is converted to ATP (adenosine triphosphate) in the mitochondria of plants, but the activation of alternative pathways can help mitigate the effects of stress on mitochondrial function and energy production. Shown are: (i) the *de novo* synthesis of purines pathway (AMP = adenosine monophosphate, GMP = guanosine monophosphate, IMP = inosine monophosphate) in the cytoplasm in response to salt stress, followed by the synthesis of JA (jasmonic acid) and NO (nitric oxide) via the octadecanoid pathway, which can be fuelled by the purine nucleotides produced in the cytoplasm; (ii) activation of the MAPK (mitogen-activated protein kinase) pathway in the nucleus in response to salt stress (cAMP = cyclic adenosine monophosphate, cGMP = cyclic guanosine monophosphate), followed by the phosphorylation of transcription factors and the activation of stress-responsive genes (*LTP*, *NHX*, *APX*, *KUP*, *SOS*) involved in the synthesis of osmoprotectants; (iii) the activation of MYC2-mediated gene expression in response to JA signalling pathway activation in the nucleus, leading to the biosynthesis of JA and its derivatives, which activate ABA (abscisic acid) biosynthesis (GA = gibberellic acid; CNG = cyclic nucleotide-gated ion channels). (Authors' own figure.)

catabolism of purines and decreasing the salvage of pyrimidines, the plant synthesizes precursors, important metabolic compounds, such as compatible solutes and amino acids, which can help minimize the effects of salts (Stasolla *et al.*, 2003). Furthermore, it is interesting that purine bases salvage does not rise proportionally to hypoxanthine's increased absorption and catabolism in salt-stressed maize roots (Akatsu *et al.*, 1996). This observation suggests that the plant prioritizes utilizing purine-derived compounds for other purposes, such as the synthesis of compatible solutes or the production of energy, over the salvage of purine bases. It also highlights plant metabolic pathways' complex and dynamic nature and ability to adapt and respond to changing environmental conditions.

Another report (Fan *et al.*, 2014) suggested a relationship between nucleotide metabolism and abiotic stress tolerance in *Hylocereus undatus*, also known as the pitahaya or dragon fruit. This cactus species is known for its ability to tolerate water stress and grow in arid regions, and it has been suggested that changes in nucleotide metabolism, such as an increase in the metabolism of purine nucleotides, may contribute to its drought tolerance. This is likely due to the role that purine-derived compounds play in helping the plant to cope with water stress, such as by serving as precursors for the synthesis of compatible solutes or by generating energy through metabolic pathways (Calder and Yaqoob, 2009). The underlying mechanisms driving these changes in purines metabolism under salt stress, and the extent to which they vary among plant species, require further investigation. A similar mechanism of purines metabolism increased the abiotic stress tolerance in *Arabidopsis* (Watanabe *et al.*, 2014). Allantoin, a purine metabolite, has been shown to enhance abiotic stress tolerance in plants through its synergistic activation of abscisic acid (ABA) metabolism (Takagi *et al.*, 2016; Zhang *et al.*, 2023). The phytohormone ABA regulates plant responses to stress, such as drought and excessive salinity. The ability of plants to withstand stress is improved by allantoin by enhancing the levels of ABA in plants (Barrero *et al.*, 2006). This activation of ABA metabolism is thought to result from the synergistic interaction between allantoin and ABA, which increases the expression of genes involved in the stress response and enhances the production of stress-tolerance compounds. Also, allantoin has been shown to activate jasmonates (JAs) signalling in an MYC2-regulated and ABA-dependent manner (Takagi *et al.*, 2016). JAs are a class of plant hormones essential for controlling stress and its damage. MYC2 is a transcription factor involved in regulating JA signalling (Delgado *et al.*, 2021). The activation of JA signalling by allantoin and ABA likely contributes to the plant's ability to cope with stress and injury by inducing the production of stress-tolerance compounds and activating stress-response pathways. The specific mechanisms by which purines and their derivatives contribute to plant stress response and the extent of these mechanisms vary among plant species.

2.6.2 Pyrimidines

Because of their function in cellular metabolism, pyrimidines help plants tolerate stress. Pyrimidines salvage pathways, which recycle pyrimidine nucleotides, play a critical role in maintaining the balance of nucleotides in cells and ensuring that adequate supplies of nucleotides are available for DNA synthesis and cell division (Peterson *et al.*, 1988). This is particularly important under stress conditions, as cells need to divide and repair damaged tissues. In addition, pyrimidine derivatives have been shown to have direct and indirect effects on stress tolerance and the activation of stress-response pathways, such as ABA (Fig. 2.3).

The use of pyrimidines-containing compounds for increasing the stress tolerance of plants is an area of ongoing research (Cansev *et al.*, 2014). Some studies have reported the potential of pyrimidine derivatives, such as uracil, to enhance plant stress tolerance (Katahira and Ashihara, 2002; Suzuki-Yamamoto *et al.*, 2006; Witz *et al.*, 2012). It has been shown to increase plant stress tolerance, likely by regulating the expression of stress-responsive genes and activating stress-response pathways, such as the ABA pathway (Fig. 2.3). In addition, research has also explored the potential of pyrimidine analogues, such as 5-fluorouracil, for modulating plant growth and stress tolerance. These compounds may inhibit cell division and alter gene expression, leading to plant growth and stress-response changes (Kafer *et al.*, 2004; Delgado-García *et al.*, 2021). According to research, uridine supplementation will improve the plant's ability to withstand stress, particularly during abiotic stresses like drought, excessive salt, and high temperatures (Moffatt and Ashihara, 2002; Nygaard and Saxild, 2009). The exact mechanisms by which uridine improves stress tolerance have to be studied; meanwhile, it has been found to be involved in the modulation of gene expression, the regulation of cellular energy balance, and the protection of

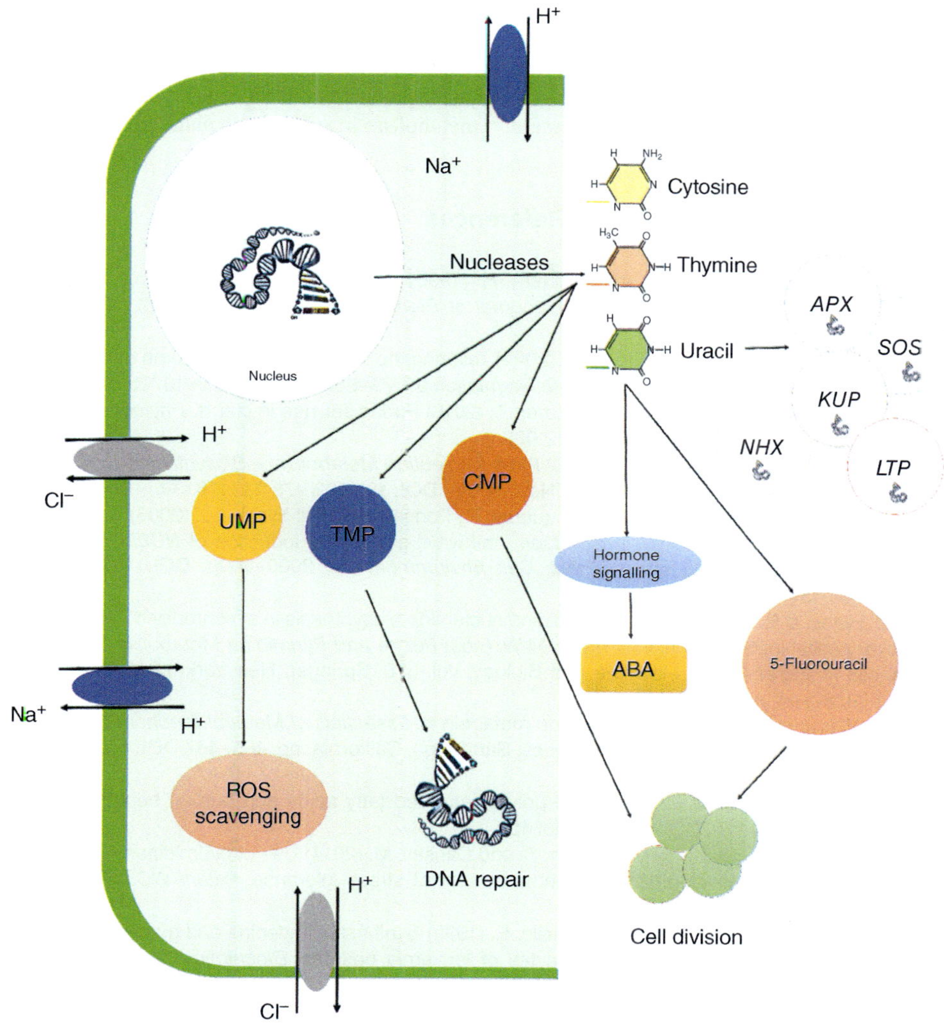

Fig. 2.3. Salvage of pyrimidines in the tolerance of salt stress in plants. The activation of stress-responsive transcription factors in the nucleus leads to increased expression of genes involved in the biosynthesis of nucleotides, including: UMP (uridine monophosphate), which can act as an antioxidant and scavenger of ROS (reactive oxygen species) in plants; TMP (thymidine monophosphate) that is used as a building block for DNA synthesis and repair, enabling plants to repair damaged DNA; CMP (cytidine monophosphate), a critical building block for nucleic acid synthesis during cell division, enabling the formation of new cells and supporting plant growth; and uracil, which is involved in the biosynthesis of pyrimidine nucleotides and pyrimidines can also be converted to ABA (abscisic acid) through a series of biochemical reactions (*LTP*, *NHX*, *APX*, *KUP*, *SOS* = stress-responsive genes). (Authors' own figure.)

cellular membranes. However, the use of these compounds is still in its early stages and requires further investigation to understand their mechanisms of action and potential applications in agriculture.

2.7 Conclusions

In conclusion, regulating purine and pyrimidine nucleotides in plants under salt stress involves multiple mechanisms. The expression and activity of genes involved in the production and breakdown of purine and pyrimidine nucleotides are altered under salt stress. These changes help the plant to maintain optimal levels of solutes and activation of stress-responsive genes in adverse conditions and to ensure the proper functioning of various cellular processes. Further studies are necessary to understand better the detailed molecular mechanisms that regulate nucleotide metabolism in response to plant stress.

References

Akatsu, M., Hosoi, Y., Sasamoto, H. and Ashihara, H. (1996) Purine metabolism in cells of a mangrove plant, *Sonneratia alba*, in tissue culture. *Journal of Plant Physiology* 149, 133–137. DOI: 10.1016/S0176-1617(96)80185-4.

Anand, G., Goel, V., Dubey, S. and Sharma, S. (2021) Tailoring the rhizospheric microbiome of *Vigna radiata* by adaptation to salt stress. *Plant Growth Regulation* 93, 79–88. DOI: 10.1007/s10725-020-00667-4.

Ashihara, H., Stasolla, C., Fujimura, T. and Crozier, A. (2018) Purine salvage in plants. *Phytochemistry* 147, 89–124. DOI: 10.1016/j.phytochem.2017.12.008.

Ashihara, H., Ludwig, I.A. and Crozier, A. (2020) *Plant Nucleotide Metabolism – Biosynthesis, Degradation, and Alkaloid Formation*. Wiley, Hoboken, New Jersey. DOI: 10.1002/9781119476139.

Barrero, J.M., Rodriguez, P.L., Quesada, V., Piqueras, P., Ponce, M.R. and Micol, J.L. (2006) Both abscisic acid (ABA)-dependent and ABA-independent pathways govern the induction of *NCED3*, *AAO3* and *ABA1* in response to salt stress. *Plant, Cell Environment* 29, 2000–2008. DOI: 10.1111/j.1365-3040.2006.01576.x.

Bein, K. and Evans, D.R. (1995) *De novo* pyrimidine nucleotide biosynthesis in synchronized Chinese hamster ovary cells. In: Sahota, A. and Taylor, M.W. (eds) *Purine and Pyrimidine Metabolism in Man VIII*. Advances in Experimental Medicine and Biology, Vol. 370. Springer, New York, pp. 545–548. DOI: 10.1007/978-1-4615-2584-4_115.

Bhagavan, N.V. and Ha, C.-E. (2015) Nucleotide metabolism. *Essentials of Medical Biochemistry, Second Edition: With Clinical Cases*. Academic Press, San Diego, California, pp. 465–487. DOI: 10.1016/B978-0-12-416687-5.00025-7.

Calder, P.C. and Yaqoob, P. (2009) Omega-3 polyunsaturated fatty acids and human health outcomes. *BioFactors* 35, 266–272. DOI: 10.1002/biof.42.

Cansev, A., Gülen, H., Kesici Zengin, M., Ergin, S. and Cansev, M. (2014) Use of pyrimidines in stimulation of plant growth and development and enhancement of stress tolerance. *Patent WO2014129996A1* (28 August 2014).

Carlucci, F., Tabucchi, A., Pagani, R. and Marinello, E. (1999) Synthesis of adenine and guanine nucleotides at the 'inosinic branch point' in lymphocytes of leukemia patients. *Biochimica Biophysica Acta – Molecular Basis of Disease* 1454, 106–114. DOI: 10.1016/S0925-4439(99)00032-0.

Caruso, G., Cavaliere, C., Guarino, C., Gubbiotti, R., Foglia, P. and Laganà, A. (2008) Identification of changes in *Triticum durum* L. leaf proteome in response to salt stress by two-dimensional electrophoresis and MALDI-TOF mass spectrometry. *Analytical Bioanalytical Chemistry* 391, 381–390. DOI: 10.1007/s00216-008-2008-x.

Chen, H., Zhang, B., Hicks, L.M. and Xiong, L. (2011) A nucleotide metabolite controls stress-responsive gene expression and plant development. *PLoS ONE* 6, e26661. DOI: 10.1371/journal.pone.0026661.

Chen, S., Tomchick, D.R., Wolle, D., Hu, P., Smith, J.L. *et al.* (1997) Mechanism of the synergistic end-product regulation of *Bacillus subtilis* glutamine phosphoribosylpyrophosphate amidotransferase by nucleotides. *Biochemistry* 36, 10718–10726. DOI: 10.1021/bi9711893.

Chou, T.F., Bieganowski, P., Shilinski, K., Cheng, J., Brenner, C. and Wagner, C.R. (2005) ^{31}P NMR and genetic analysis establish hinT as the only *Escherchia coli* purine nucleoside phosphoramidase and as essential for growth under high salt conditions. *Journal of Biological Chemistry* 280, 15356–15361. DOI: 10.1074/JBC.M500434200.

Cui, X., Cao, Y., Zhang, H. and Zhang, L. (2023) A *Picea wilsonii* NAC transcription factor, PwNAC1, interacts with RNA-binding protein PwRBP1 and synergistically enhances drought and salt tolerance of transgenic *Arabidopsis*. *Environmental and Experimental Botany* 206, 105174. DOI: 10.1016/j.envexpbot.2022.105174.

Decros, G., Beauvoit, B., Colombié, S., Cabasson, C., Bernillon, S. *et al.* (2019) Regulation of pyridine nucleotide metabolism during tomato fruit development through transcript and protein profiling. *Frontiers in Plant Science* 10, 1201. DOI: 10.3389/fpls.2019.01201.

Delgado, C., Mora-Poblete, F., Ahmar, S., Chen, J.-T. and Figueroa, C.R. (2021) Jasmonates and plant salt stress: molecular players, physiological effects, and improving tolerance by using genome-associated tools. *International Journal of Molecular Sciences* 22, 3082. DOI: 10.3390/ijms22063082.
Delgado-García, E., Piedras, P., Gómez-Baena, G., García-Magdaleno, I.M., Pineda, M. and Gálvez-Valdivieso, G. (2021) Nucleoside metabolism is induced in common bean during early seedling development. *Frontiers in Plant Science* 12, 496. DOI: 10.3389/fpls.2021.651015.
Dooki, A.D., Mayer-Posner, F.J., Askari, H., Zaiee, A. and Salekdeh, G.H. (2006) Proteomic responses of rice young panicles to salinity. *Proteomics* 6, 6498–6507. DOI: 10.1002/pmic.200600367.
Duszyn, M., Świeżawska, B., Szmidt-Jaworska, A. and Jaworski, K. (2019) Cyclic nucleotide gated channels (CNGCs) in plant signalling – current knowledge and perspectives. *Journal of Plant Physiology* 241, 153035. DOI: 10.1016/j.jplph.2019.153035.
ElSayed, A.I., Rafudeen, M.S., Ganie, S.A., Hossain, M.S. and Gomaa, A.M. (2022) Seed priming with cypress leaf extract enhances photosynthesis and antioxidative defense in zucchini seedlings under salt stress. *Scientia Horticulturae* 293, 110707.
Engelking, L.R. (2015) Nucleotides. *Textbook of Veterinary Physiological Chemistry*, 3rd edn. Academic Press, London, pp. 78–82. DOI: 10.1016/B978-0-12-391909-0.50013-X.
Fan, Q.-J., Yan, F.-X., Qiao, G., Zhang, B.-X. and Wen, X.-P. (2014) Identification of differentially-expressed genes potentially implicated in drought response in pitaya (*Hylocereus undatus*) by suppression subtractive hybridization and cDNA microarray analysis. *Gene* 533, 322–331. DOI: 10.1016/j.gene.2013.08.098.
Frischknecht, P.M. and Baumann, T.W. (1985) Stress induced formation of purine alkaloids in plant tissue culture of *Coffea arabica*. *Phytochemistry* 24, 2255–2257. DOI: 10.1016/S0031-9422(00)83020-4.
Ganie, S.A. (2020) RNA chaperones: potential candidates for engineering salt tolerance in rice. *Crop Science* 60(2), 530–540.
Ganie, S.A. (2021) Amino acids other than proline and their participation in abiotic stress tolerance. In: Wani, S.H., Gangola, M.P. and Ramadoss, B.R. (eds) *Compatible Solutes Engineering for Crop Plants Facing Climate Change*. Springer, Cham, Switzerland, pp. 47–96.
Ganie, S.A., Debnath, A.B., Gumi, A.M. and Mondal, T.K. (2017) Comprehensive survey and evolutionary analysis of genome-wide miRNA genes from ten diploid *Oryza* species. *BMC Genomics* 18(1), 711.
Gupta, R. and Gupta, N. (2021) Nucleotide biosynthesis and regulation. *Fundamentals of Bacterial Physiology and Metabolism*. Springer, Singapore, pp. 525–554. DOI: 10.1007/978-981-16-0723-3_19.
He, J., Zhuang, Y., Li, C., Sun, X., Zhao, S. *et al.* (2021) SIMP1 modulates salt tolerance by elevating ERAD efficiency through UMP1A-mediated proteasome maturation in plants. *New Phytologist* 232, 625–641. DOI: 10.1111/nph.17628.
Hove-Jensen, B., Andersen, K.R., Kilstrup, M., Martinussen, J., Switzer, R.L. and Willemoës, M. (2017) Phosphoribosyl diphosphate (PRPP): biosynthesis, enzymology, utilization, and metabolic significance. *Microbiology and Molecular Biology Reviews* 81, e00040-16. DOI: 10.1128/MMBR.00040-16.
Jha, S.K. and Pandey, G.K. (2021) Role of cyclic nucleotide-gated channels in stress and development in plants. In: Upadhyay, S.K. (ed.) *Calcium Transport Elements in Plants*. Academic Press, London, pp. 193–213. DOI: 10.1016/B978-0-12-821792-4.00016-3.
Jiménez, A., Santos, M.A., Pompejus, M. and Revuelta, J.L. (2005) Metabolic engineering of the purine pathway for riboflavin production in *Ashbya gossypii*. *Applied and Environmental Microbiology* 71, 5743–5751. DOI: 10.1128/AEM.71.10.5743-5751.2005.
Kafer, C., Zhou, L., Santoso, D., Guirgis, A., Weers, B. *et al.* (2004) Regulation of pyrimidine metabolism in plants. *Frontiers in Bioscience* 9, 1611–1625. DOI: 10.2741/1349.
Kamran, M., Parveen, A., Ahmar, S., Malik, Z., Hussain, S. *et al.* (2019) An overview of hazardous impacts of soil salinity in crops, tolerance mechanisms, and amelioration through selenium supplementation. *International Journal of Molecular Sciences* 21, 148. DOI: 10.3390/ijms21010148.
Katahira, R. and Ashihara, H. (2002) Profiles of pyrimidine biosynthesis, salvage and degradation in disks of potato (*Solanum tuberosum* L.) tubers. *Planta* 215, 821–828. DOI: 10.1007/s00425-002-0806-5.
Lescano, C.I., Martini, C., González, C.A. and Desimone, M. (2016) Allantoin accumulation mediated by allantoinase downregulation and transport by ureide permease 5 confers salt stress tolerance to *Arabidopsis* plants. *Plant Molecular Biology* 91, 581–595. DOI: 10.1007/s11103-016-0490-7.
Li, J., Zhou, H., Zhang, Y., Li, Z., Yang, Y. and Guo, Y. (2020) The GSK3-like kinase BIN2 is a molecular switch between the salt stress response and growth recovery in *Arabidopsis thaliana*. *Developmental Cell* 55, 367–380.e6. DOI: 10.1016/j.devcel.2020.08.005.
Meek, T.D. (2022) An overview of steady-state enzyme kinetics. In: Kenakin, T. (ed.) *Comprehensive Pharmacology*. Elsevier, Amsterdam, pp. 65–117. DOI: 10.1016/B978-0-12-820472-6.00124-9.

Miras-Moreno, B., Zhang, L., Senizza, B. and Lucini, L. (2021) A metabolomics insight into the cyclic nucleotide monophosphate signaling cascade in tomato under non-stress and salinity conditions. *Plant Science* 309, 110955. DOI: 10.1016/j.plantsci.2021.110955.

Mitchell, M., Pritchard, J., Okada, S., Larroque, O., Yulia, D. *et al.* (2017) Oil accumulation in transgenic potato tubers alters starch quality and nutritional profile. *Frontiers of Plant Science* 8, 554. DOI: 10.3389/fpls.2017.00554.

Moffatt, B.A. and Ashihara, H. (2002) Purine and pyrimidine nucleotide synthesis and metabolism. *Arabidopsis Book* 1, e0018. DOI: 10.1199/tab.0018.

Mostek, A., Börner, A., Badowiec, A. and Weidner, S. (2015) Alterations in root proteome of salt-sensitive and tolerant barley lines under salt stress conditions. *Journal of Plant Physiology* 174, 166–176. DOI: 10.1016/j.jplph.2014.08.020.

Mukarram, M., Petrik, P., Mushtaq, Z., Khan, M.M.A., Gulfishan, M. and Lux, A. (2022) Silicon nanoparticles in higher plants: uptake, action, stress tolerance, and crosstalk with phytohormones, antioxidants, and other signalling molecules. *Environmental Pollution* 310, 119855. DOI: 10.1016/j.envpol.2022.119855.

Naik, P. (2010) Nucleic acid metabolism. In: *Biochemistry*. Jaypee Brothers Medical Publishers (P) Ltd, New Delhi, p. 449. Available at: https://www.jaypeedigital.com/book/9788184487077/chapter/ch19 (accessed 5 March 2024).

Nassar, R.M.A., Kamel, H.A., Ghoniem, A.E., Alarcón, J.J., Sekara, A. *et al.* (2020) Physiological and anatomical mechanisms in wheat to cope with salt stress induced by seawater. *Plants* 9, 237. DOI: 10.3390/PLANTS9020237.

Nygaard, P. and Saxild, H.H. (2009) Nucleotide metabolism. In: Schaechter, M. (ed.) *Encyclopedia of Microbiology*, 3rd edn. Academic Press, San Diego, California, pp. 296–307. DOI: 10.1016/B978-012373944-5.00082-1.

Ohta, M., Guo, Y., Halfter, U. and Zhu, J.-K. (2003) A novel domain in the protein kinase SOS2 mediates interaction with the protein phosphatase 2C ABI2. *Proceedings of the National Academy of Sciences USA* 100, 11771–11776. DOI: 10.1073/pnas.2034853100.

Peterson, T.A., Nieman, R.H. and Clark, R.A. (1987) Nucleotide metabolism in salt-stressed *Zea mays* L. root tips. *Plant Physiology* 85, 984–989. DOI: 10.1104/pp.85.4.984.

Peterson, T.A., Lovatt, C.J. and Nieman, R.H. (1988) Salt stress causes acceleration of purine catabolism and inhibition of pyrimidine salvage in *Zea mays* root tips 1. *Journal of Experimental Botany* 39, 1389–1395. DOI: 10.1093/jxb/39.10.1389.

Pietrowska-Borek, M., Wojdyła-Mamoń, A., Dobrogojski, J., Młynarska-Cieślak, A., Baranowski, M.R. *et al.* (2020) Purine and pyrimidine dinucleoside polyphosphates differentially affect the phenylpropanoid pathway in *Vitis vinifera* L. cv. Monastrell suspension cultured cells. *Plant Physiology and Biochemistry* 147, 125–132. DOI: 10.1016/j.plaphy.2019.12.015.

Qureshi, M.I., Qadir, S. and Zolla, L. (2007) Proteomics-based dissection of stress-responsive pathways in plants. *Journal of Plant Physiology* 164, 1239–1260. DOI: 10.1016/j.jplph.2007.01.013.

Rubio, R. and Berne, R.M. (1980) Localization of purine and pyrimidine nucleoside phosphorylases in heart, kidney, and liver. *American Journal of Physiology – Heart and Circulatory Physiology* 239, H721–H730. DOI: 10.1152/ajpheart.1980.239.6.H721.

Schröder, M., Giermann, N. and Zrenner, R. (2005) Functional analysis of the pyrimidine *de novo* synthesis pathway in solanaceous species. *Plant Physiology* 138, 1926–1938. DOI: 10.1104/pp.105.063693.

Senecoff, J.F., McKinney, E.C. and Meagher, R.B. (1996) *De novo* purine synthesis in *Arabidopsis thaliana*. II. The PUR7 gene encoding 5′-phosphoribosyl-4-(N-succinocarboxamide)-5-aminoimidazole synthetase is expressed in rapidly dividing tissues. *Plant Physiology* 112, 905–917. DOI: 10.1104/pp.112.3.905.

Stanbury, P.F., Whitaker, A. and Hall, S.J. (2017) The isolation and improvement of industrially important microorganisms. *Principles of Fermentation Technology*, 3rd edn. Butterworth-Heinemann, Oxford, pp. 75–211. DOI: 10.1016/B978-0-08-099953-1.00003-X.

Stasolla, C., Katahira, R., Thorpe, T.A. and Ashihara, H. (2003) Purine and pyrimidine nucleotide metabolism in higher plants. *Journal of Plant Physiology* 160, 1271–1295. DOI: 10.1078/0176-1617-01169.

Sullivan, L.B., Gui, D.Y., Hosios, A.M., Bush, L.N., Freinkman, E. and Vander Heiden, M.G. (2015) Supporting aspartate biosynthesis is an essential function of respiration in proliferating cells. *Cell* 162, 552–563. DOI: 10.1016/j.cell.2015.07.017.

Sun, R.Z., Liu, J., Wang, Y.Y. and Deng, X. (2021) DNA methylation-mediated modulation of rapid desiccation tolerance acquisition and dehydration stress memory in the resurrection plant *Boea hygrometrica*. *PLoS Genetics* 17, e1009549. DOI: 0.1371/journal.pgen.1009549.

Suzuki-Yamamoto, M., Mimura, T. and Ashihara, H. (2006) Effect of short-term salt stress on the metabolic profiles of pyrimidine, purine and pyridine nucleotides in cultured cells of the mangrove tree, *Bruguiera sexangula*. *Physiologia Plantarum* 128, 405–414. DOI: 10.1111/j.1399-3054.2006.00753.X.

Takagi, H., Ishiga, Y., Watanabe, S., Konishi, T., Egusa, M. *et al.* (2016) Allantoin, a stress-related purine metabolite, can activate jasmonate signaling in a MYC2-regulated and abscisic acid-dependent manner. *Journal of Experimental Botany* 67, 2519–2532. DOI: 10.1093/jxb/erw071.

Tan, H.C., Hsu, J.W., Tai, E.S., Chacko, S., Wu, V. *et al.* (2022) *De novo* glycine synthesis is reduced in adults with morbid obesity and increases following bariatric surgery. *Frontiers in Endocrinology (Lausanne)* 13, 1135. DOI: 10.3389/fendo.2022.900343.

Wang, W., Xing, L., Xu, K., Ji, D., Xu, Y. *et al.* (2020) Salt stress-induced H_2O_2 and Ca^{2+} mediate K^+/Na^+ homeostasis in *Pyropia haitanensis*. *Journal of Applied Phycology* 32, 4199–4210. DOI: 10.1007/s10811-020-02284-0.

Watanabe, S., Matsumoto, M., Hakomori, Y., Takagi, H., Shimada, H. and Sakamoto, A. (2014) The purine metabolite allantoin enhances abiotic stress tolerance through synergistic activation of abscisic acid metabolism. *Plant, Cell & Environment* 37, 1022–1036. DOI: 10.1111/pce.12218.

Werner, A.K. and Witte, C.P. (2011) The biochemistry of nitrogen mobilization: purine ring catabolism. *Trends in Plant Science* 16, 381–387. DOI: 10.1016/j.tplants.2011.03.012.

Witte, C.-P. and Herde, M. (2019) Nucleotide metabolism in plants. *Plant Physiology* 182, 63–78. DOI: 10.1104/pp.19.00955.

Witz, S., Jung, B., Fürst, S. and Möhlmann, T. (2012) *De novo* pyrimidine nucleotide synthesis mainly occurs outside of plastids, but a previously undiscovered nucleobase importer provides substrates for the essential salvage pathway in *Arabidopsis*. *The Plant Cell* 24, 1549–1559. DOI: 10.1105/tpc.112.096743.

Yan, H., Nie, Y., Cui, K. and Sun, J. (2022) Integrative transcriptome and metabolome profiles reveal common and unique pathways involved in seed initial imbibition under artificial and natural salt stresses during germination of halophyte quinoa. *Frontiers in Plant Science* 13, 853326. DOI: 10.3389/fpls.2022.853326.

Yang, Y., Wu, Y., Ma, L., Yang, Z., Dong, Q. *et al.* (2019) The Ca^{2+} sensor SCaBP3/CBL7 modulates plasma membrane H^+-ATPase activity and promotes alkali tolerance in *Arabidopsis*. *The Plant Cell* 31, 1367–1384. DOI: 10.1105/tpc.18.00568.

Zhang, P., Wang, X., Lu, Q., Zhang, H., Chen, J. *et al.* (2023) Allantoin, a purine metabolite, confers saline–alkaline tolerance to sugar beet by triggering a self-amplifying feedback loop comprising jasmonic acid and nitric oxide. *Environmental and Experimental Botany* 206, 105172. DOI: 10.1016/j.envexpbot.2022.105172.

Zrenner, R., Stitt, M., Sonnewald, U. and Boldt, R. (2006) Pyrimidine and purine biosynthesis and degradation in plants. *Annual Review of Plant Biology* 57, 805–836. DOI: 10.1146/annurev.arplant.57.032905.105421.

3 Role of DNA Replication Proteins in Salinity Tolerance of Plants

Abubakar Mohammad Gumi[1]* and Usman Rabi'u Bello[2]

[1]Department of Plant Science, Faculty of Chemical and Life Sciences, Usmanu Danfodiyo University, Sokoto, Nigeria; [2]Department of Biology, Khadija University, Majia, Jigawa, Nigeria

Abstract

Salinity adversely affects plant growth and development, thereby limiting crop production globally. Several salt-responsive genes and proteins have so far been reported in plants and salt stress affects their expression machinery. Since expression of genes requires cellular nucleic acid metabolism, it is possible that DNA replication proteins such as helicases and polymerases are involved in salt-stress responses. DNA replication proteins (mostly helicases and polymerases) are among the largest and most highly conserved gene families in eukaryotes, including plants. The replication assembly (replisome) comprises an 11-subunit helicase, replicative DNA polymerases, sliding clamp proliferating cell nuclear antigen (PCNA), replication factor C (RFC), and replication protein A (RPA), also called single-strand DNA-binding protein, that coats and protects the single strand excised by the helicase. In this chapter, we highlight the model of DNA replication involving replisome complex and major DNA replication proteins with emphasis on their classification and roles in salinity-stress tolerance of plants. We broadly outline the various reported plant helicases (mostly members of the DEAD-box family), polymerases, and topoisomerases that are implicated to confer salinity tolerance in different plants and their molecular function during salt stress. Presently, the exact mechanisms of DNA replication protein-mediated salinity tolerance of plants are not well understood. However, we speculate that the role of DNA replication proteins in conferring salinity tolerance to plants largely depends on the ability of the replication protein complex to ensure continuous replication and transcription of stress-responsive genes during salinity episodes which will in turn increase the expression levels of salt-tolerance genes in plants.

3.1 Introduction

Due to the sessile nature of plants, salt stress or salinity is one of the major abiotic constraints that limit their growth and productivity (Gupta *et al.*, 2013). Generally, most plants are relatively susceptible to soil salinity (commonly called glycophytes) and NaCl is the major salt that affects the physiology of plants (Ganie, 2020). In the last two decades, research efforts in the field of plant biology have critically examined the effects of salinity on genome structure and transcriptome in plants. High concentrations of Na^+ in soil have adverse effects on plants such as disruption of ion homeostasis, membrane dysfunction, and inhibition of metabolic activity (Hasegawa *et al.*, 2000; ElSayed *et al.*, 2022). Salt stress induces cellular accumulation of reactive oxygen species (ROS) such as O_2, H_2O_2, O_2^-, and OH^- that are highly reactive molecules capable of causing

*Email: gumi.abubakar@udusok.edu.ng; muhammadag@yahoo.co.uk

DOI: 10.1079/9781800623033.0003

oxidative damage to proteins, DNA, and lipids (Apel and Hirt, 2004; Mansour *et al.*, 2005; Eyidogan and Oz, 2007; Miller *et al.*, 2010). However, ROS also act as important signalling molecules that regulate many important processes in plants such as stress responses, growth, and development (Foreman *et al.*, 2003; Ganie *et al.*, 2017). Salinity along with other stresses may induce DNA damage via various mechanisms, resulting in aberrant DNA replication and loss of genome integrity, potentially affecting the growth and survival of plants (Fu *et al.*, 2016; Nisa *et al.*, 2019).

DNA encodes all the genetic information (in the form of codes) necessary for the normal growth and development of a cell. During cell division, these codes first have to be faithfully replicated correctly to be transmitted to the daughter cells. The replication of a DNA molecule begins at special sites called origins of replication, which are short stretches of DNA having a specific sequence of nucleotides (Campbell *et al.*, 2008). The process of protein synthesis in plants is vital and plays an important role in abiotic stress adaptation and tolerance mechanisms but salt stress generally represses protein synthesis (Tuteja, 2007). Studies in proteomics have shown that transcription factors (such as CRT/DRE, bHLH, BTF3, and NAC) and transcription- related proteins (like DNA polymerases, topoisomerases, and helicases) are responsive to salt stress and play a pivotal role in salinity tolerance of plants (Zhang *et al.*, 2012). However, numerous proteins are enhanced by salt treatment, indicating that normal cellular processes require the maintenance of protein synthesis activities under salt stress. Several proteins and molecules (protein complex), called the replisome, are responsible for this replication process (Waga and Stillman, 1998). Many proteins accumulate in plants in response to salinity stress. Plant tissues normally respond to salt stress by degrading proteins or producing abundant salt stress-related proteins (Wang, L *et al.*, 2015). Moreover, more proteins have been observed in salt-tolerant cultivars than salt-sensitive cultivars of many crops, including sunflower, barley, rice, and wheat. Thus the quality and type of proteins may be more important than the quantity of total soluble proteins for salt tolerance (Dissanayake *et al.*, 2022).

3.2 Salinity Response and Tolerance Mechanisms in Plants

During a salt-stress episode, Na^+ (a micronutrient in plants) accumulates in the soil and its excessive transport from roots to leaves affects the growth and development of plants. The entry of NaCl into the root cells, its symplasmic transport past the Casparian band, and its transfer into the transpiration stream are the primary steps of salt accumulation in plants. The pathways for the uptake of Na^+ into the root symplasmic space have not yet been established in much detail. It has been generally assumed that non-selective K^+ ion channels allow entry into the cell (Schatchman *et al.*, 1991). Many reports have shown that the Na^+ in the soil enters plant roots via the symplastic pathway (mediated by cation channels/transporters) and the apoplastic pathway through the transpiration stream (Gao *et al.*, 2007). The apoplastic pathway bypasses symplastic control over Na^+ transport by the transpirational bypass flow. Moreover, a major part of Na^+ influx depends on leakage of the apoplastic pathway (Yadav *et al.*, 1996; Garcia *et al.*, 1997). However, more studies suggest that Na^+ is taken up with K^+ by a high-affinity K^+ uptake carrier (Schatchman and Schoeder, 1994). Salinity imposes two main stresses on plant tissues: (i) an osmotic stress due to the relatively high solute concentration in the soil; and (ii) ion-specific stresses resulting mostly from altered K^+/Na^+ ratios and Na^+ and Cl^- concentrations that are adverse to plant physiology.

The lower threshold limit of saline soils is electrical conductivity (EC) of 4 dS/m but most plants show stress symptoms and reduced growth even at EC lower than 4 dS/m (Maas, 1990). Plants resort to adaptive strategies to cope with the consequent effects of salinity, including accumulation of compatible osmolytes (proline, glycine betaine, and other proteins) and cytotoxic-ion sequestration using antiporters (Arefian *et al.*, 2019). Numerous studies have shown that protein modification at both quantitative and qualitative levels is an essential process that critically influences salinity tolerance in plants (Seki *et al.*, 2002; Subba *et al.*, 2013). Plants are either dormant during salt stress or there must be cellular adjustment to tolerate the saline environment (Ganie, 2021). Cellular adjustment mechanisms can be categorized as those that function to minimize osmotic stress or ion disequilibrium or to alleviate the consequent secondary effect caused by these stresses. The chemical potential of the saline solution initially establishes a water potential imbalance between the apoplast and symplast that leads to a turgor decrease which, if severe enough, can cause growth reduction (Bohnert *et al.*, 1995).

In the last decade, studies in proteomics and metabolomics have revealed many important components of the protein synthesis machinery that are expressed under salt-stress conditions, including different ribosomal proteins, translation initiation factors, translation elongation factors, polyA-binding proteins, translationally controlled tumour proteins, RNA recognition motif (RRM)-containing proteins, and transfer RNA (tRNA) synthases (Jiang *et al.*, 2007; Pang *et al.*, 2010; Sobhanian *et al.*, 2010; Bandehagh *et al.*, 2011; Chattopadhyay *et al.*, 2011). However, recent studies have established that plants accumulate different types of soluble proteins as a vital strategy to combat the consequent effects of salinity (Athar *et al.*, 2022). In addition to many other physiological mechanisms employed by plants to survive under salinity, proteins are a more reliable determinant of salt tolerance than simple gene expression as they play major roles in shaping physiological traits in salt-tolerant phenotypes. However, little information is available on the salt stress-responsive replisome and its possible roles in conferring salinity tolerance in plants.

3.3 Replication Fork and Replisome Assembly – An Overview

The mechanisms of DNA replication are highly conserved across the three kingdoms of life as originally hypothesized by Watson and Crick using their semi-conservative model of DNA replication (Watson and Crick, 1953; Burgers and Kunkel, 2017). The replisome (complex replication machinery) is assembled in the nuclear replication factories to form a Y-shaped region where the DNA strands are being unwound, called a replication fork (Hozak *et al.*, 1993). In eukaryotes, the initiation of DNA replication occurs at multiple individual replication origins, involving the recruitment of the origin recognition complex (ORC), multiple other proteins, and the attachment of the helicase to the double-stranded DNA (dsDNA) molecule, and together forming the pre-replicative complex (pre-RC) of DNA replication (Bell and Labib, 2016; Bleichert *et al.*, 2017). Generally, during DNA replication, the replisome assembly takes place in four progressive stages:

1. The attachment of the double-hexamer minichromosome maintenance (MCM) helicase at replication origin sites marked by the ORC.
2. Loading of other accessory molecules to form a preinitiation complex.
3. Rearrangement of the MCM complex from inactive to active helicase.
4. Priming of DNA replication.

The replisome comprises an 11-subunit helicase, replicative DNA polymerases, sliding clamp proliferating cell nuclear antigen (PCNA), replication factor C (RFC), and replication protein A (RPA), also called single-strand DNA-binding protein, that coats and protect the single strand excised by the helicase. The process of replication begins with the unwinding activity of helicase brought about by the MCM2–7 (minichromosome maintenance proteins 2–7) hetero hexamer on the template DNA. The MCM complex needs to be associated with GINS (four proteins Sld5-Psf1-Psf2-Psf3 referred to go-ichi-ni-san in Japanese for 5-1-2-3) and CDC45. Altogether, these three complexes (CDC45, MCM, and GINS) form the CMG complex (Yang *et al.*, 2019). Despite the availability of numerous DNA polymerases (DNA pols) in eukaryotes, only three (DNA pols α, δ, and ε) are responsible for genome duplication; in addition, pol δ and pol ε perform the bulk of the replication process while pol α initiates the process via replication priming (Lujan *et al.*, 2016). After the unwinding of the replication fork by the CMG complex, the pol α initiates the leading-strand synthesis by producing an RNA/DNA primer.

Although it is highly debated, numerous reports have shown that pol δ is responsible for elongation of primers on both leading and lagging strands (Pursell *et al.*, 2007; Johnson *et al.*, 2015; Burgers *et al.*, 2016; Garbacz *et al.*, 2018) and that an occasional switch to pol ε on the leading strand could be induced by replication errors (Zhou *et al.*, 2019). The commonly accepted model is that pol δ performs initiation and termination on both strands, as well as the synthesis of Okazaki fragments, whereas pol ε elongates only the leading strand (Pedroza-Garcia *et al.*, 2019). In general, the genomes of bacteria and eukaryotes are replicated by specific replication enzymes such as helicases, polymerases, and DNA primase. However, other enzymes related to replication, like DNA topoisomerase, single-stranded DNA maintenance protein, DNA ligase, primer removal enzyme, and several DNA recombination-related enzymes, have been identified (Gualberto *et al.*, 2014).

3.4 The MCM Complex Helicase – Role in DNA Replication

During replication, the cellular DNA is double stranded and the DNA polymerases require only a single-stranded DNA (ssDNA) template to act upon, yet lack dsDNA-unwinding activity (Kawasaki and Sugino, 2001). However, the ssDNA is generated by the replication helicase, a motor complex protein that uses nucleotide triphosphate (NTP)-dependent conformational changes to unwind duplex DNA (Patel and Picha, 2000). A number of studies *in vivo* over the last 15 years have implicated the MMC complex (MCM2–7) either alone or in combination with additional activators (like CDC45 and the GIN complex) as the eukaryotic replicative helicase (Moyer *et al.*, 2006).

The MCM helicase is a part of a large complex named the pre-RC and belongs to the Superfamily 6 helicases (MCM2, MCM3, MCM4, MCM5, MCM6, and MCM7) with each subunit containing an AAA+ domain (Bell and Kaguni, 2013). Within the complex, each subunit has its individual role that helps in the process of accurate replication. The MCM2 helps in loading the complex on to the chromatin through interplay with histones, favouring their deposition on chromatin. Also in association with subunit MCM5, they function as a gate by hydrolysing ATP to switch between open and closed state to mediate loading of the entire complex. Subunits MCM4, MCM6, and MCM7 have a helicase activity that helps to unwind dsDNA in the 5′–3′ direction while the MCM3 functions as a surveillance factor for the complex integrity and MCM7 interacts with DNA damage checkpoint (Bell and Kaguni, 2013; Li and Xu, 2019).

In the pre-RC, the MCM2–7 helicase is inactive and unable to unwind the dsDNA. However, during the G1/S phase transition the DBF4-dependent kinase (DDK) and cyclin-dependent kinase (CDK) phosphorylate specific sites in the pre-RC, facilitating the recruitment of CDC45 and the GINS complex to form the CMG complex (CDC45–MCM–GINS), which has active replicative helicase. Additional factors cooperate with the active helicase to unwind the DNA and further recruit other replication factors, such as RFC, RPA, the sliding clamp PCNA, and multiple DNA polymerases, known collectively as the replisome (O'Donnell *et al.*, 2013). All the subunits use energy derived from ATP hydrolysis in order to bind each other, which allows the formation of this large multimeric complex (Fernandez-Cid *et al.*, 2013).

Generally, the replisome assembly in eukaryotes is composed of many protein complexes that associate to carry out the important task of duplicating a cell's genome. Among these proteins, the helicase MCM in association with other complexes forms the CMG complex that initiates the unwinding process and the three replicative polymerases (pols α, δ, and ε), with the help of the other complex, complete the replication process, thereby making helicase and polymerases an important component of replisome assembly.

3.5 The Helicase Unwinding Machinery that Initiates Replication

Generally, helicase enzymes act as a rotating handle that unwinds the dsDNA during replication in order to create multiple duplicate copies of the ssDNA template. They are motor proteins that catalyse stable duplex DNA and RNA unwinding using energy derived from the hydrolysis of NTPs (Patel and Donmez, 2006). They have a wide range of shapes ranging from hexameric (ring-shaped with six subunits), dimeric (two subunits) to monomeric (one subunit) with average molecular weight of 300 kDa (Takahashi *et al.*, 2019). The helicases contain conserved sequence motifs and are generally classified into six superfamilies (SF1–SF6) based on their motif sequence, with SF3–SF6 members forming a toroidal structure that is ring-like hexameric in nature while the others (SF1 and SF2) exist as monomeric (Yadav *et al.*, 2021). The monomeric SF1 and SF2 groups are known to contain the largest number of DNA and RNA helicases and are structurally characterized by the presence of a conserved helicase core consisting of two tandemly placed RecA domains (Singleton *et al.*, 2007; Jankowsky, 2011). All helicase domains have conserved amino acid residues involved in hydrolysis and NTP binding and an arginine residue that facilitates DNA-dependent ATPase energy to complete the unwinding process (Yang *et al.*, 2019).

3.5.1 Classification of helicase families

Generally, helicases share extensive structural similarity although considerable differences exist in their amino acid motif that forms the basis of their classification into six superfamilies (SF1–SF6) (Fairman-Williams *et al.*, 2010). This sequence-based classification of helicases covers DNA helicase, RNA helicase, and chromatin remodellers (Table 3.1). The biological function of each helicase is correlated with its structural assembly and motifs. The core domain of helicases has conserved amino acid residues involved in NTP binding and hydrolysis in addition to an arginine residue that aids in energy coupling for translocation and/or unwinding. Additionally, based on their structures, the superfamilies are divided into two categories: (i) those with monomeric structure (SF1 and SF2 members) consisting of two tandemly placed RecA1 and RecA2 domains; and (ii) another group (SF3–SF6) having a toroidal structure that is predominantly hexameric with RecA domains arranged in a hexagon shape (Singleton *et al.*, 2007; Jankowsky, 2011). Overall, the helicases found in eukaryotes are reported as members of the monomeric SF1 and SF2 and hexameric SF6, while those of prokaryotes (specifically bacteria and viruses) are members of SF3–SF5 (Sami *et al.*, 2021). The SF1 and SF2 groups are known to contain the largest number of DNA and RNA helicases.

Most helicases belong to the DEAD/DEAH-box protein family and share a core region (200–700 amino acids) of nine highly conserved motifs arranged in order from the N-amino terminal to the C-carboxyl terminal as Q, I, Ia, Ib, II, III, IV, V, and VI (Tuteja and Tuteja, 2004). The DEAD/H helicases contain the conserved amino acid sequence Asp-Glu-Ala-Asp (DEAD) or Asp-Glu-Ala-His (DEAH) in their motif II. These conserved motifs play a role in various activities like Mg^{2+} binding, unwinding dsDNA, nucleic acid binding, ATP binding, and hydrolysis (Table 3.2).

Table 3.1. General classification of the different helicase superfamilies with examples of some prominent members of each family.

Family	Structure	Motif	Prominent members	References
Helicase Superfamily 1 (SF1)	Monomeric; RecA1 linked to RecA2 domains	Q, I, Ia, Ib, Ic, II, III, IIIa, IV, V, Va, Vb and VI	PcrA, Pif1	Moreno-del Alamo *et al.* (2021); Yokota (2021)
Helicase Superfamily 2 (SF2)	Monomeric; RecA1 linked to RecA2 domains	Q, I, Ia, Ib, Ic, II, III, IV, IVa, V, Va, Vb and VI	DEAD-box, DEAH/RHA helicases	Yokota (2021)
Helicase Superfamily 3 (SF3)	Hexameric form of RecA motif; some members are double hexamers	I, II, B′, C and VI	Papillomavirus E1 helicases	Chojnacki and Melendy (2018); Brosh (2021)
Helicase Superfamily 4 (SF4)	Hexameric form	H1, H1a, H2, H3, H4 and VI	gp4 from bacteriophage T7	Brosh and Matson (2020)
Helicase Superfamily 5 (SF5)	Hexameric form	I, Ia, II and VI	*Escherichia coli* Rho factor	O'Donnell and Li (2018)
Helicase Superfamily 6 (SF6)	Hexameric form	I, II Sensor 1, VI and Sensor 2	MCM, RuvA, RuvB, RuvC	Mohapatra *et al.* (2023)

Table 3.2. The conserved sequences and functions of the different motifs found in the helicase superfamilies.

Class of motif	Conserved sequence	Function/role
Q	GFccPoSIQ	ATP binding and hydrolysis
I	AxxGxGKT	Mg^{2+} binding, ATP binding and hydrolysis
Ia	PTRELA	DNA/RNA binding
Ib	TPGR	DNA/RNA binding
II	DEAD/H	Mg^{2+} binding, ATP binding and hydrolysis
III	SAT	DNA/RNA unwinding
IV	FVNT	RNA binding
V	RGxD	RNA binding
VI	HRIGRxxR	ATP binding and hydrolysis, DNA/RNA unwinding

Despite reports on numerous DNA helicases present in prokaryotes and selected model organisms in eukaryotes (*Escherichia coli* – 14, bacteriophages – 6, viruses – 12, yeast – 15, humans – 34), only 11 biochemically active purified DNA helicases have been reported in plants (Tuteja and Tuteja, 2004). In addition, the exact biological roles of very few DNA helicases have been identified and little is known about DNA helicases in plants (Lohman and Bjornson, 1996; Tuteja and Tuteja, 2004). However, recent reports have speculated that the plant DEAD-box helicase plays a significant role in regulating environmental stresses.

3.5.2 Role of helicases in salinity

The cellular DNA of plants is constantly being damaged by various endogenous and exogenous factors. A wide variety of environmental stresses including salinity lead to the accumulation of endogenous ROS, thereby aggravating the genotoxic effects on plant cells and subsequently leading to oxidative damage in form of deletions, base modification, or double-strand breaks (Manova and Gruszka, 2015). Helicases have been proven to be an effective and vital component of replisome enzymes that play a major role in alleviating the consequence of oxidative damage caused by these stresses (Wiedemann *et al.*, 2018). The possible site of action for the helicase-mediated stress tolerance in plants could be at the level of transcription/translation to enhance protein synthesis or in association with DNA–multi-subunit protein complexes to alter gene expression. Reports have shown that the complex process of protein synthesis in eukaryotes is sensitive to salt stress and therefore factors and vital molecules involved in transcription and translation of proteins are potential targets of salinity stress (Tuteja *et al.*, 2012)

To date, numerous helicases have been identified in plants (mostly the *Arabidopsis* and rice models), most of which are DExD- or DEAH/RHA-box proteins. Many genes, such as *HKT* and *NHX*, encoding K^+ transporters and channels have been identified and cloned in various plant species. During salt stress, enhanced expression was found of some low-abundance transcripts that are involved in K^+ uptake (Esmaeili *et al.*, 2022). The DEAD-box helicases are members of the RNA helicase enzymes and can be considered as means to identify pathways involved in tolerance to multiple abiotic stresses (Chourasia *et al.*, 2021). Analysis of genes whose expression is induced under stress conditions is important to understand the mechanism of abiotic stress tolerance and possibly to use it for breeding stress-tolerant plants. However, the functions of helicases are poorly understood in plants (Aubourg *et al.*, 1999). Among all the sequenced genomes including those of human, fly, worm, and yeast, *Arabidopsis* has the largest number (94 stress-responsive helicases reported from TAIR, The Arabidopsis Information Resource) of DEAD/H-box helicase genes (Umate *et al.*, 2010).

The expression of stress-responsive genes, including the genes encoding for the helicases, is upregulated under the influence of various abiotic stresses. The overexpression of the helicase genes (listed in Table 3.3) has been reported to confer salt-stress tolerance in plants (Sanan-Mishra *et al.*, 2005; Kim *et al.*, 2008; Liu *et al.*, 2008; Luo *et al.*, 2009). Umate *et al.* (2010) have reported a number of gene products like MEE29, SNF2 domain-containing protein, RH55, CHR31, CHR9, EDA16, RH30, RH40, and RNA helicase DRH1 that showed high levels under salt stress. Several salinity-tolerant lines of crops like rice, maize, *Arabidopsis*, tomato, soybean, and cotton were developed using overexpression

Table 3.3. Summary of plant helicases obtained from different species and their role during salt stress.

No.	Gene	Plant species	Role in salt stress	References
1.	*Hordeum vulgare* DEAD-box 1 (*HvD1*)	Barley (*Hordeum vulgare*)	Regulates transcript(s) concerned with salt tolerance, important metabolism such as photosynthesis	Nakamura *et al.* (2004)
2.	Pea DNA helicase 45 (*PDH45*)	Pea (*Pisum sativum*)	Enhanced photosynthetic and antioxidant machinery, improved ROS scavenging, cell membrane stability	Pham *et al.* (2000); Gill *et al.* (2013)
3.	Pea DNA helicase 47 (*PDH47*)	Pea (*P. sativum*)	Reduced ROS production	Sanan-Mishra *et al.* (2005); Vashisht *et al.* (2005); Singha *et al.* (2017)
4.	Stress response repressor 1 (*STRS1*)	*Arabidopsis thaliana*	Mutations in either gene cause increased tolerance to salt, osmotic and heat stresses, suggesting that the helicases suppress responses to abiotic stress	Kant *et al.* (2007)
5.	Stress response repressor 2 (*STRS2*)	*A. thaliana*	Mutations in either gene cause increased tolerance to salt, osmotic and heat stresses, suggesting that the helicases suppress responses to abiotic stress	Kant *et al.* (2007)
6.	*Apocynum venetum* DEAD-box helicase 1 (*AvDH1*)	*Apocynum venetum*	ATP-dependent DNA helicase and ATP-independent RNA unwinding activities	Chen *et al.* (2016)
7.	*Medicago* helicase 1 (*MH1*)	Alfalfa (*Medicago sativa*)	Improved ROS scavenging under salt stress; the ectopic expression of *MH1* in *Arabidopsis* improved seed germination and plant growth under drought, salt and oxidative stress	Luo *et al.* (2009)
8.	*Glycine max* RNA helicase (*GmRH*)	Soybean (*Glycine max*)	RNA processing during salinity	Chung *et al.* (2009)
9.	*Pisum sativum* minichromosome maintenance 6 (*PsMCM6*)	Pea (*P. sativum*)	Enhanced yield stability under salinity	Tran *et al.* (2010); Dang *et al.* (2011)
10.	*Oryza sativa* suppressor of var 3 (*OsSUV3*)	Rice (*Oryza sativa*)	Improved photosynthetic and antioxidant machinery under salinity	Sahoo *et al.* (2014); Tuteja *et al.* (2013, 2014b)
11.	*Oryza sativa* HLA-B associate transcript 1 (*OsBAT1*)	Rice (*O. sativa*)	Improved ROS scavenging	Tuteja *et al.* (2015)
12.	*Pennisetum glaucum* eukaryotic initiation factor 4A (*PgeIF4A*)	*Pennisetum glaucum*	Reduced ROS accumulation and increased membrane stability	Rao *et al.* (2017)
13.	*p68*	*P. sativum*	Reduced oxidative stress, improved photosynthetic machinery, enhanced ROS scavenging	Tuteja *et al.* (2014a)

Continued

Table 3.3. Continued.

No.	Gene	Plant species	Role in salt stress	References
14.	*Solanum lycopersicum* DEAD-box 30 (*SlDEAD30*)	Tomato (*Solanum lycopersicum*)	Improved ROS scavenging	Zhu *et al*. (2015)
15.	*Solanum lycopersicum* DEAD-box 31 (*SlDEAD31*)	Tomato (*S. lycopersicum*)	Improved ROS scavenging	Zhu *et al*. (2015)
16.	*Arabidopsis thaliana* RNA helicase 9 (*AtRH9*)	*A. thaliana*	RNA chaperone activity	Kim *et al*. (2008)
17.	*Arabidopsis thaliana* RNA helicase 17 (*AtRH17*)	*A. thaliana*	RNA chaperone activity	Nguyen *et al*. (2018); Seok *et al*. (2020)
18.	*Arabidopsis thaliana* RNA helicase 25 (*AtRH25*)	*A. thaliana*	RNA chaperone activity	Kim *et al*. (2008)
19.	*Arachis hypogaea* RNA helicase 47 (*AhRH47*)	*Arachis hypogaea*	Enhanced protein synthesis	Umate *et al*. (2010)
20.	*Brassica rapa* RNA helicase 22 (*BrRH22*)	*Brassica rapa*	RNA chaperone activity within chloroplasts	Nawaz *et al*. (2018)
21.	*Arabidopsis thaliana* suppressor of var 3 (*AtSUV3*)	*A. thaliana*	Improved photosynthetic and antioxidant machinery under salinity	Gagliardi *et al*. (1999)
22.	*Arabidopsis* DEVH-box RNA helicase (*AtHELPS*)	*A. thaliana*	Active under K^+ deprivation stress	Xu *et al*. (2011)
23.	*Arabidopsis thaliana* eukaryotic initiation factor 4A-III (*AtEIF4A-III*)	*A. thaliana*	Enhanced photosynthetic and antioxidant machinery, improved ROS scavenging	Sami *et al*. (2021)
24.	*OsRuvBL1a*	*O. sativa*	Enhanced tolerance under *in vivo* salinity stress	Saifi *et al*. (2023)
25.	Translation initiation factor 2 (*TIF2*)	*Saccharomyces cerevisiae*	Involved in translation initiation	Montero-Lomeli *et al*. (2002)
26.	Rice *OsPDH45* similar to pea DNA helicase 45	*O. sativa*	Improved the antioxidant system and protected the photosynthetic machinery during salt stress	Gill *et al*. (2013)
27.	Wheat RNA helicase *TaDEAD-box57–3B*	*Triticum aestivum*	Enhanced germination rates, root lengths, fresh weights, and survival rates under salinity	Ru *et al*. (2021)

of selected RNA helicase genes. In rice, the suppressor of var-3 helicase gene (*SUV3*) was shown to improve the photosynthetic and antioxidant machinery of rice under salt stress (Tuteja *et al.*, 2013, 2014b) while the HLA-B associate transcript 1 gene (*OsBAT1*) increases ROS scavenging in salt-stress episodes, thereby conferring salinity tolerance (Tuteja *et al.*, 2015). The *Hordeum vulgare* DEAD-box protein-1 gene (*HvD1*) encoding a putative ATP-dependent DEAD-box RNA helicase was reported to accumulate its transcript under a salt-stress episode (Nakamura *et al.*, 2004). The HvD1 protein has a conserved helicase domain, five repeats of an RNA recognition motif (RGG), a hydrophilic C terminus, and is localized in the chloroplast of the cell.

Since MCM proteins are present in excess relative to origins of replication, it is therefore possible that these proteins may also serve other functions. Recently, the MCM6 single subunit of pea plant has been shown to play an important role in salinity tolerance through upregulation of its transcript driven by CAM-V 35S promoter in tobacco plants (Dang *et al.*, 2011). In the last few years, the role of DNA helicase in response to stress has been studied extensively both in plants and other non-plant systems (Briolat and Reysset, 2002; Liu *et al.*, 2002; Westermarck *et al.*, 2002). In addition, multiple stress-induced DNA helicases have been isolated from plants and characterized using molecular and biochemical approaches (Tuteja, 2003; Tuteja and Tuteja; 2004). The report by Dang *et al.* (2011) provides the first direct evidence that overexpression of single MCM6 subunit confers salinity tolerance in pea plant without any yield loss, although the exact mechanism by which MCM6 confers stress tolerance of the plant towards salinity is not known yet.

Tran *et al.* (2007) speculate that MCM6 in pea might function as a helicase that helps in unwinding the secondary structure of mRNA of stress-responsive genes and also acts as a transcription factor of the zinc finger-homeodomain (Zn-HD) family and interacts with other proteins (including ion transporters) that may be directly or indirectly involved in stress tolerance. It has been reported that reduction of MCM levels in cells could cause defects in genome stability (Liang *et al.*, 1999; Fitch *et al.*, 2003). Therefore, overexpression of MCM6 might help the DNA replication to work normally in the salinity-stress condition. It is interesting to note that MCM6 subunit forms an oligomer (without the involvement of other subunits) and functions as DNA helicase (Tran *et al.*, 2010). Thus in the stress condition, MCM6 subunit may also function separately by forming oligomers.

3.6 DNA Polymerases

DNA polymerases are widespread in almost all organisms and in conjunction with other proteins are involved in cell division, genome replication, and the maintenance of genetic integrity (Johansson and Dixon, 2013). The polymerases catalyse the phosphodiester bond formation between the terminal 3′-OH of the primer and the α-phosphate of the incoming triphosphate to extend the DNA strand based on the template DNA sequence (Daimon *et al.*, 2018). Living organisms (whether prokaryotes or eukaryotes) encode several DNA polymerases in their genome; to date, humans have the highest number (up to 16) of characterized DNA polymerase enzymes while *E. coli* has five and yeast (*Saccharomyces cerevisae*) has eight (Jain *et al.*, 2018). In addition, the model plant *Arabidopsis thaliana* was reported to possess ten DNA polymerase genes that encode enzymes which are orthologous to human polymerases (Garcia-Diaz and Bebenek, 2007).

On the basis of their tertiary structures and conserved sequences, DNA polymerases are generally classified into seven subfamilies, namely A, B, C, D, X, Y, and reverse transcriptase (RT) (Rothwell and Waksman, 2005). Despite the availability of numerous DNA polymerases, there exist only three replicative polymerases (responsible for DNA replication or genome duplication), namely DNA pols α, δ, and ε, all belonging to the subfamily B category. The non-replicative polymerases are widespread across the seven subfamilies and are involved in different DNA repair pathways. To date, six members of the subfamily X polymerases (pol β, pol λ, pol μ, terminal deoxynucleotidyl transferase (Tdt), pol σ_1, and pol σ_2) have been identified from eukaryotes and are believed to be highly conserved among most organisms from bacteria to humans (Ramadan *et al.*, 2004; Uchiyama *et al.*, 2009). Despite their conserved nature, pol λ is the only member of the subfamily X present in plant systems (Amoroso *et al.*, 2011; Roy *et al.*, 2011).

The replicative polymerases are characterized by conserved amino acid motifs within their catalytic sites and exonuclease domain. In the model plant *Arabidopsis* and most eukaryotes, the pol α is responsible for DNA replication priming while pols δ and ε serve as the main replicative DNA polymerases and together perform the bulk of the replication process (Lujan *et al.*, 2016). Replicative polymerases are associated into a large protein complex (an 11-subunit helicase complex) called the replisome that encompasses all the core activities required for high-fidelity DNA replication (Garg and Burgers, 2005). Recently, reports on plant DNA polymerases demonstrate the conservation of their function compared to what is known in mammals or yeast. However, available data so far have revealed the paucity of biochemical studies that precludes the clear identification of each polymerase's activity (Pedroza-Garcia *et al.*, 2019).

3.6.1 Plant replicative DNA polymerases

As described above, the replicative polymerases are members of the B subfamily of DNA polymerases with characteristic conserved amino acid motifs and high-fidelity replication due to the tight conformation of their active site that allows the incorporation of mismatched nucleotides with a frequency of $\sim 10^{-7}$, as well as their proof-reading activity that improves fidelity by about 100–1000 times (Ganai and Johansson, 2016; Bebenek and Ziuzia-Graczyk, 2018). In addition to their role in DNA synthesis, replicative DNA polymerases are involved in the surveillance of stalled forks that may be generated by constraints on the replication machinery such as G-quadruplexes, DNA lesions, or collisions between transcription and replication. In plants (particularly *Arabidopsis* and rice), the pol α forms a complex with primase and consists of four subunits (POLA1/POL1, POLA2/POL12, PRI1, and PR2) that are essential for cell survival and are highly conserved among plants (Collins *et al.*, 1993). POLA1/POL1 (also known as *INCURVATA2/ICU2*) is the largest subunit that is responsible for the polymerase activity while the POLA2/POL12 subunit has a regulatory role of linking pol α to the replication fork with no known enzymatic activity. The PRI1 and 2 are the smallest subunits that harbour the DNA primase activity (Bryant and Aves, 2011).

Additionally, the pol ε is a four-subunit complex consisting of a larger catalytic (POLE1/POL2A) subunit and other three accessory DNA Pol II subunit B, namely DPB2, DPB3, and DPB4 (Jaszczur *et al.*, 2008). Among the accessory subunits, only DPB2 is required for cell viability while the rest (DPB3 and DPB4) are dispensable for cell viability although they may play a crucial role in the fidelity of pol ε (Aksenova *et al.*, 2010; Ganai *et al.*, 2015). The pol δ complex is composed of POLD monomeric subunits and their number varies greatly among prokaryotes and eukaryotes. In the model plant *Arabidopsis* and other eukaryotes, it is composed of four POLD subunits (POLD1–POLD4) in the form of a heterotetramer complex while in rice it is made up of two POLD4 subunits (Johnson and O'Donnell, 2005; Shultz *et al.*, 2007). The POLD1 subunit acts as the catalytic component and serves both polymerase and exonuclease functions while the other subunits (POLD2–4) are involved in the stabilization of the complex and interaction with PCNA (Sanchez *et al.*, 2012).

Despite the widespread occurrence of DNA polymerases in all organisms, their role in DNA replication by extending the new strand using template DNA has made them the talking point of the replication fork. However, the generalized concept of replication using only polymerases is far from complete as they never work alone but form part of a much larger and more dynamic nucleoprotein complex (the replisome). Additionally, among the numerous polymerase enzymes in plants, only three members of the B subfamily (pols α, δ, and ε) are replicative (involved in new strand synthesis) while the vast majority (distributed among the seven subfamilies) are involved in other DNA repair-related pathways.

3.6.2 Role of DNA polymerases in salinity tolerance

Generally, most enzymes (including polymerases) lose activity in high-salt conditions (Yamagami *et al.*, 2014; Wang, J. *et al.*, 2015; Baba *et al.*, 2017). It is a known hypothesis that the survival of living organisms (plants included) depends on accurate and faithful transmission of genetic information involving many enzymes of DNA replication and that salt concentration has substantial effects on the catalytic properties as well as the structural stability of these enzymes. To date, very scanty or little information is available on the exact roles of DNA polymerases in the abiotic stress tolerance of plants. The non-replicative DNA polymerase pol λ (a member of the X subfamily) showed enhanced level of protein and enzyme activity in model plants *Arabidopsis* and rice treated with different concentrations of salt (Sihi *et al.*, 2017).

Studies on Phi29 DNA polymerase (Phi29 pol) obtained from *Bacillus* phage Phi29 have reported the enzyme to be a good candidate with enhanced activity under salt-stress episodes (Gao *et al.*, 2021). Although there is limit to the tolerance level of Phi29 pol, it is widely used and commercially available for *in vitro* DNA duplication. Additionally, reports by Nisa *et al.* (2021) showed that another DNA polymerase, pol θ, a non-replicative DNA polymerase encoded by the *POLQ* gene in *Arabidopsis tebichi*, plays a significant role in the salinity tolerance of plants and provides evidence for a link between abiotic stress (salt stress included) response and DNA damage. Similarly, identification and analysis of POLA1 subunit in *A. thaliana* encoded by *INCURVATA2* (ICU2: At5g67100) gene revealed the role of pol α in maintaining replication ability under abiotic stress (Barrero *et al.*, 2007).

In summary, the DNA replication process in plants is initiated by a molecular complex consisting of DNA primase (its two subunits, POLA3 and POLA4) and DNA pol α (both its catalytic and non-catalytic subunits). After initiation, pol ε and pol δ replace pol α to continue the synthesis of leading and lagging strands, respectively (Perera *et al.*, 2013). Conclusively, the role of DNA polymerase in conferring salinity tolerance to plants largely depends on the ability of the polymerases to ensure continuous replication and transcription of stress-responsive genes during salinity episodes, which will in turn increase the expression levels of salt-tolerance genes in plants.

3.7 Topoisomerases

Topoisomerases are enzymes of DNA replication that play a crucial role in untangling and relaxing the interwound DNA supercoiled duplex. Unwinding of the DNA double helix is either transient as in transcription and recombination or stable as during DNA replication (Champoux, 2001; Vos *et al.*, 2011; Ashour *et al.*, 2015). Toposiomerases are divided into type I (TopoI; ATP-independent monomeric proteins) and type II (TopoII; ATP-dependent monomeric proteins) where type I nicks and seals only one strand while type II nicks both strands of DNA (Singh *et al.*, 2003, 2004; Vrielynck *et al.*, 2016). In addition, type II plays a role in removing positive supercoils ahead of replication forks and moving transcription complexes in archaea and bacteria, since the TopoIA present in these two domains cannot relax positive supercoils (Gadelle *et al.*, 2003). To date, all eukaryotic type II topoisomerases characterized show similarity in their primary structure, comprising 1429–1530 amino acid residues with a predictable molecular mass of 160–170 kDa (Wang, 2002; Corbett and Berger, 2004). In the recent past, there has been a surge in research on the biochemical and biological aspects of TopoII and various processes underlying its functions and reactions have been established (Nitiss, 2009; Blattner, 2016) but studies are mostly restricted to animal systems.

3.7.1 The role of topoisomerase in salinity tolerance

Although a large number of genes have been identified in plants in response to abiotic stress, unravelling the relationship between DNA replication enzymes and abiotic stress is crucial for developing abiotic stress-tolerant transgenic crops (Hettiarachchi *et al.*, 2005). Singh *et al.* (2003) isolated and characterized TopoII from *Nicotiana tabacum* and proposed that TopoII under GAL1 promoter functionally complements a temperature-sensitive topoisomerase II in a yeast mutant and its transcript levels are upregulated in response to cold and salinity stress in tobacco and pea. Additionally, reports have shown that overexpression of TopoII in *N. tabacum* leads to enhanced salinity-stress tolerance (Hettiarachchi *et al.*, 2005).

3.8 Summary and Concluding Remarks

The DNA replication proteins (helicases, polymerases, and topoisomerases) are highly conserved among all domains of life and play a critical role in different biological pathways including stress physiology. They collectively interact to ensure faithful duplication of cellular DNA. In plants, salinity causes accumulation of endogenous ROS which in turn leads to oxidative damage of DNA and other vital functions. DNA replication proteins play a vital role in salinity tolerance of plants by ensuring sustained transient replication into transcription and translation of salinity-responsive genes during salt stress. The understanding of the roles of DNA replication proteins in salinity tolerance of plants is getting broader, mostly with regard to helicases and polymerases. Although a large number of DNA replication proteins are identified there is still a handful with no known function. The involvement of replication proteins in various metabolic pathways of plants might have general implications and their role in salinity response is not well understood (or beginning to be understood). Recently, overexpression of salt-induced helicases and polymerases has proven effective and paved the way for engineering salt-tolerant crop plants. Recently, the DNA replication proteins (mostly helicase and polymerase) have emerged as new molecules to study stress signalling in plants. A few studies of salinity-induced helicase and polymerase suggested salinity affects the stability of nucleic acid base pairing in plants. With very little knowledge of helicase and polymerase involvement in salinity responses of plants, in this chapter we have attempted to outline some of the reported helicase and polymerase enzymes of different plant species and their possible role or activity during salinity episodes. Overall, this provides a framework to explore the true capacity of helicases, polymerases, and topoisomerases as multifunctional modules that can be the focus of further experimental validation to better understand plant responses to salinity. Finally, replication proteins offer vast opportunities for the development of salt-tolerant crops.

References

Aksenova, A., Volkov, K., Maceluch, J., Pursell, Z.F., Rogozin, I.B. *et al.* (2010) Mismatch repair-independent increase in spontaneous mutagenesis in yeast lacking non-essential subunits of DNA polymerase epsilon. *PLoS Genetics* 6, e1001209.

Amoroso, A., Concia, L., Maggio, C., Raynaud, C., Bergounioux, C. *et al.* (2011) Oxidative DNA damage bypass in *Arabidopsis thaliana* requires DNA polymerase λ and proliferating cell nuclear antigen 2. *The Plant Cell* 23, 806–822. DOI: 10.1105/tpc.110.081455.

Apel, K. and Hirt, H. (2004) Reactive oxygen species: metabolism, oxidative stress, and signal transduction. *Annual Review of Plant Biology* 55, 379–399. DOI: 10.1146/annurev.arplant.55.031903.141701.

Arefian, M., Vessal, S., Malekzadeh-Shafaroudi, S., Siddique, K.H.M. and Bagheri, A. (2019) Comparative proteomics and gene expression analyses revealed responsive proteins and mechanisms for salt tolerance in chickpea genotypes. *BMC Plant Biology* 19, 300. DOI: 10.1186/s12870-019-1793-z.

Ashour, M.E., Atteya, R. and El-Khamisy, S.F. (2015) Topoisomerase-mediated chromosomal break repair: an emerging player in many games. *Nature Reviews Cancer* 15, 137–151. DOI: 10.1038/nrc3892.

Athar, H.-U.-R., Zulfiqar, F., Moosa, A., Ashraf, M., Zafar, Z.U. *et al.* (2022) Salt stress proteins in plants: an overview. *Frontiers in Plant Science* 13, 999058. DOI: 10.3389/fpls.2022.999058.

Aubourg, S., Kreis, M. and Lecharny A. (1999) The DEAD box RNA helicase family in *Arabidopsis thaliana*. *Nucleic Acids Research* 27, 628–636.

Baba, M., Kakue, R., Leucht, C., Rasor, P., Walch, H. *et al.* (2017) Further increase in thermostability of moloney murine leukemia virus reverse transcriptase by mutational combination. *Protein Engineering, Design & Selection* 30(8), 551–557. DOI: 10.1093/protein/gzx046.

Bandehagh, A., Salekdeh, G.H., Toorchi, M., Mohammadi, A. and Komatsu, S. (2011) Comparative proteomic analysis of canola leaves under salinity stress. *Proteomics* 11(10), 1965–1975.

Barrero, J.M., Gonzalez-Bayon, R., del Pozo, J.C., Ponce, M.R. and Micol, J.L. (2007) *INCURVATA2* encodes the catalytic subunit of DNA polymerase alpha and interacts with genes involved in chromatin-mediated cellular memory in *Arabidopsis thaliana*. *The Plant Cell* 19(9), 2822–2838. DOI: 10.1105/tpc.107.054130.

Bebenek, A. and Ziuzia-Graczyk, I. (2018) Fidelity of DNA replication – a matter of proof reading. *Current Genetics* 64, 985–996.

Bell, S.P. and Kaguni, J.M. (2013) Helicase loading at chromosomal origins of replication. *Cold Spring Harbor Perspectives in Biology* 5(6), a010124.

Bell, S.P. and Labib, K. (2016) Chromosome duplication in *Saccharomyces cerevisiae*. *Genetics* 203(3), 1027–1067.

Blattner, F.R. (2016) TOPO6: a nuclear single-copy gene for plant phylogenetic inference. *Plant Systematics and Evolution* 302, 239–244. DOI: 10.1007/s00606-015-1259-1.

Bleichert, F., Botchan, M.R. and Berger, J.M. (2017) Mechanisms for initiating cellular DNA replication. *Science* 355, eaah6317.

Bohnert, H.J., Nelson, D.E. and Jensen, R.G. (1995) Adaptation to environmental stresses. *The Plant Cell* 7, 1099–1111.

Briolat, V. and Reysset, G. (2002) Identification of the *Clostridium perfringens* genes involved in the adaptive response to oxidative stress. *Journal of Bacteriology* 184, 2333–2343.

Brosh, R.M. (2021) DNA helicases: mechanisms, biological pathways, and disease relevance. *Genes (Basel)* 12, 356–360.

Brosh, R.M. and Matson, S.W. (2020) History of DNA helicases. *Genes (Basel)* 11, 255–300.

Bryant, J.A. and Aves, S.J. (2011) Initiation of DNA replication: functional and evolutionary aspects. *Annals of Botany* 107, 1119–1126.

Burgers, P.M.J. and Kunkel, T.A. (2017) Eukaryotic DNA replication fork. *Annual Review of Biochemistry* 86, 417–438. DOI: 10.1146/annurev-biochem-061516-044709.

Burgers, P.M.J., Gordenin, D. and Kunkel, T.A. (2016) Who is leading the replication fork, Pol δ or Pol ε? *Molecular Cell* 61, 492–493.

Campbell, N.A., Reece, J.B., Urry, L.A., Cain, M.L., Wasserman, S.A. *et al.* (2008) *Biology*, 8th edn. Pearson, New York.

Champoux, J.J. (2001) DNA topoisomerases: structure, function, and mechanism. *Annual Review of Biochemistry* 70, 369–413. DOI: 10.1146/annurev.biochem.70.1.369.

Chattopadhyay, A., Subba, P., Pandey, A., Bhushan, D., Kumar, R. *et al.* (2011) Analysis of the grasspea proteome and identification of stress-responsive proteins upon exposure to high salinity, low temperature, and abscisic acid treatment. *Phytochemistry* 72(10), 1293–1307.

Chen, J., Wan, S., Liu, H., Fan, S., Zhang, Y. *et al.* (2016) Overexpression of an *Apocynum venetum* DEAD-box helicase gene (AvDH1) in cotton confers salinity tolerance and increases yield in a saline field. *Frontiers in Plant Science* 6, 1227. DOI: 10.3389/fpls.2015.01227.

Chojnacki, M. and Melendy, T. (2018) The human papillomavirus DNA helicase E1 binds, stimulates, and confers processivity to cellular DNA polymerase epsilon. *Nucleic Acids Research* 46, 229–241.

Chourasia, K.N., Lal, M.K., Tiwari, R.K., Dev, D., Kardile, H.B. *et al.* (2021) Salinity stress in potato: understanding physiological, biochemical and molecular responses. *Life* 11(6), 545–569. DOI: 10.3390/life11060545.

Chung, E., Cho, C.W., Yun, B.H., Choi, H.K., So, H.A. *et al.* (2009) Molecular cloning and characterization of the soybean DEAD-box RNA helicase gene induced by low temperature and high salinity stress. *Gene* 443, 91–99.

Collins, K.L., Russo, A.A., Tseng, B.Y. and Kelly, T.J. (1993) The role of the 70 kDa subunit of human DNA polymerase alpha in DNA replication. *EMBO Journal* 12, 4555.

Corbett, K.D. and Berger, J.M. (2004) Structure, molecular mechanisms, and evolutionary relationships in DNA topoisomerases. *Annual Review of Biophysics and Biomolecular Structure* 33, 95–118. DOI: 10.1146/annurev.biophys.33.110502.140357.

Daimon, K., Ishino, S., Imai, N., Nagumo, S., Yamagami, T. *et al.* (2018) Two family B DNA polymerases from *Aeropyrum pernix*, based on revised translational frames. *Frontiers in Molecular Biosciences* 5, 37. DOI: 10.3389/fmolb.2018.00037.

Dang, H.Q., Tran, N.Q., Gill, S.S., Tuteja, R. and Tuteja, N. (2011) A single subunit MCM6 from pea promotes salinity stress tolerance without affecting yield. *Plant Molecular Biology* 76(1–2), 19–34. DOI: 10.1007/s11103-011-9758-0.

Dissanayake, B.M., Staudinger, C., Munns, R., Taylor, N.L. and Millar, A.H. (2022) Distinct salinity-induced changes in wheat metabolic machinery in different root tissue types. *Journal of Proteomics* 256, 104502. DOI: 10.1016/j.jprot.2022.104502.

ElSayed, A.I., Rafudeen, M.S., Ganie, S.A., Hossain, M.S. and Gomaa, A.M. (2022) Seed priming with cypress leaf extract enhances photosynthesis and antioxidative defense in zucchini seedlings under salt stress. *Scientia Horticulturae* 293, 110707.

Esmaeili, N., Shen, G. and Zhang, H. (2022) Genetic manipulation for abiotic stress resistance traits in crops. *Frontiers in Plant Science* 13, 1011985. DOI: 10.3389/fpls.2022.1011985.

Eyidogan, F. and Oz, T. (2007) Effect of salinity on antioxidant responses of chickpea seedlings. *Acta Physiologiae Plantarum* 29, 485–493. DOI: 10.1007/s11738-007-0059-9.

Fairman-Williams, M.E., Guenther, U.P. and Jankowsky, E. (2010) SF1 and SF2 helicases: family matters. *Current Opinion in Structural Biology* 20, 313–324. DOI: 10.1016/j.sbi.2010.03.011.

Fernandez-Cid, A., Reira, A., Tognetti, S., Herrera, M.C., Samel, S. *et al.* (2013) An ORC/Cdc6/MCM2–7 complex is formed in a multistep reaction to serve as a platform for MCM double-hexamer assembly. *Molecular Cell* 50(4), 577–588. DOI: 10.1016/j.molcel.2013.03.026.

Fitch, M.J., Donato, J.J. and Tye, B.K. (2003) MCM7, a subunit of the presumptive MCM helicase, modulates its own expression in conjunction with MCM1. *Journal of Biological Chemistry* 278, 25408–25416. DOI: 10.1074/jbc.M300699200.

Foreman, J., Demidchik, V., Bothwell, J.H., Mylona, P., Miedema, H. *et al.* (2003) Reactive oxygen species produced by NADPH oxidase regulate plant cell growth. *Nature* 422(6930), 442–446. DOI: 10.1038/nature01485.

Gadelle, D., Filée, J., Buhler, C. and Forterre P. (2003) Phylogenomics of type II DNA topoisomerases. *Bioessays* 25, 232–242. DOI: 10.1002/bies.10245.

Gagliardi, D., Kuhn, J., Spadinger, U., Brennicke, A., Leaver, C.J. *et al.* (1999) An RNA helicase (AtSUV3) is present in *Arabidopsis thaliana* mitochondria. *FEBS Letters* 458, 337–342. DOI: 10.1016/s0014-5793(99)01168-0.

Ganai, R.A. and Johansson, E. (2016) DNA replication – a matter of fidelity. *Molecular Cell* 62, 745–755.

Ganai, R.A., Osterman, P. and Johansson, E. (2015) Yeast DNA polymerase ε, catalytic core and holoenzyme have comparable catalytic rates. *Journal of Biological Chemistry* 290, 3825–3835.

Ganie, S.A. (2020) RNA chaperones: potential candidates for engineering salt tolerance in rice. *Crop Science* 60(2), 530–540.

Ganie, S.A. (2021) Amino acids other than proline and their participation in abiotic stress tolerance. In: Wani, S.H., Gangola, M.P. and Ramadoss, B.R. (eds) *Compatible Solutes Engineering for Crop Plants Facing Climate Change*. Springer, Cham, Switzerland, pp. 47–96.

Ganie, S.A., Debnath, A.B., Gumi, A.M. and Mondal, T.K. (2017) Comprehensive survey and evolutionary analysis of genome-wide miRNA genes from ten diploid *Oryza* species. *BMC Genomics* 18(1), 711.

Gao, J.P., Chao, D.Y. and Lin H.X. (2007) Understanding abiotic stress tolerance mechanisms: recent studies on stress response in rice. *Journal of Integrative Plant Biology* 49(6), 742–750.

Gao, Y., He, Y., Chen, L., Liu, X., Ivanov, I. *et al.* (2021) Chimeric Phi29 DNA polymerase with helix-hairpin-helix motifs shows enhanced salt tolerance and replication performance. *Microbial Biotechnology* 14, 1642–1656. DOI: 10.1111/1751-7915.13830.

Garbacz, M.A., Lujan, S.A., Burkholder, A.B., Cox, P.B., Wu, Q. *et al.* (2018) Evidence that DNA polymerase δ contributes to initiating leading strand DNA replication in *Saccharomyces cerevisiae*. *Nature Communications* 9, 858.

Garcia, A., Rizzo, C.A., Ud-Din, J., Bartos, S.L., Senadhira, D. *et al.* (1997) Sodium and potassium transport to the xylem are inherited independently in rice, and the mechanisms of sodium:potassium selectivity differs between rice and wheat. *Plant, Cell & Environment* 20, 1167–1174.

Garcia-Diaz, M. and Bebenek, K. (2007) Multiple functions of DNA polymerases. *Critical Reviews in Plant Sciences* 26, 105–122.

Garg, P. and Burgers, P.M.J. (2005) DNA polymerases that propagate the eukaryotic DNA replication fork. *Critical Reviews in Biochemistry and Molecular Biology* 40, 115–128.

Gill, S.S., Tajrishi, M., Madan, M. and Tuteja, N. (2013) A DESD-box helicase functions in salinity stress tolerance by improving photosynthesis and antioxidant machinery in rice (*Oryza sativa* L. cv. PB1). *Plant Molecular Biology* 82, 1–22. DOI: 10.1007/s11103-013-0031-6.

Gualberto, J.M., Mileshina, D., Wallet, C., Niazi, A.K., Weber-Lotfi, F. and Dietrich, A. (2014) The plant mitochondrial genome: dynamics and maintenance. *Biochimie* 100, 107–120. DOI: 10.1016/j.biochi.2013.09.016.

Gupta, B., Sengupta, A., Saha, J. and Gupta, K. (2013) Plant abiotic stress: 'omics' approach. *Journal of Plant Biochemistry & Physiology* 1(3), 108–109.

Hasegawa, P.M., Bressan, R.A., Zhu, J-K. and Bohnert, J. (2000) Plant cellular and molecular responses to high salinity. *Annual Review of Plant Physiology and Plant Molecular Biology* 51(1), 463–499.

Hettiarachchi, G.H., Reddy, M.K., Sopory, S.K and Chattopadhyay, S. (2005) Regulation of *TOP2* by various abiotic stresses including cold and salinity in pea and transgenic tobacco plants. *Plant & Cell Physiology* 46, 1154–1160. DOI: 10.1093/pcp/pci114.

Hozak, P., Hassan, A.B., Jackson, D.A. and Cook, P.R. (1993) Visualization of replication factories attached to a nucleoskeleton. *Cell* 73, 361–373.

Hu, Z., Cools, T. and De Veylder, L. (2016) Mechanisms used by plants to cope with DNA damage. *Annual Review of Plant Biology* 67, 439–462.

Jain, R., Aggarwal, A.K. and Rechkoblit, O. (2018) Eukaryotic DNA polymerases. *Current Opinion in Structural Biology* 53, 77–87.

Jankowsky, E. (2011) RNA helicases at work: binding and rearranging. *Trends in Biochemical Sciences* 36, 19–29. DOI: 10.1016/j.tibs.2010.07.008.

Jaszczur, M., Flis, K., Rudzka, J., Kraszewska, J., Budd, M.E. *et al.* (2008) Dpb2p, a noncatalytic subunit of DNA polymerase epsilon, contributes to the fidelity of DNA replication in *Saccharomyces cerevisiae*. *Genetics* 178, 633–647.

Jiang, Y., Yang, B., Harris, N.S. and Deyholos, M.K. (2007) Comparative proteomic analysis of NaCl stress-responsive proteins in *Arabidopsis* roots. *Journal of Experimental Botany* 58(13), 3591–3607.

Johansson, E. and Dixon, N. (2013) Replicative DNA polymerases. *Cold Spring Harbor Perspectives in Biology* 5, a012799.

Johnson, A. and O'Donnell, M. (2005) Cellular DNA replicases: components and dynamics at the replication fork. *Annual Review of Biochemistry* 74, 283–315.

Johnson, R.E., Klassen, R., Prakash, L. and Prakash, S.A. (2015) Major role of DNA polymerase δ in replication of both the leading and lagging DNA strands. *Molecular Cell* 59, 163–175.

Kant, P., Kant, S., Gordon, M., Shaked, R. and Barak, S. (2007) *STRESS RESPONSE SUPPRESSOR1* and *STRESS RESPONSE SUPPRESSOR2*: two DEAD-box RNA helicases that attenuate *Arabidopsis* responses to multiple abiotic stresses. *Plant Physiology* 145, 814–830.

Kawasaki, Y. and Sugino, A. (2001) Yeast replicative DNA polymerases and their role at the replication fork. *Molecular Cell* 12, 277–285.

Kim, J.S., Kim, K.A., Oh, T.R., Park, C.M. and Kang, H. (2008) Functional characterization of DEAD-box RNA helicases in *Arabidopsis thaliana* under abiotic stress conditions. *Plant & Cell Physiology* 49, 1563–1571.

Li, Z. and Xu, X. (2019) Post-translational modifications of the mini-chromosome maintenance proteins in DNA replication. *Genes (Basel)* 10(5), 331. DOI: 10.3390/genes10050331.

Liang, D.T., Hodson, J.A. and Forsburg, S.L. (1999) Reduced dosage of a single fission yeast MCM protein causes genetic instability and S phase delay. *Journal of Cell Science* 112, 559–567.

Liu, H.H., Liu, J., Fan, S.L., Song, M.Z., Han, X.L. *et al.* (2008) Molecular cloning and characterization of a salinity stress induced gene encoding DEAD-box helicase from the halophyte *Apocynum venetum*. *Journal of Experimental Botany* 59, 633–644.

Liu, H.Y., Nefsky, B.S. and Walworth, N.C. (2002) The Ded1 DEAD-box helicase interacts with Chk1 and Cdc2. *Journal of Biological Chemistry* 277, 2637–2643.

Lohman, T.M. and Bjornson, K.P. (1996) Mechanism of helicase-catalyzed DNA unwinding. *Annual Review of Biochemistry* 65, 169–214.

Lujan, S.A., Williams, J.S. and Kunkel, T.A. (2016) DNA polymerases divide the labor of genome replication. *Trends in Cell Biology* 26, 640–654.

Luo, Y., Liu, Y.B., Dong, Y.X., Gao, X.Q. and Zhang, X.S. (2009) Expression of a putative alfalfa helicase increases tolerance to abiotic stress in *Arabidopsis* by enhancing the capacities for ROS scavenging and osmotic adjustment. *Journal of Plant Physiology* 166, 385–394.

Maas, E.V. (1990) Crop salt tolerance. In: Tanji, K.K. (ed.) *Agricultural Salinity Assessment and Management*. ASCE Manual Reports on Engineering Practice, Vol. 71. ASCE, New York, pp. 262–304.

Manova, V. and Gruszka, D. (2015) DNA damage and repair in plants – from models to crops. *Frontiers in Plant Science* 6, 885. DOI: 10.3389/fpls.2015.00885.

Mansour, M.M.F., Salama, K.H.A., Ali, F.Z.M. and Abou Hadid, A.F. (2005) Cell and plant responses to NaCl in *Zea mays* L. cultivars differing in salt tolerance. *General and Applied Plant Physiology* 31(1–2), 29–41.

Miller, G., Suzuki, N., Ciftci-Yilmaz, S. and Mittler, R. (2010) Reactive oxygen species homeostasis and signalling during drought and salinity stresses. *Plant, Cell & Environment* 33, 453–467.

Mohapatra, M.D., Poosapati, S., Sahoo, R.K. and Swain, D.M. (2023) Helicase: a genetic tool for providing stress tolerance in plants. *Plant Stress* 9, e100171.

Montero-Lomeli, M., Morais, B.L., Figueiredo, D.L., Neto, D.C., Martins, J.R. *et al.* (2002) The initiation factor eIF4A is involved in the response to lithium stress in *Saccharomyces cerevisiae*. *Journal of Biological Chemistry* 277, 21542–21548.

Moreno-del Alamo, M., Carrasco, B., Torres, E. and Alonso, J.C. (2021) *Bacillus subtilis* PcrA helicase removes trafficking barriers. *Cells* 10, 935–959.

Moyer, S.E., Lewis, P.W. and Botchan, M.R. (2006) Isolation of the Cdc45/Mcm2–7/GINS (CMG) complex, a candidate for the eukaryotic DNA replication fork helicase. *Proceedings of the National Academy of Sciences USA* 103, 10236–10241.

Nakamura, T., Muramoto, Y., Yokota, S., Ueda, A. and Takabe, T. (2004) Structural and transcriptional characterization of a salt-responsive gene encoding putative ATP-dependent RNA helicase in barley. *Plant Science* 167, 63–70. DOI: 10.1016/j.plantsci.2004.03.001.

Nawaz, G., Lee, K., Park, S.J., Kim, Y.O. and Kang, H. (2018) A chloroplast-targeted cabbage DEAD-box RNA helicase BrRH22 confers abiotic stress tolerance to transgenic *Arabidopsis* plants by affecting translation of chloroplast transcripts. *Plant Physiology and Biochemistry* 127, 336–342. DOI: 10.1016/j.plaphy.2018.04.007.

Nguyen, L., Seok, H.-Y., Woo, D.-H., Lee, S.-Y. and Moon, Y.-H. (2018) Overexpression of the DEAD-box RNA helicase gene *AtRH17* confers tolerance to salt stress in *Arabidopsis*. *International Journal of Molecular Science* 19, 3777. DOI: 10.3390/ijms19123777.

Nisa, M., Bergis, C., Pedroza-Garcia, J.A., Drouin-Wahbi, J., Mazubert, C. *et al.* (2021) The plant DNA polymerase theta is essential for the repair of replication-associated DNA damage. *The Plant Journal* 106(5), 1197–1207. DOI: 10.1111/tpj.15295.

Nisa, M.U., Huang, Y., Benhamed, M. and Raynaud, C. (2019) The plant DNA damage response: signaling pathways leading to growth inhibition and putative role in response to stress conditions. *Frontiers in Plant Science* 10, 653.

Nitiss, J.L. (2009) Targeting DNA topoisomerase II in cancer chemotherapy. *Nature Reviews Cancer* 9, 338–350. DOI: 10.1038/nrc2607.

O'Donnell, M., Langston, L. and Stillman, B. (2013) Principles and concepts of DNA replication in bacteria, archaea, and eukarya. *Cold Spring Harbor Perspectives in Biology* 5, 7.

O'Donnell, M.E. and Li, H. (2018) The ring-shaped hexameric helicases that function at DNA replication forks. *Nature Structural & Molecular Biology* 25, 122–130.

Pang, Q.Y., Chen, S.X., Dai, S.J., Chen, Y.Z., Wang, Y. *et al.* (2010) Comparative proteomics of salt tolerance in *Arabidopsis thaliana* and *Thellungiella halophila*. *Journal of Proteome Research* 9(5), 2584–2599.

Patel, S.S. and Donmez, I. (2006) Mechanisms of helicases. *Journal of Biological Chemistry* 281, 18265–18268. DOI: 10.1074/jbc.R600008200.

Patel, S.S. and Picha, K.M. (2000) Structure and function of hexameric helicases. *Annual Review of Biochemistry* 69, 651–697.

Pedroza-Garcia, J., De Veylder, L. and Raynaud, C. (2019) Plant DNA polymerases. *International Journal of Molecular Sciences* 20, 4814–4835.

Perera, R.L., Torella, R., Klinge, S., Kilkenny, M.L., Maman, J.D. and Pellegrini, L. (2013) Mechanism for priming DNA synthesis by yeast DNA polymerase α. *eLlife* 2, e00482. DOI: 10.7554/eLife.00482.

Pham, X.H., Reddy, M.K., Ehtesham, N.Z., Matta, B. and Tuteja, N. (2000) A DNA helicase from *Pisum sativum* is homologous to translation initiation factor and stimulates topoisomerase I activity. *The Plant Journal* 24, 219–229.

Pursell, Z.F., Isoz, I., Lundström, E.B., Johansson, E. and Kunkel, T.A. (2007) Yeast DNA polymerase epsilon participates in leading strand DNA replication. *Science* 317, 127–130.

Ramadan, K., Shevelev, I.V., Maga, G. and Hübscher, U. (2004) *De novo* DNA synthesis by human DNA polymerase λ, DNA polymerase μ and terminal deoxyribonucleotidyl transferase. *Journal of Molecular Biology* 339(2), 395–404. DOI: 10.1016/j.jmb.2004.03.056.

Rao, T.S.R.B., Naresh, J.V., Reddy, P.S., Reddy, M.K. and Mallikarjuna, G. (2017) Expression of *Pennisetum glaucum* eukaryotic translational initiation factor 4a (PgEIF4a) confers improved drought, salinity, and oxidative stress tolerance in groundnut. *Frontiers in Plant Science* 8, 3389. DOI: 10.3389/fpls.2017.00453.

Rothwell, P.J. and Waksman, G. (2005) Structure and mechanism of DNA polymerases. *Advances in Protein Chemistry* 71, 401–440.

Roy, S., Choudhury, S.R., Singh, S.K. and Das, K.P. (2011) AtPolλ, a homolog of mamalian DNA polymerase λ in *Arabidopsis thaliana,* is involved in the repair of UV-B induced DNA damage through the dark repair pathway. *Plant & Cell Physiology* 52, 448–467. DOI: 10.1093/pcp/pcr002.

Ru, J.N., Hou, Z.H., Zheng, L., Zhao, Q., Wang, F.Z. *et al.* (2021) Genome-wide analysis of DEAD-box RNA helicase family in wheat (*Triticum aestivum*) and functional identification of TaDEAD-box57 in abiotic stress responses. *Frontiers in Plant Science* 12, 797276. DOI: 10.3389/fpls.2021.797276.

Sahoo, R.K., Ansari, M.W., Tuteja, R. and Tuteja, N. (2014) OsSUV3 transgenic rice maintains higher endogenous levels of plant hormones that mitigates adverse effects of salinity and sustains crop productivity. *Rice* 7, 17. DOI: 10.1186/s12284-014-0017-2.

Saifi, S.K., Passricha, N., Tuteja, R., Nath, M., Gill, R. *et al.* (2023) *OsRuvBL1a* DNA helicase boost salinity and drought tolerance in transgenic indica rice raised by *in planta* transformation. *Plant Science* 335, 111786. DOI: 10.1016/j.plantsci.2023.111786.

Sami, A.A., Arabia, S., Sarker, R.H. and Islam, T. (2021) Deciphering the role of helicases and translocases: a multifunctional gene family safeguarding plants from diverse environmental adversities. *Current Plant Biology* 26, 100–204.

Sanan-Mishra, N., Pham, X.H., Sopory, S.K. and Tuteja, N. (2005) Pea DNA helicase 45 overexpression in tobacco confers high salinity tolerance without affecting yield. *Proceedings of the National Academy of Sciences USA* 102, 509–514.

Sanchez, M.d.I.P., Costas, C., Sequeira-Mendes, J. and Gutierrez, C. (2012) Regulating DNA replication in plants. *Cold Spring Harbor Perspectives in Biology* 4, a010140.

Schatchman, D.P. and Schroeder, J.I. (1994) Structure and transport mechanism of a high-affinity potassium uptake transporter from higher plants. *Nature* 370, 655–658.

Schatchman, D.P., Munns, R. and Whitecross, M.I. (1991) Variation in sodium exclusion and salt tolerance in *Triticum tauschii. Crop Science* 31, 992–997.

Seki, M., Narusaka, M., Ishida, J., Nanjo, T., Fujita, M. *et al.* (2002) Monitoring the expression profiles of 7000 *Arabidopsis* genes under drought, cold and high-salinity stresses using a full-length cDNA microarray. *The Plant Journal* 31(3), 279–292.

Seok, H.Y., Nguyen, L.V., Nguyen, D.V., Lee, S.Y. and Moon, Y.H. (2020) Investigation of a novel salt stress-responsive pathway mediated by *Arabidopsis* dead-box RNA helicase gene *AtRH17* using RNA-seq analysis, *International Journal of Molecular Science* 21, 1595. DOI: 10.3390/ijms21051595.

Shultz, R.W., Tatineni, V.M., Hanley-Bowdoin, L. and Thompson, W.F. (2007) Genome-wide analysis of the core DNA replication machinery in the higher plants *Arabidopsis* and rice. *Plant Physiology* 144, 1697–1714.

Sihi, S., Maiti, S., Bakshi, S., Nayak, A., Chaudhuri, S. and Sengupta, D.N. (2017) Understanding the role of DNA polymerase λ gene in different growth and developmental stages of *Oryza sativa* L. indica rice cultivars. *Plant Physiology and Biochemistry* 120, 156–168.

Singh, B.N., Mudgil, Y., Sopory, S.K. and Reddy, M.K. (2003) Molecular characterization of a nuclear topoisomerase II from *Nicotiana tabacum* that functionally complements a temperature-sensitive topoisomerase II yeast mutant. *Plant Molecular Biology* 52, 1063–1076. DOI: 10.1023/A:1025427700337.

Singh, B.N., Sopory, S.K. and Reddy, M.K. (2004) Plant DNA topoisomerases: structure, function and cellular roles in plant development. *Critical Reviews in Plant Sciences* 23, 251–269. DOI: 10.1080/07352680490452816.

Singha, D.L., Tuteja, N., Boro, D., Hazarika, G.N. and Singh, S. (2017) Heterologous expression of *PDH47* confers drought tolerance in *indica* rice. *Plant Cell, Tissue and Organ Culture (PCTOC)* 130, 577–589. DOI: 10.1007/s11240-017-1248-x.

Singleton, M.R., Dillingham, M.S. and Wigley, D.B. (2007) Structure and mechanism of helicases and nucleic acid translocases. *Annual Review of Biochemistry* 76, 23–50. DOI: 10.1146/annurev.biochem.76.052305.115300.

Sobhanian, H., Razavizadeh, R., Nanjo, Y., Ehsanpour, A.A., Jazii, F.R. *et al.* (2010) Proteome analysis of soybean leaves, hypocotyls and roots under salt stress. *Proteome Science* 8, 19.

Subba, P., Kumar, R., Gayali, S., Shekhar, S., Parveen, S. *et al.* (2013) Characterisation of the nuclear proteome of a dehydration-sensitive cultivar of chickpea and comparative proteomic analysis with a tolerant cultivar. *Proteomics* 13(12–13), 1973–1992.

Takahashi, N., Ogita, N., Takahashi, T., Taniguchi, S., Tanaka, M. *et al.* (2019) A regulatory module controlling stress-induced cell cycle arrest in *Arabidopsis*. *eLife* 9(8), e43944. DOI: 10.7554/eLife.43944.

Tran, L.S., Nakashima, K., Sakuma, Y., Osakabe, Y., Qin, F. *et al.* (2007) Co-expression of the stress-inducible zinc finger homeodomain ZFHD1 and NAC transcription factors enhances expression of the *ERD1* gene in *Arabidopsis. The Plant Journal* 49(1), 46–63. DOI: 10.1111/j.1365-313X.2006.02932.x.

Tran, N.Q., Dang, H.Q., Tuteja, R. and Tuteja, N. (2010) A single subunit MCM6 from pea forms homohexamer and functions as DNA helicase. *Plant Molecular Biology* 74, 327–336.

Tuteja, N. (2003) Plant DNA helicases: the long unwinding road. *Journal of Experimental Botany* 54, 2201–2214.

Tuteja, N. (2007) Mechanisms of high salinity tolerance in plants. *Methods in Enzymology* 428, 419–438.

Tuteja, N. and Tuteja, R. (2004) Prokaryotic and eukaryotic DNA helicases: essential molecular motor proteins for cellular machinery. *European Journal of Biochemistry* 271, 1835–1848.

Tuteja, N., Singh, S. and Tuteja, R. (2012) Helicases in improving abiotic stress tolerance, in crop plants. In: Tuteja, N., Gill, S.S., Tiburcio, A.F. and Tuteja, R. (eds) *Improving Crop Resistance to Abiotic Stress*. Wiley-VCH, Weinheim, Germany, pp. 435–449.

Tuteja, N., Sahoo, R.K., Garg, B. and Tuteja, R. (2013) OsSUV3 dual helicase functions in salinity stress tolerance by maintaining photosynthesis and antioxidant machinery in rice (*Oryza sativa* L. cv. IR64). *The Plant Journal* 76(1), 115–127. DOI: 10.1111/tpj.12277.

Tuteja, N., Tarique, M., Banu, M.S.A., Ahmad, M. and Tuteja, R. (2014a) *Pisum sativum* p68 DEAD-box protein is ATP-dependent RNA helicase and unique bipolar DNA helicase. *Plant Molecular Biology* 85, 639–651. DOI: 10.1007/s11103-014-0209-6.

Tuteja, N., Tarique, M. and Tuteja, R. (2014b) Rice SUV3 is a bidirectional helicase that binds both DNA and RNA. *BMC Plant Biology* 14, 283. DOI: 10.1186/s12870-014-0283-6.

Tuteja, N., Tarique, M., Trivedi, D.K., Sahoo, R.K. and Tuteja, R. (2015) Stress-induced *Oryza sativa* BAT1 dual helicase exhibits unique bipolar translocation. *Protoplasma* 252, 1563–1574. DOI: 10.1007/s00709-015-0791-8.

Uchiyama, Y., Takeuchi, R., Kodera, H. and Sakaguchi, K. (2009) Distributions and roles of X-family DNA polymerases in eukaryotes. *Biochimie* 91(2), 165–170. DOI: 10.1016/j.biochi.2008.07.005.

Umate, P., Tuteja, R. and Tuteja, N. (2010) Genome wide analysis of helicase gene family from rice and *Arabidopsis*: a comparison with yeast and human. *Plant Molecular Biology* 73, 449–465.

Vashisht, A., Pradhan, A., Tuteja, R. and Tuteja, N. (2005) Cold- and salinity stress-induced pea bipolar pea DNA helicase 47 is involved in protein synthesis and stimulated by phosphorylation with protein kinase C. *The Plant Journal* 44, 76–87.

Vos, S.M., Tretter, E.M., Schmidt, B.H. and Berger, J.M. (2011) All tangled up: how cells direct, manage and exploit topoisomerase function. *Nature Reviews Molecular Cell Biology* 12, 827–841. DOI: 10.1038/nrm3228.

Vrielynck, N., Chambon, A., Vezon, D., Pereira, L., Chelysheva, L. *et al.* (2016) A DNA topoisomerase VI-like complex initiates meiotic recombination. *Science* 351, 939–943. DOI: 10.1126/science.aad5196.

Waga, S. and Stillman, B. (1998) The DNA replication fork in eukaryotic cells. *Annual Review of Biochemistry* 67, 721–751. DOI: 10.1146/annurev.biochem.67.1.721.

Wang, J., Meng, Y., Li, B., Ma, X., Lai, Y. *et al.* (2015) Salt stress response in *Hologeton glomeratus*. *Plant, Cell & Environment* 38, 655–669. DOI: 10.1111/pce.12428.

Wang, J.C. (2002) Cellular roles of DNA topoisomerases: a molecular perspective. *Nature Reviews Molecular Cell Biology* 3, 430–440. DOI: 10.1038/nrm831.

Wang, L., Pan, D., Li, J., Tan, F., Hoffmann-Benning, S. *et al.* (2015) Proteomic analysis of changes in the *Kandelia candel* chloroplast proteins reveals pathways associated with salt tolerance. *Plant Science* 231, 159–172. DOI: 10.1016/j.plantsci.2014.11.013.

Watson, J.D. and Crick, F.H. (1953) Genetical implications of the structure of deoxyribonucleic acid. *Nature* 171, 964–967.

Westermarck, J., Weiss, C., Saffrich, R., Kast, J., Musti, A.M. *et al.* (2002) The DEXD/H-box RNA helicase RHII/Gu is a co-factor for c-Jun-activated transcription. *EMBO Journal* 21, 451–460.

Wiedemann, G., Van Gessel, N., Ochl, F.K., Hunn, L., Schulze, K. *et al.* (2018) RecQ helicases function in development, DNA repair, and gene targeting in *Physcomitrella patens*. *The Plant Cell* 30, 717–736. DOI: 10.1105/tpc.17.00632.

Xu, R.R., Qi, S.D., Lu, L.T., Chen, C.T., Wu, C.A. *et al.* (2011) A DExD/H box RNA helicase is important for K^+ deprivation responses and tolerance in *Arabidopsis thaliana*: analysis of an *Arabidopsis* DExD/H box RNA helicase. *FEBS Journal* 278, 2296–2306. DOI: 10.1111/j.1742-4658.2011.08147.x.

Yadav, M., Singh, R.S., Hogan, D., Dmitriev, O.Y., Cygler, M. and Wu, Y. (2021) The KH domain facilitates the substrate specificity and unwinding processivity of DDX43 helicase. *Journal of Biological Chemistry* 296, 100085. DOI: 10.1074/jbc.RA120.015824.

Yadav, R., Flowers, T.J. and Yeo, A.R. (1996) The involvement of the transpirational bypass flow in sodium uptake by high- and low-sodium-transporting lines of rice developed through intravarietal selection. *Plant, Cell & Environment* 19(3), 329–336.

Yamagami, T., Ishino, S., Kawarabayasi, Y. and Ishino, Y. (2014) Mutant *Taq* DNA polymerases with improved elongation ability as a useful reagent for genetic engineering. *Frontiers in Microbiology* 5 461.

Yang, W., Seidman, M.M., Rupp, W.D. and Gao, Y. (2019) Replisome structure suggests mechanism for continuous fork progression and post-replication repair. *DNA Repair* 81, 102658.

Yokota, H. (2021) Roles of the C-terminal amino acids of non-hexameric helicases: insights from *Escherichia coli* UvrD. *International Journal of Molecular Sciences* 22, 1018.

Zhang, H., Han, B., Wang, T., Chen, S., Li, H. *et al.* (2012) Mechanisms of plant salt response: Insights from proteomics. *Journal of Proteome Research* 11, 49–67.

Zhou, Z.X., Lujan, S.A., Burkholder, A.B., Garbacz, M.A. and Kunkel, T.A. (2019) Roles for DNA polymerase δ in initiating and terminating leading strand DNA replication. *Nature Communications* 10, 3992.

Zhu, M., Chen, G., Dong, T., Wang, L., Zhang, J. *et al.* (2015) *SlDEAD31*, a putative DEAD-box RNA helicase gene, regulates salt and drought tolerance and stress-related genes in tomato. *PLoS ONE* 10, e0133849. DOI: 10.1371/journal.pone.0133849.

4 Chromatin Architecture: Role of Epigenetic Modifications and Nucleosome Occupancy in Modulating Plant Responses to Salt Stress

Md Sazzad Hossain[1,4*], Mst Ishrat Zahan[2], Md Zillur Rahman[3,4] and Britta Pitann[1]

[1]Institute of Plant Nutrition and Soil Science, Kiel University, Kiel, Germany; [2]Breeding Division, Bangladesh Sugarcrop Research Institute, Ishurdi, Pabna, Bangladesh; [3]Centre for Carbon, Water, and Food, The University of Sydney, Camden, New South Wales, Australia; [4]Department of Agronomy and Haor Agriculture, Sylhet Agricultural University, Sylhet, Bangladesh

Abstract

Crop performance is severely affected by high salt concentrations in soils. To engineer salt-tolerant plants, it is crucial to unravel the main components of the plant salt-tolerance network. Growing evidence shows that regulation of chromatin architecture through epigenetic mechanisms and nucleosome occupancy controls complex plant traits, with implications across many fields of biology, especially molecular stress physiology research. Epigenetic regulation in plants can be mediated through a dynamic interplay between DNA methylation, histone modifications, non-coding RNAs (ncRNAs), and nucleosome occupancy, which modulate transcriptional responses together. These modifications have emerged as essential regulators of plant responses to salt stress by facilitating the recruitment of more transcription factors and co-activators on the mRNAs to synthesize many stress-tolerance proteins under salt stress. This chapter highlights the pivotal role of chromatin remodelling through epigenetic processes involving changes in DNA methylation, histone modification, ncRNAs, and nucleosome occupancy in plant responses to salt stress.

4.1 Introduction

Agriculture, the ultimate source of feeding the global population, is facing severe challenges in terms of sustainable production, mainly due to global climate variability. At all stages of development, crop plants encounter multiple environmental challenges. Extreme climate events like salinity, drought, and heat lower the average yield and worsen global food security issues (Ganie, 2021). According to estimates, the yield reduction brought about by these stresses might be as high as (or even higher than) 50% for several of the world's major crops (Fahad *et al.*, 2017; Shew *et al.*, 2020). Among different abiotic stresses, salinity has become a major concern that reduces overall production, and ~20% of the world's agricultural lands are badly affected

*Email: shossain@plantnutrition.uni-kiel.de; sazzadmh.aha@sau.ac.bd

 The Genetics of Salt Tolerance in Plants: A Central Dogma Perspective and Strategies for Enhancement (S. Ganie and S.H. Wani eds)
DOI: 10.1079/9781800623033.0004

by salinity (Ganie, 2020; Devkota *et al.*, 2022). The salinization of coastal farming land is one of the effects of climate change. Over the last 25 years, the increase in sea levels has raised soil salinity from 1 to 33% (Rahman *et al.*, 2018). The injudicious use of irrigation water has resulted in a salinity increase which negatively impacts crop growth and causes substantial yield losses (Jha *et al.*, 2019). Salinity alters plants' regular morpho-physiological, biochemical, and molecular attributes and reduces crop growth and yield (Hossain *et al.*, 2017a,b; ElSayed *et al.*, 2022). Depending on the growth stage, plants show varying responses to salt stress. Salt induces oxidative, osmotic, and ionic stress, resulting in the activation of several physiological, biochemical, and molecular responses to ameliorate the salt-induced damage (Hossain and Dietz, 2016; Ganie *et al.*, 2017; Hossain *et al.*, 2017a,b). Among the other responses, epigenetic regulation at the transcriptional level through modifications in histones and nucleosome occupancy, and at post-transcriptional level through non-coding RNAs (ncRNAs), are of great significance for plant stress tolerance (Cheng *et al.*, 2018; Marondedze *et al.*, 2021). All these mechanisms work to bring about a robust and multilayered epigenetic circuitry (Pikaard and Scheid, 2014). Epigenetics provides us the molecular basis for understanding how heritable genomic alterations other than those encoded in the DNA sequence can alter patterns of gene expression and influence phenotypic variation. Epigenetics is responsible for chromosome-associated heritable changes to gene expression which are independent of changes in DNA sequence. Epigenetic marks such as DNA methylation, histone methylation, and other histone marks such as phosphorylation, ADP-ribosylation, glycosylation, acetylation, and SUMOylation have been reported in the regulation of environmental stress responses (Simon and Meyers, 2011). DNA methylation is mainly due to the methylation of cytosines, while histone modifications involve acetylation, phosphorylation, methylation, ubiquitination, and ribosylation of different histone proteins (Lämke and Bäurle, 2017). Besides, various small RNAs (sRNAs) produced by plant cells are also involved in the methylation of DNA, histone modifications, and chromatin remodelling in the targeted DNA sequence (Simon and Meyers, 2011). Nucleosome occupancy and location are variable and have a significant influence on eukaryotic gene expression, regulation, and evolution (Lämke and Bäurle, 2017). Stress-induced changes in plant histone modification and nucleosome occupancy correspondingly modulate the expression of stress-responsive genes. To deal with salt stress, several plant species have developed a variety of mechanisms at different levels. Till now, significant progress has been made in understanding the mechanisms and processes underlying salt-stress tolerance of plants. However, very little is known about the dynamics of salinity-induced chromatin architecture and its role in plant salt tolerance. In this chapter, we highlight the current state of knowledge about the involvement of epigenetic processes including DNA methylation, histone modifications, RNA-directed changes, and nucleosome occupancy in plant responses to salt stress.

4.2 Epigenetic Modifications in Modulating Plant Responses to Salt Stress

Several epigenetic modifications such as DNA methylation, histone modifications (e.g. histone methylation and other histone marks such as acetylation, phosphorylation, deamination, β-*N*-acetylglucosamine, ADP-ribosylation, ubiquitination, and SUMOylation), and RNA interference have been reported to play important roles in the regulation of salt stress through gene expression regulation (Figs 4.1 and 4.2).

4.2.1 DNA methylation

One strategy for successful adaptation to stress is the epigenetic modification of the genome by DNA methylation, thereby changing the activity of DNA segments without changing the sequence (Chen *et al.*, 2013). The covalent methylation of cytosine at position 5 of the pyrimidine ring (5-methylcytosine (5mC)) is the most predominant reversible DNA alteration known in plants (Kress *et al.*, 2001), where its occurrence ranges from 6 to 25% (Steward *et al.*, 2002). Apart from 5mC, plants also have evolved non-canonical base modifications such as the methylation of

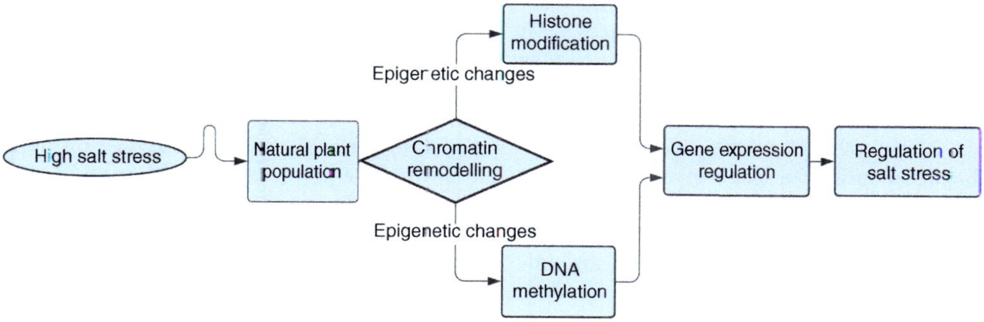

Fig. 4.1. A simplified schematic diagram of regulation through epigenetic modifications in plants in response to high salt stress.

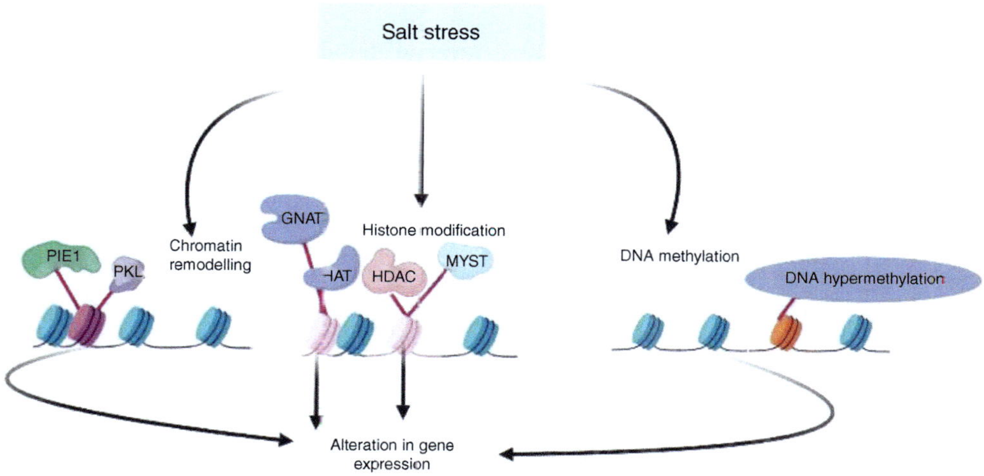

Fig. 4.2. An illustration of the nucleosome-level epigenetic mechanisms involved in plant response to salt stress. Changes in gene expression result from the chromatin-remodelling factors PIE1 and PKL modulating the accessibility of DNA to other transcriptional regulators under salt stress. HAT, HDAC, MYST, and GNAT are types of histone modifiers that control the acetylation of histone protein. DNA hypermethylation also regulates changes in gene expression and helps to reduce the effects of salt stress. Alteration in gene expression involves changes in the activity of genes that are regulated by various transcription factors and signalling pathways.

position 6 of the purine ring of adenine (N^6-methyladenine (6mA)) or position 7 of guanine (7-methylguanine (7mG)), but both only occur at a very low rate (Rashid *et al.*, 2022). Plant DNA methylation generally occurs in all cytosine sequence contexts, i.e. the symmetric CG and CxG context and the asymmetric Cxx context (x represents A, T, or C) (Zhang *et al.*, 2006; Lister *et al.*, 2008) and mostly occurs in gene promoter regions, while being less prominent at other transcriptional regulatory sites (Ficz *et al.*, 2013). This way, the methylated DNA base provides a binding site for various proteins, which, in combination with histone modifications and non-histone proteins, modifies the chromatin scaffold and thus accessibility. This in turn leads to the subsequent modulation of, for example, gene expression, replication, trait inheritance, transposon silencing, and epigenetic memory (Zhang *et al.*, 2018; Karanthamalai *et al.*, 2020)

(Fig. 4.2). In the past, it was a widely held belief that gene silencing is a consequence of DNA methylation, while demethylation leads to gene activation (Bilichak *et al.*, 2012). Although lots of progress has been made in understanding the contribution of DNA methylation in salt-stress adaptation during the last few years, the knowledge is still incomplete and only fragmented, especially for important crops; comprehensive methylation profiling for salt stress has only been reported for rice (Garg *et al.*, 2015), wheat (Wang *et al.*, 2014), and *Medicago truncatula* (Yaish *et al.*, 2018). Nevertheless, more recent studies contradict the assumption of unidirectional DNA modification and prove a more diverse association between hyper/hypomethylation and gene expression for various crops, depending on plant species, genotype, tissue, and stress duration (Garg *et al.*, 2015; Fan *et al.*, 2020; Shams *et al.*, 2020; Skorupa *et al.*, 2021). Therefore, we aim to review differences in DNA methylation between mono- and dicotyledonous crops.

Different modification streams can be observed in plants under salt stress, such as (i) DNA methylation mediated by conserved methyltransferases, for which *S*-adenosyl-l-methionine serves as methyl group donor (Niehrs, 2009); (ii) demethylation, involving enzymes to actively (glycolases) or passively (non-enzymatic) control the methylation status of DNA (Niehrs, 2009). In addition to the reversible modification of DNA, there is also an inheritable RNA-directed DNA methylation (RdDM) pathway (iii). This pathway (iii) is important for *de novo* methylation observed in plants (Matzke and Mosher, 2014), possibly conferring the appearance of novel traits (Boyko *et al.*, 2010). In this context, Fan *et al.* (2020) identified 22 genes coding for DNA methylases and six coding for DNA demethylases in **rapeseed** which are salt responsive. Both gene families contained a high number of introns, the main sites important for the regulation of gene expression (Monteuuis *et al.*, 2019), a phenomenon also described for *Arabidopsis* and rice. Fan *et al.* (2020) showed that high salt stress resulted in the downregulation of methylase gene subfamilies *BnaCMT* and *BnaMET*, while subfamily *BnaDNM* was expressed more under salt stress. Also, demethylase gene subfamilies *BnaDME* and *BnaDML* were upregulated by saline conditions, while *BnaROS1* was not. The role of demethylases in salt-stress adaptation was also proven by Bharti *et al.* (2015) for **tobacco**. In their study, they showed that transgenic tobacco which overexpressed *AtROS1*, decreasing the 5mC level, displayed higher tolerance. The authors referred to this as hypomethylation of genes encoding for enzymes involved in flavonoid and antioxidant biosynthesis. It is thus reasonable that variation in the methylation of key genes is essential for the salt-stress response. As already proposed by Fan *et al.* (2020), the altered methylation status might (i) lead to more alternatively spliced transcripts or (ii) provide more protection against the mutation frequency of the coding genes. Methylation in general can occur genome-wide or at specific loci, which may be related to transcriptional changes of genes involved in salt-stress response (Marconi *et al.*, 2013; Zhang *et al.*, 2018). For **lucerne (alfalfa)**, Al-Lawati *et al.* (2016) showed in a comparison of different landraces that high salinity (20 dS/m) not only increased the number of methylated loci (10% non-methylated, 20% semi-methylated, 70% methylated), but also promoted global DNA methylation by approximately 26%. Since demethylation by 5-azacytidine, an DNA methylation inhibitor, reduced total plant performance, they concluded that, unlike other abiotic stressors, methylation is a prerequisite to maintaining salt tolerance at least in lucerne. This is in line with Wang, B. *et al.* (2015) who found that susceptible **cotton** cultivars showed a higher hypomethylation level, whereas hypermethylation dominated in tolerant cultivars. On the contrary, an opposing relationship was found for **soybean**, generally classified as more susceptible to salt stress: Chen *et al.*'s (2019) study confirmed the general belief that hypomethylation confers acclimatization of plants to salinity. When comparing non-treated with stressed plants (80 mM NaCl), soybean showed a slightly increased methylation rate of all cytosine sequence contexts under control conditions, which is in line with an increased number of expressed genomic loci (+50%) in response to salt stress. Thus, hypomethylation helped in reshaping the soybean transcriptome to withstand the adverse stress effects. However, Shams *et al.* (2020) showed an opposite reaction of different **pepper** cultivars to salt stress. In 'Maras', a more tolerant variety, methylation was decreased under a salinity level of 7 dS/m, accompanied by an increase in K^+/Na^+ ratio and antioxidant enzyme activity (peroxidase, superoxide dismutase, ascorbate peroxidase). In contrast, hypermethylation and subsequent gene silencing were triggered in 'Carliston' and 'Dolmalik' cultivars, decreasing salt tolerance as reflected by increased malondialdehyde (MDA) and H_2O_2 content together with a decrease in K^+/Na^+ ratio. Similar effects were also observed for **rapeseed** (Marconi *et al.*, 2013). Comparing salt-sensitive 'Toccata' with salt-tolerant 'Exagone', methylation was increased in 'Toccata' while decreased in 'Exagone'. This indicated that hypermethylation led to gene silencing and increased susceptibility without the ability to adapt to the stress. Among all the mentioned dicot plants, **sugarbeet** can be classified as the most salt-tolerant crop. Skorupa *et al.* (2021) observed that under salt stress relatively more genes are upregulated compared to its wild ancestor *Beta maritima*. To achieve such transcriptional mobilization, hypomethylation of certain functionally inactive but salt-responsive genes (e.g. aquaporins, expansins, osmotin, peroxidases) also seems to be the key driver for stress adaption in sugarbeet. Such diverse reactions upon salt treatment are also described for monocot plants. Zhong and Wang (2014) described an increase in salt tolerance due to increased DNA methylation in **wheat**. Also, Kumar *et al.* (2017) observed that salt-tolerant wheat genotypes showed an approximately 12% higher methylation level compared to salt-sensitive genotypes, although salt stress overall increased the methylation level. Hypermethylation in the coding region of salt-responsive high-affinity K^+ transporters (*HKTs*) correlated with downregulation of *HKT2* genes in roots of both genotypes while upregulation occurred in shoots of susceptible variants (Kumar *et al.*, 2017). Contrasting results by Zhong *et al.* (2009) found that hypomethylation was the dominant effect under salt stress both in salt-tolerant wheat cultivar 'Dekang-961' and salt-sensitive 'Lumai-15', while hypermethylation was only present at a low extent. Here, they argued that the formation of reactive oxygen species (ROS), a typical response to salt stress, may be a direct inhibitor of DNA methylation due to the formation of 8-hydroxyguanosine after ROS attack on DNA (Cerda and Weitzman, 1997). In the presence of 8-hydroxyguanosine, especially in CG sequences, the methylation of adjacent cytosine residues is strongly inhibited (Wachsman, 1997). Likewise, a general downward trend in methylation upon salt treatment was also observed for a salt-tolerant **maize** inbred line (Sun *et al.*, 2018), although different genes showed a diverse pattern. While a total of 828 genes were hypomethylated under 100 mM NaCl, 566 genes appeared to be hypermethylated, the numbers even increasing at 200 mM NaCl (939 versus 783 genes), covering all functions in the plant. By quantitative real-time reverse-transcription PCR (qRT-PCR), they further investigated the outcome of methylation and concluded that DNA methylation might affect the transcription of genes, while others may not be affected only by DNA methylation. DNA methylation is also obvious from Tan (2010) who observed that intron methylation of protein phosphatases (PP2C), which act as negative regulators in the abscisic acid (ABA)-induced antioxidative defence system of plants, repressed *zmPPC2* transcription. On the other hand, salt-induced demethylation of glutathione-*S*-transferase (GST), playing a key role in ROS detoxification, led to an upregulated expression of *zmGST*, stimulation of the antioxidant system, and acclimatization of maize to salt stress (100 mM NaCl). Another important crop showing a high dynamic in DNA methylation is **rice** (Ferreira *et al.*, 2015). In phenotypically contrasting rice genotypes total hypomethylation dominated under salinity, which was well linked to an increased expression of DNA demethylases in salt-tolerant 'Pokkali' (Ferreira *et al.*, 2015). In contrast, in salt-sensitive variety 'IR29' both methylases as well as demethylases were induced at a similar rate, leading to the conclusion that this resulted in lower plasticity and thus susceptibility. Similarly, it was further shown for rice that salt stress led to rapid demethylation of *OsMYB91* and subsequent changes in histone modification (Zhu *et al.*, 2015). This consequently increased proline levels and enhanced the ROS scavenging capacity. Consistent with Ferreira *et al.* (2015), Wang, W. *et al.* (2015) also demonstrated hypomethylation of two contrasting rice genotypes ('IL177-103', tolerant; 'IR64', sensitive), but pointed to tissue-specific alterations, showing a high degree of demethylation in roots, while shoots were only slightly affected. Tissue-specific differences in methylation patterns upon salt-stress perception were also described by Karan *et al.* (2012). They found for salt-sensitive rice varieties ('IR29') a predominance of hypomethylation in shoots, but hypermethylation in salt-tolerant 'Pokkali' and 'Nipponbare', while in roots hypomethylation dominated in all these genotypes. Further expression analysis revealed that these (de)methylation processes were not necessarily linked to a unidirectional response in expression pattern. For example, while demethylation of *Os02g44330* led to an increased expression in the shoot of both salt-sensitive ('IR29') and salt-tolerant ('Nipponbare', 'Pokkali') varieties, a reduced expression was found in salt-sensitive 'Geumgangbyeo' and salt-tolerant 'Nonabokra' under salt stress. From these results, it can be concluded that DNA methylation is a key molecular switch for crops to adapt to salt stress. However, the relationship between the DNA methylation pattern and the susceptibility of the crop to salt stress is highly dynamic and needs to be more investigated to elucidate the effects of salt stress on crop-specific epigenetic regulations.

4.2.2 Histone modifications

Chromatin is an instructive DNA scaffold of ~2 m in size with histones inside a nucleus, forming a DNA–protein complex that responds to external cues (Li *et al.*, 2014). The nucleosome in eukaryotes is the basic unit of chromatin, which is composed of approximately 146 bp of DNA-wrapped protein octamers (i.e. duplicates of four core histones such as H2A, H2B, H3, and H4) (Mariño-Ramírez *et al.*, 2005; Lämke and Bäurle, 2017). The histone structure gives several genetic variabilities through many enzymatic modifications *in vivo* which play fundamental roles in most biological processes (Gujral *et al.*, 2020). Histone modifications alter the structural and functional capabilities of chromatin by undergoing extensive post-translational modifications (PTMs), such as acetylation, phosphorylation, lysine methylation, arginine methylation, deamination, ADP-ribosylation, proline isomerization, β-*N*-acetylglucosamine, ubiquitylation, and SUMOylation, in response to environmental changes (Strahl and Allis, 2000; Mariño-Ramírez *et al.*, 2005; Yung *et al.*, 2021) (Fig. 4.2). Besides protruding tails, PTMs of histone also occur in the core histone domain amino acids such as lysine or arginine as well as serine, threonine, tyrosine, etc. (Zhou *et al.*, 2013). Among those modifications, histone methylation (H3K79me, H4K59me, and H2AK99me), histone acetylation (H3K56ac, H3K122ac, and H4K91ac), and histone phosphorylation H3T118 are mentionable (Valencia-Sánchez *et al.*, 2021). Histone modifications regulate chromatin structures and are responsible for recruiting remodelling enzymes (Bannister and Kouzarides, 2011). Together with DNA methylation, histone PTMs represent the main level of epigenetic regulation to control chromatin structure and gene transcription, impacting various important cellular phenotypes to counter unfavourable environmental conditions for better survival. The involvement of histone modifications in influencing the salt-stress response in plants is gradually being explored (Song *et al.*, 2012; Li *et al.*, 2014; Shen *et al.*, 2014; Yung *et al.*, 2021) (Fig. 4.2). Differential regulation of gene expression occurs due to the significant differences in histone modifications under salt stress (Sokol *et al.*, 2007; Chen *et al.*, 2010; Pikaard and Scheid, 2014; Paul *et al.*, 2017) (Fig. 4.2). Through histone modifications, researchers can uncover novel epigenetic insights into cellular processes. Here, we describe the known histone modifications and discuss their functional consequences under salinity stress in plants.

Histone methylation regulates the transfer of methyl groups to histone proteins via histone methyltransferases (HMTs) and histone demethylases (HDMs). Histone methylation and demethylation lead to modifications in chromatin structure that cause transcriptional activation as well as silencing (Liu *et al.*, 2016). N-terminal tails, which protrude from the histone proteins and are exposed on the surface of the nucleosome, act as sites for the methyl groups and are the most common location for PTMs (Strahl and Allis, 2000). The side chains of lysine (K) and arginine (R) are the sites for histone methylation and do not alter the charge of the histone protein. Furthermore, lysine may be mono-, di-, or trimethylated, whereas arginine may be mono-, symmetrically, or asymmetrically dimethylated (Bannister and Kouzarides, 2011). In *Arabidopsis*, histone methylation represents repressive (H4R3me2, H3K9me2/3, and H3K27me3) and active marks (H4R3me2, H3K4me3, and H3K36me2/3) (Liu *et al.*, 2016). In *Arabidopsis* (Song *et al.*, 2012) and soybean (Bilichak *et al.*,

2012), salt stress increases methylation at the fourth lysine of H3 (H3K4me3) and/or decreases histone H3 lysine 9 dimethylation (H3K9me2) associated with salt-responsive genes. Mild salt stress induced the *HKT1* gene by the change in the H3K27me3 level, which was due to the physiological effects of the priming treatment (Sani *et al.*, 2013). HD2C regulates gene expression through histone modifications under salt stress and is functionally associated with HDA6 (Luo *et al.*, 2012). DNA methylation at H3K4 and H3K27 regulates transcriptional expression in response to salt stress in castor bean and rice, respectively (Karan *et al.*, 2012; Han *et al.*, 2020). A similar observation has been reported with castor bean where methylation of a key salt-response regulator *RSM1* (RADIALIS-LIKE SANT, a myeloblastosis (MYB) transcription factor (TF)) also occurred at H3K4 and H3K27 (Han *et al.*, 2020). In another study on soybean, it was confirmed that the transcription of some salt-tolerance genes was regulated by the H3K4me0/1/2 code reader (GmPHD6) (Wei *et al.*, 2017). Variation in methylation levels at H3K4me3 and H3K27me3 has been reported in rice tolerant ('Nonabokra') and sensitive ('IR64') varieties that impart salt tolerance through differential expression levels of the salt-responsive *OsBZ8* gene (Paul *et al.*, 2017). Furthermore, the *JMJ15* gene (coding for H3K4 demethylase) overexpression in *Arabidopsis* improved salt tolerance under salt stress (Shen *et al.*, 2014).

Histone acetylation is a primary and critical epigenetic modification to regulate gene expression. It changes chromatin architecture by opening or closing the chromatin structure through reducing the affinity between DNA and histone protein, and enhancing DNA's accessibility by the transcriptional machinery resulting in transcriptional activation and/or repression (Zheng *et al.*, 2016; Onufriev and Schiessel, 2019). Acetylation of lysine residues of histone H3 (e.g. H3K9ac) by histone acetyltransferases (HATs) leads to transcriptionally active euchromatin, and deacetylation by histone deacetylases (HDAs) leads to inactive and condensed heterochromatin in plants under salt stress (Zheng *et al.*, 2016; Kim *et al.*, 2019; Onufriev and Schiessel, 2019). Histone acetylation at H3K9/H3K14 increased salt tolerance through elevated expression levels of *GCN5* (*general control non-derepressible* 5) and activated chitinase-like (CTL) protein involved in cell-wall biosynthesis under salt conditions (Zheng *et al.*, 2021). Salt stress-induced histone acetylation triggers transcriptional activation of salt stress-responsive genes in tobacco (Sokol *et al.*, 2007) and maize (Li *et al.*, 2014). Histone deacetylase shows both positive and negative relation to salinity tolerance in rice (Zhao *et al.*, 2015; Zhao *et al.*, 2016). TaHAG1 plays a decisive role in strengthening the salt tolerance in bread wheat (Zheng *et al.*, 2021). In maize roots, acetylation at H3K9 and H4K5 in the promoter and coding regions upregulates cell wall-related genes such as *ZmEXPA1*, *ZmEXPA3*, *ZmEXPA5*, *ZmEXPB1*, *ZmEXPB2*, and *ZmXET1* which are necessary for high salinity response (Li *et al.*, 2014). *Arabidopsis* mutants *HD2C*, *HDA6*, and *HDA19* showed hypersensitivity to salt due to increased levels of histone acetylation (Chen *et al.*, 2010; Luo *et al.*, 2012). Deposition of active histone marks (e.g. H3K9ac and H3K4ac) was demonstrated by Yolcu *et al.* (2016) resulting in peroxidase gene activation in *Beta vulgaris* and *B. maritima*. Overexpression of *AtHD2* in transgenic *Arabidopsis* plants enhanced tolerance to salt stress, suggesting the involvement of HD2 proteins in plant response to salt stresses (Sridha and Wu, 2006). Treatment with Ky-2, a histone deacetylase inhibitor, enhanced high-salinity stress tolerance by increased histone acetylation of AtSOS1 and AtSOS3 under salinity stress (Sako *et al.*, 2016). By locus-specific acetylation of histones H3 and H4, ADA2b positively regulates salt-induced gene expression (Kaldis *et al.*, 2011). In cotton, suberoylanilide hydroxamic acid (SAHA) pretreatment induces global hyperacetylation of histone H3 at lysine 9 (H3K9) and histone H4 at lysine 5 (H4K5) under salt stress, indicating that SAHA could act as the HDAC inhibitor for conferring plant salt tolerance (He *et al.*, 2020). Overexpression of histone deacetylase in transgenic poplar plants reduced tolerance to salt stress (Ma *et al.*, 2019). In *Hibiscus cannabinus*, the levels of acetylation at H3K9ac, H3K27ac, and H4K5ac were upregulated under salt-stress conditions (Wei *et al.*, 2019). OsHDA1 was reported to negatively affect the transcriptional activation of OsSOS1 in rice (Cheng *et al.*, 2018). Altering the levels of H4 acetylation (H4K5 and H4K16) in their promoters HDA710/OsHDAC2, an HDA RPD3/HDA1 family member, contributes to controlling the rice salt-stress genes (Ullah *et al.*, 2021).

The addition of a phosphoryl group to any of the four nucleosomal histone tails, i.e. **histone phosphorylation**, can play an important role in gene expression, DNA damage repair, and chromatin remodelling. Serine, threonine, and/or tyrosine residues at H3 are the sites for this PTM. Although this PTM is highly conserved and confers a negative charge to the histone, resulting in a more open chromatin conformation (Wang, Z. *et al.*, 2015), in *Arabidopsis* phosphorylation of H3T3 in pericentromeric/knob regions involving MUT9-like kinase 1 and 2 (repressive epigenetic mark) is important for maintaining proper heterochromatic organization and, possibly, chromosome function(s) to salt stress (Wang, Z. *et al.*, 2015). Interestingly, rapid transient upregulation of histone H3 serine 10 phosphorylation, H3 phosphoacetylation, and histone H4 acetylation follow immediately in tobacco BY-2 and *Arabidopsis* T87 cell cultures under high salinity stress, indicating histone phosphorylation along with other histone modification (Paulson and Taylor, 1982; Sokol *et al.*, 2007).

Histone deamination is a PTM involving the conversion of an arginine to a citrulline, i.e. arginine citrullination, by L-arginine deiminases (e.g. Ca^{2+}-dependent peptidyl arginine deiminases (PADs)) (Witalison *et al.*, 2015; Marondedze *et al.*, 2021). As a PTM, arginine citrullination in higher plants assigns its role in abiotic stress responses in *Arabidopsis* (Marondedze *et al.*, 2021).

β-*N*-Acetylglucosamine (O-GlcNAc) sugar residues modify both non-histone and histone proteins via modification of their serine (Ser) and threonine (Thr) side chains (Sakabe *et al.*, 2010). Addition or removal of the sugar is catalysed by O-GlcNAc transferase (OGT) and *N*-acetylglucosaminidase (O-GlcNAcase), respectively. The covalent attachment of O-GlcNAc to Ser/Thr residues of proteins acts as not only a PTM but also a nutritional sensor in the nucleus and cytoplasm, which directly regulates the expression of genes and multiple crucial signal transduction pathways (Liu *et al.*, 2015). *O*-linked β-*N*-acetylglucosamine (O-GlcNAc) modification of protein abundances in *Suaeda maritima* under NaCl stress resulted from overexpression of the O-GlcNAc transferase gene (*OGT*) (Sahu and Shaw, 2009). *O*-linked β-*N*-acetylglucosamine transferase SPINDLY acts as a negative regulator of gibberellic acid (GA) signalling as well as a negative regulator of salt-stress response in *Arabidopsis* (Kim *et al.*, 2013).

ADP-ribosylation is an important and reversible PTM carried out by mono-ADP-ribosyltransferases (MARTs), poly-ADP-ribose polymerases (PARPs), and poly-ADP-ribose glycohydrolases (PARGs), where histones are mono- and poly-ADP-ribosylated on glutamate and arginine residues (Lamb *et al.*, 2012). Moreover, PARP-mediated ribosylation inhibits the demethylase and excludes it from chromatin, thereby making target promoters more accessible (Muthamilarasan *et al.*, 2016). A class-one ADP-ribosylation factor gene (*SaARF1*) from the halophyte *Spartina alterniflora* imparts salinity-stress tolerance when it is expressed in both mono- and dicot systems (Joshi *et al.*, 2014). By regulating genes involved in proline biosynthesis, three ADP-ribosylation factors (*PvArf1*, *PvArf-B1C*, and *PvArf*-related) contribute to salinity tolerance in transgenic switchgrass (Guan *et al.*, 2020). A higher amount of ADP-ribosylation factor 1 (*ARF1*) transcripts was accumulated in response to salinity stress in *M. truncatula* (de Lorenzo *et al.*, 2007).

Ubiquitination is a larger covalent modification that involves the sequential action of E1-ubiquitin-activating, ubiquitin-conjugating, and E3-ubiquitin-ligating enzymes (Hershko, 1998). The enzyme complexes determine both substrate specificity as well as the degree of ubiquitination. However, two well-characterized sites lie within H2A and H2B involved in gene silencing (Wang *et al.*, 2004) as well as transcriptional initiation and elongation (Kim *et al.*, 2009). The modification is removed via the action of isopeptidases (de-ubiquitin enzymes) which is important for both gene activity and silencing (Xhemalce *et al.*, 2011). The ubiquitination (U-box and RING) of proteins affected the protective mechanisms in rice seedlings at the initial phase against salt stress (Liu, C.W. *et al.*, 2012). Gene expression assays indicated that several salt stress-related genes were significantly upregulated in transgenic rice and *Arabidopsis* (Zhang, H. *et al.*, 2015). In brassinosteroid (BR)-mediated salt-stress response and BR signalling, U-box (UBC32) plays a role necessary for the plant to tolerate salt in *Arabidopsis* (Cui *et al.*, 2012). Plant U-box (PUB) E3 ubiquitin ligase PnSAG1 plays a negative role in plant response to ABA and salt stress (Wang *et al.*, 2019). F-box proteins, as key components of the

E3 ubiquitin ligase in the Skp1/Cullin or CDC53/F-box (SCF) complex, are negative regulatory proteins that improve tolerance to salt stress in *Arabidopsis* (Yu *et al.*, 2020). The U-box E3 ubiquitin ligase TaPUB1 regulates salt-stress tolerance in wheat positively (Wang *et al.*, 2020), whereas GmPUB21 shows the opposite effect in soybean (Yang *et al.*, 2022).

SUMOylation mirrors the modification of ubiquitylation and covalently attaches small ubiquitin-like modifier molecules to histone lysine residues via the sequential action of three enzymes E1, E2, and E3 (Bannister and Kouzarides, 2011). SUMOylation has been detected on all core histones and antagonized acetylation and ubiquitylation (Nathan *et al.*, 2006). Salt directly affects OTS1-dependent signalling in *Arabidopsis* by inducing OTS1 protein degradation, indicating the requirement of OTS1 deSUMOylation activity in plant salt-tolerance responses (Conti *et al.*, 2009).

4.2.3 Influence of ncRNAs in modulating salt tolerance of plants

Plants' response to salinity may vary from avoidance, tolerance, and escape to recovery from stress. Plant salt-tolerance mechanisms aim for genes associated with the photosynthesis process, TFs controlling growth, activating genes associated with salt-responsive hormone signalling, genes that reduce the toxicity by minimizing take up of toxic substances like Na^+, and genes that minimize a function of ROS (Wang, W. *et al.*, 2015; Sun *et al.*, 2016; Khandal *et al.*, 2017). To tackle salinity stress plants adopt a variety of biochemical and molecular processes regulated by genes and proteins, where non-coding RNA molecules (ncRNAs) – including short-chain or small non-coding RNA molecules, e.g. microRNAs (miRNAs), small interference RNAs (siRNAs), piwi-interacting RNAs (piRNAs), circular RNAs (circRNAs), and long non-coding RNAs (lncRNAs) – control gene expression at the transcription, post-transcription, and translation levels (Wang, T.Z. *et al.*, 2015; Sun *et al.*, 2016; Wei *et al.*, 2017; Chen *et al.*, 2019; Jha *et al.*, 2019, 2020; Tian *et al.*, 2020). A large number of studies have shown a significant role of non-encoding RNAs in epigenetic modification and their ability to regulate gene expression (Xia *et al.*, 2012; Wei *et al.*, 2017; Ma *et al.*, 2021; Tang *et al.*, 2022). In the regulation of the salt stress-induced alterations in plants, ncRNAs play critical roles in fine-tuning the PTMs for modulating plant responses to salt stress as potential targets to engineer salinity tolerance in crops (Banerjee *et al.*, 2017). Many studies have revealed that stresses induce aberrant expression of ncRNAs, which could be used as a tool for plant improvement through enhanced tolerance to salinity (Tables 4.1 and 4.2). Many ncRNAs and their target genes participate in salinity-stress response in plants by their regulatory roles whereby their expressions are up- or downregulated under salt stress. In groundnut (peanut), the expression of plant miR160 and its ARF target gene under salt stress and the miR160 overexpression facilitated ARF18 pathway checks on seedling growth against the impacts of ROS by salinity stress (Tang *et al.*, 2022). Additionally, overexpression of miR156a weakens salt resistance in apple but its target gene, i.e. *MdSPL13*, provides strength for salt resistance (Ma *et al.*, 2021). Moreover, the similar miRNA or different miRNAs from the identical miRNA family results in diverse promoting or inhibitory impacts on the salt-tolerance mechanism of diverse crop types. For instance, salt tolerance in rice and *Arabidopsis* decreases from overexpression of *Osa-miR393* (Xia *et al.*, 2012) whereas overexpression of the miR393-resistant *mTIR1* form augments salt tolerance in *Arabidopsis* (Chen *et al.*, 2015). By contrast, overexpression of *Osa-miR393a* improves the K^+ uptake of creeping bentgrass and enhances the tolerance to salinity stress (Zhao *et al.*, 2019). Moreover, overexpression of *Osa-miR396c* and its growth-regulating factor (GRF) target gene reduces salt- and alkali-stress tolerance in rice and *Arabidopsis* (Gao *et al.*, 2010) but improves the salt-tolerance mechanism of transgenic creeping bentgrass via salinity overly sensitive 1 (As-SOS1)-facilitated Na^+ elimination (Yuan *et al.*, 2019). The overexpression of miR164-targeted gene *PeNAC070* results in increased sensitivity to salt stresses in *Arabidopsis* (Lu *et al.*, 2017) and so does the overexpression of miR169 and its target gene *GmNFYA3* (Ni *et al.*, 2013). Another miR414c and its target gene *GhFSD1* (iron superoxide dismutase 1) play roles in developing a response to salinity-stress tolerance in cotton. Overexpression of miR414c can reduce the expression of *GhFSD1* and improve the response to

Table 4.1. List of several published ncRNAs in different crops controlling salt-stress responses.

Number of ncRNAs	Crop species	References
Drought induced lncRNA (DRIR)	*Arabidopsis thaliana*	Qin *et al.* (2017)
1117 unique lncRNAs identified and 44 lncRNAs expressed under stress	*Gossypium hirsutum*	Deng *et al.* (2018)
44,698 lncRNAs identified and 2278 lncRNAs expressed under stress	*Triticum aestivum*	Shumayla *et al.* (2017)
10,646 lncRNAs identified and 8592 expressed under stress	*Populus deltoides*	Jha *et al.* (2020)
CNT0018772 and CNT0031477	*Hordeum vulgare*	Karlik and Gözükırmızı (2018)
48,345 lncRNAs identified and 1710 lncRNAs expressed under stress	*Zea mays*	Huanca-Mamani *et al.* (2018)
385 conserved miRNAs and 68 novel miRNAs	*Medicago truncatula* *Medicago sativa*	Long *et al.* (2015)
876 miRNAs	*M. truncatula*	Cao *et al.* (2018)
3030 long intergenic non-coding RNAs (lincRNAs), 275 long non-coding natural antisense transcripts (lncNATs)	*Glycine max*	Chen *et al.* (2019)
71 miRNAs	*G. max*	Sun *et al.* (2016)
59 novel miRNAs	*Cicer arietinum*	Kohli *et al.* (2014)
3457 high-confidence lncRNAs	*C. arietinum*	Singh *et al.* (2017)
284 unique miRNAs	*C. arietinum*	Khandal *et al.* (2017)
18 miRNAs	*Vigna unguiculata*	Paul *et al.* (2011)
Among the differentially expressed miRNAs: 298 upregulated and 395 downregulated in IC-1 and IS-4 combination; 284 upregulated and 243 downregulated in HC-4 and HSA combination	*Vicia faba*	Alzahrani *et al.* (2019)

salt stress by regulating ROS metabolism (Wang *et al.*, 2019). Similarly, another study found that the expression of miRNVL5 shows susceptibility to salt stress by suppressing CBF (C-repeat binding factor), ERF (ethylene-responsive component binding factor), etc. (Gao *et al.*, 2016). Functional validation by quantitative real-time PCR showed the upregulatory activity of miRNAs under stress coming from soil salinization. In a study on the role of miRNAs in regulating salinity stress in legumes, 18 conserved miRNAs were recovered from root tissue of legume crops and 15 corresponding target gene(s) were detected as TFs that are involved in managing different developmental processes (Paul *et al.*, 2011).

4.3 Nucleosome Occupancy in Modulating Plant Responses to Salt Stress

Transcriptional activity often correlates inversely with nucleosome occupancy. When transcription is activated in a genomic area, the number of nucleosomes drops and the chromatin structure relaxes (Eberharter and Becker, 2002). Nucleosome occupancy modulates salt-stress tolerance through different epigenetic mechanisms occurring at the nucleosome level in response to salt stress. The nucleosome-regulated responses to salt stress and the accessibility of TFs and cofactors for DNA-tempered processes are influenced by chromatin structure. Numerous epigenetic mechanisms control the structure and function of chromatin, such as DNA methylation, histone modifications, ATP-dependent chromatin remodelling, positioning of histone variations, and regulation by ncRNA (Berger, 2007). The dynamic competition between nucleosomes and TFs regulates the activity of the promoter in gene transcription regulated by RNA polymerase. Consequently, nucleosome removal or replacement with other nucleosomes of different compositions occurs after gene transcription (Zhang, T. *et al.*, 2015). For example, the substitution of H2A with the histone variation H2A.Z improves variable gene expression without altering gene DNA methylation (To and Kim, 2014). In yeast, H2A/H2B dimers are substituted for H2A.Z/H2B dimers by the chromatin-remodelling SWR1 complex. Based on these findings, it was hypothesized that salt stress encourages the eviction of nucleosomes that contain H2A.Z. This indicates that salt stress causes H2A.Z-containing nucleosomes to be ejected from the promoter region. As a result, H2A.Z-containing nucleosomes at gene-body regions wrap the DNA more tightly and reduce transcription by changing the binding affinity of the AtMYB44 protein, which acts as a repressor of the protein's transcription (Coleman-Derr and Zilberman, 2012). Dai *et al.* (2017) reported that H2A.Z preferentially binds to H3K4me3 at promoters and suppresses gene expression by altering nucleosome structure in a genome-wide mapping of nucleosome occupancy in *Arabidopsis* inflorescences. Additionally, Chen *et al.* (2010) demonstrated that salt stress might raise the amounts of H3K4me3 on DREB2A and RD29B and that this was associated with increased expression. Osmotic stress or salinity stress is often related to chromatin-level response. According to Yu *et al.* (2018), ZmCHB101, a homolog of *Arabidopsis thaliana* SWI3D in maize, is crucial for the osmotic and dehydration stress responses. Osmotic and salt stress-sensitive phenotypes were present in ZmCHB101-RNA interference (RNAi) transgenic plants. Genome-wide RNA-sequencing investigation revealed that ZmCHB101 affects the transcriptional expression landscape of genes that respond to osmotic stress. Nucleosome concentrations around the transcription start sites of crucial stress-responsive genes are controlled by ZmCHB101. H2A.Z and CENH3 are two histone variations that play a distinct and precise role in regulating gene activity and genome organization (Coleman-Derr and Zilberman, 2012). Under salt stress, it is expected that protein factors or complexes will be able to access packaged DNA containing stress-responsive genes and thus facilitate transcription. In contrast, a recent study found that most chromatin had little to no change when *A. thaliana* seedlings were exposed to the abiotic stresses of heat, cold, salt, and drought (Raxwal *et al.*, 2020). Furthermore, extreme temperatures increased chromatin accessibility whereas drought and salt stressors did not affect the outcome (Raxwal *et al.*, 2020). Numerous researchers have found that epigenetic pathways have a role in the plant stress response. For example, DNA methylation, histone modifications, chromatin-remodelling factors, and remodelling proteins have been reported to play their part in modulating the gene expression of various genes influencing salt stress through chromatin modification during stress conditions in various plant species (Fig. 4.1; Table 4.3).

High-salinity stress can change how genes are expressed by modifying DNA methylation in plants. Therefore, salt stress influences genome-wide DNA methylation and histone changes, and both mechanisms are interconnected to resist salt stress (Pandey *et al.*, 2016) (Fig. 4.2). Additionally, the SNF2/Brahma-type chromatin-remodelling gene *CHR12* is essential for controlling how salt stress causes temporary growth stopping in *Arabidopsis* (Mlynárová *et al.*, 2007). Several genes involved in chromatin control and RNA processing have been identified as part of the salt-stress signal transduction (Nishimura *et al.*, 2005). By replacing the H2A–H2B dimer with H2A.Z–H2B or directly ejecting the H2A–H2B dimer to loosen the nucleosome wrapping and trigger transcription, chromatin remodelling under stress may be accomplished (Jiang and Berger, 2017). The responsiveness of gene expression under stress conditions in *A. thaliana* is considerably positively linked with H2A.Z occupancy in the gene body (Sura *et al.*, 2017). It has been shown that the *hta9* and *hta11* double mutant's H2A.Z depletion caused slightly slower germination rates but slightly faster cotyledon greening when exposed to salt (Sura *et al.*, 2017) (Table 4.3).

In rice, the miR156–squamosa promoter binding-like (SPL) pathway functions indicate that miR156 plays a vital role in modulating plant development and responses to abiotic stress, e.g. salt stress (Cui *et al.*, 2014). One of the two documented natural antisense short interfering RNAs (nat-siRNAs) is induced by salt stress (Katiyar-Agarwal *et al.*, 2006). DiRNAs are formed from both the sense and the antisense strands near the sites of double-strand breaks, and they can act as guides for chromatin changes or the recruitment of protein complexes to source sites to aid in repair (Wei *et al.*, 2012). Abiotic or biotic variables can affect DNA methylation and histone changes, which make plants more resilient to harsh environmental circumstances. For example, in an experiment with a wheat genotype, salt stress greatly raised the methylation level in

Table 4.2. Role of ncRNAs controlling salinity stress and other growth and developmental processes in different crops, with the potential molecular mechanisms involved.

Name of ncRNA	Crop	Target gene(s)/protein-coding gene(s)	Function	References
miR172a	*Glycine max*	Glyma.10G116600, Glyma.02G087400, Glyma.13G329700, Glyma.12G073300, Glyma.15G044400, Glyma.11G053800, AP2/EREBP-type TF gene *SSAC1*, thiamine biosynthesis gene *THI1*	Induction cleaves mRNA transcripts of salt-suppressed AP2 domain-containing genes, increasing expression of thiamine biosynthesis gene *THI1* and resulting in salinity tolerance	Pan *et al.* (2016)
18 conserved miRNAs	*Vigna unguiculata*	15 target genes	Regulate root growth and plant developmental process under stresses by targeting various genes encoding TFs, i.e. SBP, ARF, SPL, TCP, NFY, and AP2	Paul *et al.* (2011)
miR156_1, miR156_10, car-miR008, car-miR011, car-miR015	*Cicer arietinum*	Squamosa promoter-binding protein	Target protein-encoding gene to control salinity stress	Kohli *et al.* (2014)
lncRNAs TCONS_00097188, TCONS_00046739, TCONS_00100258, TCONS_00118328, TCONS_00047650, TCONS_00020253, TCONS_00116877	*Medicago truncatula*	*Medtr6g006990*, *cytochromeP450*, *Medtr3g069280*, *Medtr1g081900*, *Medtr7g094600*	Upregulate diverse gene expressions influencing salt-stress adaptation	Wang, T.Z. *et al.* (2015)
lncRNAs TCONS_00292946, TCONS_00176941, TCONS_00011551	*Arachis hypogaea*	–	Regulate salt-stress tolerance	Tian *et al.* (2020)
pvu-miR159.2	*Phaseolus vulgaris*	–	–	Arenas-Huertero *et al.* (2009)
miR160, miR156/157, miR159, miR169, miR172, miR408	*V. unguiculata*	Auxin response factor (ARF), squamosa promoter-binding protein (SBP), TCP family transcription factor, CCAAT-binding transcription factor (CBF), PHAP2B protein, APETALA2 protein (AP2), Basic blue copper protein/Plantacyanin	Target TFs and regulate salt stress	Paul *et al.* (2011)

Table 4.3. Histone modifications, DNA methylation, chromatin-remodelling factors, and remodelling proteins involved in salt-stress regulation.

Type	Target gene/genomic region	Plant species	References
GNAT family of histone acetyltransferases	*OsHAG703*	Rice	Liu, X. *et al.* (2012); Fang *et al.* (2014)
MYST	*OsHAM701*	Rice	
CBP/p300	*OsHAC703*	Rice	
	OsHAC704	Rice	Liu, C.W. *et al.* (2012)
RPD3/HDA1	*AtHDA19*	*Arabidopsis*	Zhou *et al.* (2005); Choi *et al.* (2012)
RPD3/HDA1	*OsHDA714*	Rice	Fu *et al.* (2007)
HD2-like	*AtHD2C*	*Arabidopsis*	
SIR2	*OsSIRT701*	Rice	
SIR2	*OsSIRT702*	Rice	
Arginine	*AtPRMT5*	*Arabidopsis*	Zhang *et al.* (2011)
Remodellers	*AtCHR12*	*Arabidopsis*	Mlynárová *et al.* (2007)
CGH hypermethylation	Nuclear genome	*Mesembryanthemum crystallinum*	Dyachenko *et al.* (2006)
HAT complex	*Ada2b*	*Arabidopsis*	Kaldis *et al.* (2011)
	SGF29a	*Arabidopsis*	
HDAC	*HDA6*	*Arabidopsis*	Chen *et al.* (2010)
	HDA19	*Arabidopsis*	
HDAC complex	*AtSAP18*	*Arabidopsis*	Song and Galbraith (2006)
Arginine methyltransferase	*SKB1*	*Arabidopsis*	Zhang *et al.* (2011)
Histone variants	*HTA9*, *HTA11*	*Arabidopsis thaliana*	Sura *et al.* (2017)
Histone chaperones	*DEK3*	*A. thaliana*	Waidmann *et al.* (2014)
	MSI1	*A. thaliana*	Mehdi *et al.* (2016)
	NAP1;1	*A. thaliana*	Liu *et al.* (2009)
	OsNAPL6	*Oryza sativa*	Tripathi *et al.* (2016)
Chromatin-remodelling factors	*ARP6A*	*A. thaliana*	Sura *et al.* (2017)
	CHR12A	*A. thaliana*	Mlynárová *et al.* (2007)
	PIE1A	*A. thaliana*	Sura *et al.* (2017)
	PKL (*CHR6*)	*A. thaliana*	Yang *et al.* (2019)
	EEN (*IES6*), *EIN6* (*REF6*), *IED6*	*A. thaliana*	Zander *et al.* (2019)

the shoot of a salt-sensitive genotype (Kumar *et al.*, 2017). Furthermore, the downregulated expression of *HKT2* genes in the investigated wheat genotype was linked with increased methylation brought on by salt stress.

4.4 Future Perspectives

To address the ultimate constraints on crop production, it is important to understand and utilize natural variation in plants as a genetic reservoir for adaptation strategies. Early evolution of terrestrial plants occurred under dry conditions with extreme temperatures while the domestication of the crops we use today occurred under favourable conditions. Consequently, in the strong selection for important production traits, genetic variation has been increasingly limited. As a result, our current crop plants are no longer able to respond adequately to various stress factors. This is compounded by the genetic complexity of stress responses, which has limited the success of previous breeding efforts to restore plant tolerance to, for example, salinity. Plants respond to salinity differently via modulating plant osmoticum, reducing ion toxicity, and scavenging ROS (Hossain and Dietz, 2016). However, the aim must be to develop resistant, climate-smart varieties that can withstand climate change and its consequences in terms of productivity. Besides natural salinization, improper irrigation techniques in arid and semi-arid climates lead to an accumulation of soluble salts in the soil, affecting plant growth and development. Today, approximately 20% of all irrigated land systems worldwide are already adversely affected by soil salinization (Qadir *et al.*, 2014), with an increasing trend. As most crops are glycophytes, high soil salinity not only hampers growth but also seriously limits crop productivity and yield. Considering the increase of the world's population to above 9 billion by 2050 (Obaid, 2001), unravelling the underlying mechanisms of salt tolerance is crucial to guarantee food security. This becomes even more challenging as salt tolerance is a multigenic trait, which makes conventional breeding more challenging. Therefore, also the understanding of epigenetics should be focused upon. Current studies of the role of epigenetic modifications and nucleosome occupancy in modulating salt stress are incomplete. The model plant *A. thaliana* is still the most frequently used to study histone changes, DNA methylation, and chromatin remodelling in plant responses to salt stress. However, the genomes of crops are often substantially more complex than that of *A. thaliana*. This relationship must be properly understood to grasp the role played by chromatin remodelling, histone modification, and DNA methylation impacted by epigenetic alterations in response to salt stress. Furthermore, in most crop species, it is frequently difficult to find particular mutants of histone-modifying enzymes, histone variations, histone chaperones, and chromatin-remodelling proteins. For basic research and practical application, it is thus still difficult to transfer knowledge from *A. thaliana* to crops. Moreover, relatively little research has been done on the precise DNA sequences or protein factors that control the histone modifications or chromatin remodelling under salt stress. Contrarily, crops grown in salt-affected fields are stressed throughout their whole life cycle and into the next generations. However, more research must be done by subjecting different plant species to continuous salt exposure throughout their whole life cycle to determine whether alterations in histone modifications can also be found in genes involved in salt tolerance and passed down to the next generation. To do this, one must look back at the role of genetics and evolution propagated by Darwin (Ruse, 1975). He represented the view that life forms can acquire information from their environment and pass it on in their genes. Today, epigenetic studies together with the complexity of current genomic technologies show the accuracy of Darwin's assessment. To withstand abiotic stress factors, plants can make specific changes that are controlled by a sophisticated biological system. For instance, they use 'molecular switches' to e.g. modify the chromatin architecture to facilitate gene expression and thus produce stress-tolerant phenotypes with only a small loss in productivity under stress. While some of the stress-driven changes in plants are temporary, other modifications are stable and inheritable (Akhter *et al.*, 2021). Thus, the epigenome of the plant indeed passes on information to the offspring not being encoded in the DNA and further provides a mechanism for acclimatization and adaptation to stress (Avramova, 2015). Therefore, it can be rightly concluded that permanent plant–stress interactions do have a significant epigenetic impact on a plant's behaviour, which presents a new challenge to speed up breeding programmes enhancing abiotic stress tolerance. Combined investigations by molecular geneticists and bioinformaticians in deciphering epigenetic modifications may boost our knowledge about different epigenetic marks and their exploitation to breed climate-smart crops.

Acknowledgements

M.S.H. greatly acknowledges the support of the Alexander von Humboldt Foundation (AvH), Germany.

References

Akhter, Z., Bi, Z., Ali, K., Sun, C., Fiaz, S. *et al.* (2021) In response to abiotic stress, DNA methylation confers epigenetic changes in plants. *Plants* 10(6), 1096.

Al-Lawati, A., Al-Bahry, S., Victor, R., Al-Lawati, A.H. and Yaish, M.W. (2016) Salt stress alters DNA methylation levels in alfalfa (*Medicago* spp). *Genetics and Molecular Research* 15(1), 15018299.

Alzahrani, S.M., Alaraidh, I.A., Khan, M.A., Migdadi, H.M., Alghamdi, S.S. and Alsahli, A.A. (2019) Identification and characterization of salt-responsive microRNAs in *Vicia faba* by high-throughput sequencing. *Genes (Basel)* 10(4), 303.

Arenas-Huertero, C., Pérez, B., Rabanal, F., Blanco-Melo, D., De la Rosa, C. *et al.* (2009) Conserved and novel miRNAs in the legume *Phaseolus vulgaris* in response to stress. *Plant Molecular Biology* 70, 385–401.

Avramova, Z. (2015) Transcriptional 'memory' of a stress: transient chromatin and memory (epigenetic) marks at stress-response genes. *The Plant Journal* 83(1), 149–159.

Banerjee, S., Sirohi, A., Ansari, A.A. and Gill, S.S. (2017) Role of small RNAs in abiotic stress responses in plants. *Plant Gene* 11, 180–189.

Bannister, A.J. and Kouzarides, T. (2011) Regulation of chromatin by histone modifications. *Cell Research* 21(3), 381–395.

Berger, S.L. (2007) The complex language of chromatin regulation during transcription. *Nature* 447(7143), 407–412.

Bharti, P., Mahajan, M., Vishwakarma, A.K., Bhardwaj, J. and Yadav, S.K. (2015) *AtROS1* overexpression provides evidence for epigenetic regulation of genes encoding enzymes of flavonoid biosynthesis and antioxidant pathways during salt stress in transgenic tobacco. *Journal of Experimental Botany* 66(19), 5959–5969.

Bilichak, A., Ilnystkyy, Y., Hollunder, J. and Kovalchuk, I. (2012) The progeny of *Arabidopsis thaliana* plants exposed to salt exhibit changes in DNA methylation, histone modifications and gene expression. *PLoS ONE* 7(1), e30515.

Boyko, A., Blevins, T., Yao, Y., Golubov, A., Bilichak, A. *et al.* (2010) Transgenerational adaptation of *Arabidopsis* to stress requires DNA methylation and the function of Dicer-like proteins. *PLoS ONE* 5(3), e9514.

Cao, C., Long, R., Zhang, T., Kang, J., Wang, Z. *et al.* (2018) Genome-wide identification of microRNAs in response to salt/alkali stress in *Medicago truncatula* through high-throughput sequencing. *International Journal of Molecular Sciences* 19(12), 4076.

Cerda, S. and Weitzman, S.A. (1997) Influence of oxygen radical injury on DNA methylation. *Mutation Research/Reviews in Mutation Research* 386(2), 141–152.

Chen, L.J., Wei, S.Y. and Chiu, J.J. (2013) Mechanical regulation of epigenetics in vascular biology and pathobiology. *Journal of Cellular and Molecular Medicine* 17(4), 437–448.

Chen, L.T., Luo, M., Wang, Y.Y. and Wu, K. (2010) Involvement of *Arabidopsis* histone deacetylase HDA6 in ABA and salt stress response. *Journal of Experimental Botany* 61(12), 3345–3353.

Chen, R., Li, M., Zhang, H., Duan, L., Sun, X. *et al.* (2019) Continuous salt stress-induced long non-coding RNAs and DNA methylation patterns in soybean roots. *BMC Genomics* 20(1), 730.

Chen, Z., Hu, L., Han, N., Hu, J., Yang, Y. *et al.* (2015) Overexpression of a miR393-resistant form of transport inhibitor response protein 1 (mTIR1) enhances salt tolerance by increased osmoregulation and Na^+ exclusion in *Arabidopsis thaliana*. *Plant & Cell Physiology* 56(1), 73–83.

Cheng, X., Zhang, S., Tao, W., Zhang, X., Liu, J. *et al.* (2018) INDETERMINATE SPIKELET1 recruits histone deacetylase and a transcriptional repression complex to regulate rice salt tolerance. *Plant Physiology* 178(2), 824–837.

Choi, S.M., Song, H.R., Han, S.K., Han, M., Kim, C.Y. *et al.* (2012) HDA19 is required for the repression of salicylic acid biosynthesis and salicylic acid-mediated defense responses in *Arabidopsis*. *The Plant Journal* 71(1), 135–146.

Coleman-Derr, D. and Zilberman, D. (2012) Deposition of histone variant H2A.Z within gene bodies regulates responsive genes. *PLoS Genetics* 8(10), e1002988.

Conti, L., Kioumourtzoglou, D., O'Donnell, E. Dominy, P. and Sadanandom, A. (2009) OTS1 and OTS2 SUMO proteases link plant development and survival under salt stress. *Plant Signaling & Behavior* 4(3), 225–227.

Cui, F., Liu, L., Zhao, Q., Zhang, Z., Li, Q. *et al.* (2012) *Arabidopsis* ubiquitin conjugase UBC32 is an ERAD component that functions in brassinosteroid-mediated salt stress tolerance. *The Plant Cell* 24(1), 233–244.

Cui, L.G., Shan, J.X., Shi, M., Gao, J.P. and Lin, H.X. (2014) The miR156–SPL 9–DFR pathway coordinates the relationship between development and abiotic stress tolerance in plants. *The Plant Journal* 80(6), 1108–1117.

Dai, X., Bai, Y., Zhao, L., Dou, X., Liu, Y. *et al.* (2017) H2A.Z represses gene expression by modulating promoter nucleosome structure and enhancer histone modifications in *Arabidopsis*. *Molecular Plant* 10(10), 1274–1292.

de Lorenzo, L., Merchan, F., Blanchet, S., Megías, M., Frugier, F. *et al.* (2007) Differential expression of the TFIIIA regulatory pathway in response to salt stress between *Medicago truncatula* genotypes. *Plant Physiology* 145(4), 1521–1532.

Deng, F., Zhang, X., Wang, W., Yuan, R. and Shen, F. (2018) Identification of *Gossypium hirsutum* long non-coding RNAs (lncRNAs) under salt stress. *BMC Plant Biology* 18, 23.

Devkota, K.P., Devkota, M., Rezaei, M. and Oosterbaan, R. (2022) Managing salinity for sustainable agricultural production in salt-affected soils of irrigated drylands. *Agricultural Systems* 198, 103390.

Dyachenko, O.V., Zakharchenko, N.S., Shevchuk, T.V., Bohnert, H.J., Cushman, J.C. *et al.* (2006) Effect of hypermethylation of CCWGG sequences in DNA of *Mesembryanthemum crystallinum* plants on their adaptation to salt stress. *Biochemistry (Moscow)* 71, 461–465.

Eberharter, A. and Becker, P.B. (2002) Histone acetylation: a switch between repressive and permissive chromatin. *EMBO Reports* 3(3), 224–229.

ElSayed, A.I., Rafudeen, M.S., Ganie, S.A., Hossain, M.S. and Gomaa, A.M. (2022) Seed priming with cypress leaf extract enhances photosynthesis and antioxidative defense in zucchini seedlings under salt stress. *Scientia Horticulturae* 293, 110707.

Fahad, S., Bajwa, A.A., Nazir, U., Anjum, S.A., Farooq, A. *et al.* (2017) Crop production under drought and heat stress: plant responses and management options. *Frontiers in Plant Science* 8, 1147.

Fan, S., Liu, H., Liu, J., Hua, W., Xu, S. and Li, J. (2020) Systematic analysis of the DNA methylase and demethylase gene families in rapeseed (*Brassica napus* L.) and their expression variations after salt and heat stresses. *International Journal of Molecular Sciences* 21(3), 953.

Fang, H., Liu, X., Thorn, G., Duan, J. and Tian, L. (2014) Expression analysis of histone acetyltransferases in rice under drought stress. *Biochemical and Biophysical Research Communications* 443(2), 400–405.

Ferreira, L.J., Azevedo, V., Maroco, J., Oliveira, M.M. and Santos, A.P. (2015) Salt tolerant and sensitive rice varieties display differential methylome flexibility under salt stress. *PLoS ONE* 10(5), e0124060.

Ficz, G., Hore, T.A., Santos, F., Lee, H.J., Dean, W. *et al.* (2013) FGF signaling inhibition in ESCs drives rapid genome-wide demethylation to the epigenetic ground state of pluripotency. *Cell Stem Cell* 13(3), 351–359.

Fu, W., Wu, K. and Duan, J., (2007) Sequence and expression analysis of histone deacetylases in rice. *Biochemical and Biophysical Research Communications* 356(4), 843–850.

Ganie, S.A. (2020) RNA chaperones: potential candidates for engineering salt tolerance in rice. *Crop Science* 60(2), 530–540.

Ganie, S.A. (2021) Amino acids other than proline and their participation in abiotic stress tolerance. In: Wani, S.H., Gangola, M.P. and Ramadoss, B.R. (eds) *Compatible Solutes Engineering for Crop Plants Facing Climate Change*. Springer, Cham, Switzerland, pp. 47–96.

Ganie, S.A., Debnath, A.B., Gumi, A.M. and Mondal, T.K. (2017) Comprehensive survey and evolutionary analysis of genome-wide miRNA genes from ten diploid *Oryza* species. *BMC Genomics* 18(1), 711.

Gao, P., Bai, X., Yang, L., Lv, D., Li, Y. *et al.* (2010) Over-expression of osa-MIR396c decreases salt and alkali stress tolerance. *Planta* 231, 991–1001.

Gao, S., Yang, L., Zeng, H.Q., Zhou, Z.S. *et al.* (2016) A cotton miRNA is involved in regulation of plant response to salt stress. *Scientific Reports* 6(1), 19736.

Garg, R., Narayana Chevala, V.V.S., Shankar, R. and Jain, M. (2015) Divergent DNA methylation patterns associated with gene expression in rice cultivars with contrasting drought and salinity stress response. *Scientific Reports* 5(1), 14922.

Guan, C., Li, X., Tian, D.Y., Liu, H.Y., Cen, H.F. *et al.* (2020) ADP-ribosylation factors improve biomass yield and salinity tolerance in transgenic switchgrass (*Panicum virgatum* L.). *Plant Cell Reports* 39, 1623–1638.

Gujral, P., Mahajan, V., Lissaman, A.C. and Ponnampalam, A.P. (2020) Histone acetylation and the role of histone deacetylases in normal cyclic endometrium. *Reproductive Biology and Endocrinology* 18, 84.

Han, B., Xu, W., Ahmed, N., Yu, A., Wang, Z. and Liu, A. (2020) Changes and associations of genomic transcription and histone methylation with salt stress in castor bean. *Plant & Cell Physiology* 61(6), 1120–1133.

He, S., Hao, Y., Zhang, Q., Zhang, P., Ji, F. *et al.* (2020) Histone deacetylase inhibitor SAHA improves high salinity tolerance associated with hyperacetylation-enhancing expression of ion homeostasis-related genes in cotton. *International Journal of Molecular Sciences* 21(19), 7105.

Hershko, A. (1998) The ubiquitin system: past, present, and future perspectives. In: Peters, J.M., Harris, J.R. and Finley, D. (eds) *Ubiquitin and the Biology of the Cell*. Springer, Boston, Massachusetts, pp. 1–17.

Hossain, M.S. and Dietz, K.J. (2016) Tuning of redox regulatory mechanisms, reactive oxygen species and redox homeostasis under salinity stress. *Frontiers in Plant Science* 7, 548.

Hossain, M.S., ElSayed, A.I., Moore, M. and Dietz, K.J. (2017a) Redox and reactive oxygen species network in acclimation for salinity tolerance in sugar beet. *Journal of Experimental Botany* 68(5), 1283–1298.

Hossain, M.S., Persicke, M., ElSayed, A.I., Kalinowski, J. and Dietz, K.J. (2017b) Metabolite profiling at the cellular and subcellular level reveals metabolites associated with salinity tolerance in sugar beet. *Journal of Experimental Botany* 68(21–22), 5961–5976.

Huanca-Mamani, W., Arias-Carrasco, R., Cárdenas-Ninasivincha, S., Rojas-Herrera, M., Sepúlveda-Hermosilla, G. *et al.* (2018) Long non-coding RNAs responsive to salt and boron stress in the hyper-arid Lluteno maize from Atacama Desert. *Genes* 9(3), 170.

Jha, U.C., Bohra, A., Jha, R. and Parida, S.K. (2019) Salinity stress response and 'omics' approaches for improving salinity stress tolerance in major grain legumes. *Plant Cell Reports* 38, 255–277.

Jha, U.C., Nayyar, H., Jha, R., Khurshid, M., Zhou, M. *et al.* (2020) Long non-coding RNAs: emerging players regulating plant abiotic stress response and adaptation. *BMC Plant Biology* 20(1), 466.

Jiang, D. and Berger, F. (2017) Histone variants in plant transcriptional regulation. *Biochimica et Biophysica Acta (BBA) – Gene Regulatory Mechanisms* 1860(1), 123–130.

Joshi, R., Ramanarao, M.V., Lee, S., Kato, N. and Baisakh, N. (2014) Ectopic expression of ADP ribosylation factor 1 (*SaARF1*) from smooth cordgrass (*Spartina alterniflora* Loisel) confers drought and salt tolerance in transgenic rice and *Arabidopsis*. *Plant Cell, Tissue, and Organ Culture (PCTOC)* 117, 17–30.

Kaldis, A., Tsementzi, D., Tanriverdi, O. and Vlachonasios, K.E. (2011) *Arabidopsis thaliana* transcriptional co-activators ADA2b and SGF29a are implicated in salt stress responses. *Planta* 233, 749–762.

Karan, R., DeLeon, T., Biradar, H. and Subudhi, P.K. (2012) Salt stress induced variation in DNA methylation pattern and its influence on gene expression in contrasting rice genotypes. *PLoS ONE* 7(6), e40203.

Karanthamalai, J., Chodon, A., Chauhan, S. and Pandi, G. (2020) DNA N^6-methyladenine modification in plant genomes – a glimpse into emerging epigenetic code. *Plants* 9(2), 247.

Karlik, E. and Gözükırmızı, N. (2018) Evaluation of barley lncRNAs expression analysis in salinity stress. *Russian Journal of Genetics* 54, 198–204.

Katiyar-Agarwal, S., Morgan, R., Dahlbeck, D., Borsani, O., Villegas, A. Jr *et al.* (2006) A pathogen-inducible endogenous siRNA in plant immunity. *Proceedings of the National Academy of Sciences USA* 103(47), 18002–18007.

Khandal, H., Parween, S., Roy, R., Meena, M.K. and Chattopadhyay, D. (2017) MicroRNA profiling provides insights into post-transcriptional regulation of gene expression in chickpea root apex under salinity and water deficiency. *Scientific Reports* 7(1), 4632.

Kim, J., Guermah, M., McGinty, R.K., Lee, J.S., Tang, Z. *et al.* (2009) RAD6-mediated transcription-coupled H2B ubiquitylation directly stimulates H3K4 methylation in human cells. *Cell* 137(3), 459–471.

Kim, M., Park, C., Jung, J. and Yeo, S.G. (2019) The histone deacetylase class I, II inhibitor trichostatin A delays peripheral neurodegeneration. *Journal of Molecular Histology* 50(2), 167–178.

Kim, W.Y., Ali, Z., Park, H.J., Park, S.J., Cha, J.Y. *et al.* (2013) Release of SOS2 kinase from sequestration with GIGANTEA determines salt tolerance in *Arabidopsis*. *Nature Communications* 4(1), 1352.

Kohli, D., Joshi, G., Deokar, A.A., Bhardwaj, A.R., Agarwal, M. *et al.* (2014) Identification and characterization of wilt and salt stress-responsive microRNAs in chickpea through high-throughput sequencing. *PLoS ONE* 9(10), e108851.

Kress, C., Thomassin, H. and Grange, T. (2001) Local DNA demethylation in vertebrates: how could it be performed and targeted? *FEBS Letters* 494(3), 135–140.

Kumar, S., Beena, A.S., Awana, M. and Singh, A. (2017) Physiological, biochemical, epigenetic and molecular analyses of wheat (*Triticum aestivum*) genotypes with contrasting salt tolerance. *Frontiers in Plant Science* 8, 1151.

Lamb, R.S., Citarelli, M. and Teotia, S. (2012) Functions of the poly (ADP-ribose) polymerase superfamily in plants. *Cellular and Molecular Life Sciences* 69, 175–189.

Lämke, J. and Bäurle, I. (2017) Epigenetic and chromatin-based mechanisms in environmental stress adaptation and stress memory in plants. *Genome Biology* 18(1), 124.

Li, H., Yan, S., Zhao, L., Tan, J., Zhang, Q. *et al.* (2014) Histone acetylation associated up-regulation of the cell wall related genes is involved in salt stress induced maize root swelling. *BMC Plant Biology* 14, 105.

Lister, R., O'Malley, R.C., Tonti-Filippini, J., Gregory, B.D., Berry, C.C. *et al.* (2008) Highly integrated single-base resolution maps of the epigenome in *Arabidopsis*. *Cell* 133(3), 523–536.

Liu, C., Wang, C., Wang, G., Becker, C., Zaidem, M. and Weigel, D. (2016) Genome-wide analysis of chromatin packing in *Arabidopsis thaliana* at single-gene resolution. *Genome Research* 26(8), 1057–1068.

Liu, C.W., Hsu, Y.K., Cheng, Y.H., Yen, H.C., Wu, Y.P. *et al.* (2012) Proteomic analysis of salt-responsive ubiquitin-related proteins in rice roots. *Rapid Communications in Mass Spectrometry* 26(15), 1649–1660.

Liu, X., Luo, M., Zhang, W., Zhao, J., Zhang, J. *et al.* (2012) Histone acetyltransferases in rice (*Oryza sativa* L.): phylogenetic analysis, subcellular localization and expression. *BMC Plant Biology* 12(1), 145.

Liu, Y., Dai, S., Xing, L., Xu, Y. and Chong, K. (2015) *O*-linked β-*N*-acetylglucosamine modification and its biological functions. *Science Bulletin* 60, 1055–1061.

Liu, Z., Zhu, Y., Gao, J., Yu, F., Dong, A. and Shen, W.H. (2009) Molecular and reverse genetic characterization of nucleosome assembly protein1 (NAP1) genes unravels their function in transcription and nucleotide excision repair in *Arabidopsis thaliana*. *The Plant Journal* 59(1), 27–38.

Long, R.C., Li, M.N., Kang, J.M., Zhang, T.J., Sun, Y. and Yang, Q.C. (2015) Small RNA deep sequencing identifies novel and salt-stress-regulated microRNAs from roots of *Medicago sativa* and *Medicago truncatula*. *Physiologia Plantarum* 154(1), 13–27.

Lu, X., Dun, H., Lian, C., Zhang, X., Yin, W. and Xia, X. (2017) The role of peu-miR164 and its target PeNAC genes in response to abiotic stress in *Populus euphratica*. *Plant Physiology and Biochemistry* 115, 418–438.

Luo, M., Wang, Y.Y., Liu, X., Yang, S., Lu, Q. *et al.* (2012) HD2C interacts with HDA6 and is involved in ABA and salt stress response in *Arabidopsis*. *Journal of Experimental Botany* 63(8), 3297–3306.

Ma, J., Bai, X., Luo, W., Feng, Y., Shao, X. *et al.* (2019) Genome-wide identification of long noncoding RNAs and their responses to salt stress in two closely related poplars. *Frontiers in Genetics* 10, 777.

Ma, Y., Xue, H., Zhang, F., Jiang, Q., Yang, S. *et al.* (2021) The miR156/SPL module regulates apple salt stress tolerance by activating MdWRKY100 expression. *Plant Biotechnology Journal* 19(2), 311–323.

Marconi, G., Pace, R., Traini, A., Raggi, L., Lutts, S. *et al.* (2013) Use of MSAP markers to analyse the effects of salt stress on DNA methylation in rapeseed (*Brassica napus* var. *oleifera*). *PLoS ONE* 8(9), e75597.

Mariño-Ramírez, L., Kann, M.G., Shoemaker, B.A. and Landsman, D. (2005) Histone structure and nucleosome stability. *Expert Review of Proteomics* 2(5), 719–729.

Marondedze, C., Elia, G., Thomas, L., Wong, A. and Gehring, C. (2021) Citrullination of proteins as a specific response mechanism in plants. *Frontiers in Plant Science* 12, 638392.

Matzke, M.A. and Mosher, R.A. (2014) RNA-directed DNA methylation: an epigenetic pathway of increasing complexity. *Nature Reviews Genetics* 15(6), 394–408.

Mehdi, S., Derkacheva, M., Ramström, M., Kralemann, L., Bergquist, J. and Hennig, L. (2016) The WD40 domain protein MSI1 functions in a histone deacetylase complex to fine-tune abscisic acid signaling. *The Plant Cell* 28(1), 42–54.

Mlynárová, L., Nap, J.P. and Bisseling, T. (2007) The SWI/SNF chromatin-remodeling gene AtCHR12 mediates temporary growth arrest in *Arabidopsis thaliana* upon perceiving environmental stress. *The Plant Journal* 51(5), 874–885.

Monteuuis, G., Wong, J.J., Bailey, C.G., Schmitz, U. and Rasko, J.E. (2019) The changing paradigm of intron retention: regulation, ramifications and recipes. *Nucleic Acids Research* 47(22), 11497–11513.

Muthamilarasan, M., Mangu, V.R., Zandkarimi, H., Prasad, M. and Baisakh, N. (2016) Structure, organization and evolution of ADP-ribosylation factors in rice and foxtail millet and their expression in rice. *Scientific Reports* 6(1), 24008.

Nathan, D., Ingvarsdottir, K., Sterner, D.E., Bylebyl, G.R., Dokmanovic, M. *et al.* (2006) Histone sumoylation is a negative regulator in *Saccharomyces cerevisiae* and shows dynamic interplay with positive-acting histone modifications. *Genes & Development* 20(8), 966–976.

Ni, Z., Hu, Z., Jiang, Q. and Zhang, H. (2013) GmNFYA3, a target gene of miR169, is a positive regulator of plant tolerance to drought stress. *Plant Molecular Biology* 82, 113–129.

Niehrs, C. (2009) Active DNA demethylation and DNA repair. *Differentiation* 77(1), 1–11.

Nishimura, N., Kitahata, N., Seki, M., Narusaka, Y., Narusaka, M. *et al.* (2005) Analysis of ABA hypersensitive germination2 revealed the pivotal functions of PARN in stress response in *Arabidopsis*. *The Plant Journal* 44(6), 972–984.

Obaid, T.A. (2001) *Footprints and Milestones: Population and Environmental Change*. United Nations Population Fund.

Onufriev, A.V. and Schiessel, H. (2019). The nucleosome: from structure to function through physics. *Current Opinion in Structural Biology* 56, 119–130.

Pan, W.J., Tao, J.J., Cheng, T., Bian, X.H., Wei, W. *et al.* (2016) Soybean miR172a improves salt tolerance and can function as a long-distance signal. *Molecular Plant* 9(9), 1337–1340.

Pandey, G., Sharma, N., Pankaj Sahu, P. and Prasad, M. (2016) Chromatin-based epigenetic regulation of plant abiotic stress response. *Current Genomics* 17(6), 490–498.

Paul, A., Dasgupta, P., Roy, D. and Chaudhuri, S. (2017) Comparative analysis of histone modifications and DNA methylation at CsBZ8 locus under salinity stress in IR64 and Nonabokra rice varieties. *Plant Molecular Biology* 95, 63–88.

Paul, S., Kundu, A. and Pal, A. (2011) Identification and validation of conserved microRNAs along with their differential expression in roots of *Vigna unguiculata* grown under salt stress. *Plant Cell, Tissue, and Organ Culture (PCTOC)* 105, 233–242.

Paulson, J.R. and Taylor, S.S. (1982) Phosphorylation of histones 1 and 3 and nonhistone high mobility group 14 by an endogenous kinase in HeLa metaphase chromosomes. *Journal of Biological Chemistry* 257(11), 6064–6072.

Pikaard, C.S. and Scheid, O.M. (2014) Epigenetic regulation in plants. *Cold Spring Harbor Perspectives in Biology* 6(12), a019315.

Qadir, M., Quillérou, E., Nangia, V., Murtaza, G., Singh, M. *et al.* (2014) Economics of salt-induced land degradation and restoration. *Natural Resources Forum* 38(4), 282–295.

Qin, T., Zhao, H., Cui, P., Albesher, N. and Xiong, L. (2017) A nucleus-localized long non-coding RNA enhances drought and salt stress tolerance. *Plant Physiology* 175(3), 1321–1336.

Rahman, A.K.M.M., Ahmed, K.M., Butler, A.P. and Hoque, M.A. (2018) Influence of surface geology and micro-scale land use on the shallow subsurface salinity in deltaic coastal areas: a case from southwest Bangladesh. *Environmental Earth Sciences* 77, 423.

Rashid, M.M., Vaishnav, A., Verma, R.K., Sharma, P., Suprasanna, P. and Gaur, R.K. (2022) Epigenetic regulation of salinity stress responses in cereals. *Molecular Biology Reports* 49, 761–772.

Raxwal, V.K., Ghosh, S., Singh, S., Katiyar-Agarwal, S., Goel, S. *et al.* (2020) Abiotic stress-mediated modulation of the chromatin landscape in *Arabidopsis thaliana*. *Journal of Experimental Botany* 71(17), 5280–5293.

Ruse, M. (1975) Darwin's debt to philosophy: an examination of the influence of the philosophical ideas of John FW Herschel and William Whewell on the development of Charles Darwin's theory of evolution. *Studies in History and Philosophy of Science Part A* 6(2), 159–181.

Sahu, B.B. and Shaw, B.P. (2009) Isolation, identification and expression analysis of salt-induced genes in *Suaeda maritima*, a natural halophyte, using PCR-based suppression subtractive hybridization. *BMC Plant Biology* 9, 69.

Sakabe, K., Wang, Z. and Hart, G.W. (2010) β-*N*-acetylglucosamine (O-GlcNAc) is part of the histone code. *Proceedings of the National Academy of Sciences USA* 107(46), 19915–19920.

Sako, K., Kim, J.M., Matsui, A., Nakamura, K., Tanaka, M. *et al.* (2016) Ky-2, a histone deacetylase inhibitor, enhances high-salinity stress tolerance in *Arabidopsis thaliana*. *Plant & Cell Physiology* 57(4), 776–783.

Sani, E., Herzyk, P., Perrella, G., Colot, V. and Amtmann, A. (2013) Hyperosmotic priming of *Arabidopsis* seedlings establishes a long-term somatic memory accompanied by specific changes of the epigenome. *Genome Biology* 14(6), R59.

Shams, M., Yildirim, E., Arslan, E. and Agar, G. (2020) Salinity induced alteration in DNA methylation pattern, enzyme activity, nutrient uptake and H_2O_2 content in pepper (*Capsicum annuum* L.) cultivars. *Acta Physiologiae Plantarum* 42, 59.

Shen, Y., Conde e Silva, N., Audonnet, L., Servet, C., Wei, W. and Zhou, D.X. (2014) Over-expression of histone H3K4 demethylase gene JMJ15 enhances salt tolerance in *Arabidopsis*. *Frontiers in Plant Science* 5, 290.

Shew, A.M., Tack, J.B., Nalley, L.L. and Chaminuka, P. (2020) Yield reduction under climate warming varies among wheat cultivars in South Africa. *Nature Communications* 11(1), 4408.

Shumayla, Sharma, S., Taneja, M., Tyagi, S., Singh, K. and Upadhyay, S.K. (2017) Survey of high throughput RNA-Seq data reveals potential roles for lncRNAs during development and stress response in bread wheat. *Frontiers in Plant Science* 8, 1019.

Simon, S.A. and Meyers, B.C. (2011) Small RNA-mediated epigenetic modifications in plants. *Current Opinion in Plant Biology* 14(2), 148–155.

Singh, U., Khemka, N., Rajkumar, M.S., Garg, R. and Jain, M. (2017) PLncPRO for prediction of long non-coding RNAs (lncRNAs) in plants and its application for discovery of abiotic stress-responsive lncRNAs in rice and chickpea. *Nucleic Acids Research* 45(22), e183.

Skorupa, M., Szczepanek, J., Mazur, J., Domagalski, K., Tretyn, A. and Tyburski, J. (2021) Salt stress and salt shock differently affect DNA methylation in salt-responsive genes in sugar beet and its wild, halophytic ancestor. *PLoS ONE* 16(5), e0251675.

Sokol, A., Kwiatkowska, A., Jerzmanowski, A. and Prymakowska-Bosak, M. (2007) Up-regulation of stress-inducible genes in tobacco and *Arabidopsis* cells in response to abiotic stresses and ABA treatment correlates with dynamic changes in histone H3 and H4 modifications. *Planta* 227, 245–254.

Song, C.P. and Galbraith, D.W. (2006) AtSAP18, an orthologue of human SAP18, is involved in the regulation of salt stress and mediates transcriptional repression in *Arabidopsis*. *Plant Molecular Biology* 60, 241–257.

Song, Y., Ji, D., Li, S., Wang, P., Li, Q. and Xiang, F. (2012) The dynamic changes of DNA methylation and histone modifications of salt responsive transcription factor genes in soybean. *PLoS ONE* 7(7), e41274.

Sridha, S. and Wu, K. (2006) Identification of AtHD2C as a novel regulator of abscisic acid responses in *Arabidopsis*. *The Plant Journal* 46(1), 124–133.

Steward, N., Ito, M., Yamaguchi, Y., Koizumi, N. and Sano, H. (2002) Periodic DNA methylation in maize nucleosomes and demethylation by environmental stress. *Journal of Biological Chemistry* 277(40), 37741–37746.

Strahl, B.D. and Allis, C.D. (2000) The language of covalent histone modifications. *Nature* 403(6765), 41–45.

Sun, L., Miao, X., Cui, J., Deng, J., Wang, X. *et al.* (2018) Genome-wide high-resolution mapping of DNA methylation identifies epigenetic variation across different salt stress in maize (*Zea mays* L.). *Euphytica* 214, 25.

Sun, Z., Wang, Y., Mou, F., Tian, Y., Chen, L. *et al.* (2016) Genome-wide small RNA analysis of soybean reveals auxin-responsive microRNAs that are differentially expressed in response to salt stress in root apex. *Frontiers in Plant Science* 6, 1273.

Sura, W., Kabza, M., Karlowski, W.M., Bieluszewski, T., Kus-Slowinska, M. *et al.* (2017) Dual role of the histone variant H2A.Z in transcriptional regulation of stress-response genes. *The Plant Cell* 29(4), 791–807.

Tan, M.P. (2010) Analysis of DNA methylation of maize in response to osmotic and salt stress based on methylation-sensitive amplified polymorphism. *Plant Physiology and Biochemistry* 48(1), 21–26.

Tang, Y., Du, G., Xiang, J., Hu, C., Li, X. *et al.* (2022) Genome-wide identification of auxin response factor (ARF) gene family and the miR160-ARF18-mediated response to salt stress in peanut (*Arachis hypogaea* L.). *Genomics* 114(1), 171–184.

Tian, H., Guo, F., Zhang, Z., Ding, H., Meng, J. *et al.* (2020) Discovery, identification, and functional characterization of long noncoding RNAs in *Arachis hypogaea* L. *BMC Plant Biology* 20, 308.

To, T.K. and Kim, J.M. (2014) Epigenetic regulation of gene responsiveness in *Arabidopsis*. *Frontiers in Plant Science* 4, 548.

Tripathi, A.K., Pareek, A. and Singla-Pareek, S.L. (2016) A NAP-family histone chaperone functions in abiotic stress response and adaptation. *Plant Physiology* 171(4), 2854–2868.

Ullah, F., Xu, Q., Zhao, Y. and Zhou, D.X. (2021) Histone deacetylase HDA710 controls salt tolerance by regulating ABA signaling in rice. *Journal of Integrative Plant Biology* 63(3), 451–467.

Valencia-Sánchez, M.I., De Ioannes, P., Wang, M., Truong, D.M., Lee, R. *et al.* (2021) Regulation of the Dot1 histone H3K79 methyltransferase by histone H4K16 acetylation. *Science* 371(6527), eabc6663.

Wachsman, J.T. (1997) DNA methylation and the association between genetic and epigenetic changes: relation to carcinogenesis. *Mutation Research/Fundamental and Molecular Mechanisms of Mutagenesis* 375(1), 1–8.

Waidmann, S., Kusenda, B., Mayerhofer, J., Mechtler, K. and Jonak, C. (2014) A DEK domain-containing protein modulates chromatin structure and function in *Arabidopsis*. *The Plant Cell* 26(11), 4328–4344.

Wang, B., Fu, R., Zhang, M., Ding, Z., Chang, L. *et al.* (2015) Analysis of methylation-sensitive amplified polymorphism in different cotton accessions under salt stress based on capillary electrophoresis. *Genes & Genomics* 37, 713–724.

Wang, H., Wang, L., Erdjument-Bromage, H., Vidal, M., Tempst, P. *et al.* (2004) Role of histone H2A ubiquitination in Polycomb silencing. *Nature* 431(7010), 873–878.

Wang, M., Qin, L., Xie, C., Li, W., Yuan, J. *et al.* (2014) Induced and constitutive DNA methylation in a salinity-tolerant wheat introgression line. *Plant & Cell Physiology* 55(7), 1354–1365.

Wang, T.Z., Liu, M., Zhao, M.G., Chen, R. and Zhang, W.H. (2015) Identification and characterization of long non-coding RNAs involved in osmotic and salt stress in *Medicago truncatula* using genome-wide high-throughput sequencing. *BMC Plant Biology* 15(1), 131.

Wang, W., Huang, F., Qin, Q., Zhao, X., Li, Z. and Fu, B. (2015) Comparative analysis of DNA methylation changes in two rice genotypes under salt stress and subsequent recovery. *Biochemical and Biophysical Research Communications* 465(4), 790–796.

Wang, W., Liu, D., Chen, D., Cheng, Y., Zhang, X. *et al.* (2019) MicroRNA414c affects salt tolerance of cotton by regulating reactive oxygen species metabolism under salinity stress. *RNA Biology* 16(3), 362–375.

Wang, W., Wang, W., Wu, Y., Li, Q., Zhang, G. *et al.* (2020) The involvement of wheat U-box E3 ubiquitin ligase TaPUB1 in salt stress tolerance. *Journal of Integrative Plant Biology* 62(5), 631–651.

Wang, Z., Casas-Mollano, J.A., Xu, J., Riethoven, J.J.M., Zhang, C. and Cerutti, H. (2015) Osmotic stress induces phosphorylation of histone H3 at threonine 3 in pericentromeric regions of *Arabidopsis thaliana*. *Proceedings of the National Academy of Sciences USA* 112(27), 8487–8492.

Wei, F., Tang, D., Li, Z., Kashif, M.H., Khan, A. *et al.* (2019) Molecular cloning and subcellular localization of six HDACs and their roles in response to salt and drought stress in kenaf (*Hibiscus cannabinus* L.). *Biological Research* 52, 20.

Wei, W., Ba, Z., Gao, M., Wu, Y., Ma, Y. *et al.* (2012) A role for small RNAs in DNA double-strand break repair. *Cell* 149(1), 101–112.

Wei, W., Tao, J.J., Chen, H.W., Li, Q.T., Zhang, W.K. *et al.* (2017) A histone code reader and a transcriptional activator interact to regulate genes for salt tolerance. *Plant Physiology* 175(3), 1304–1320.

Witalison, E.E., Thompson, P.R. and Hofseth, L.J. (2015) Protein arginine deiminases and associated citrullination: physiological functions and diseases associated with dysregulation. *Current Drug Targets* 16(7), 700–710.

Xhemalce, B., Dawson, M.A. and Bannister, A.J. (2011) Histone modifications. In: Meyers, R.A. (ed.) *Encyclopaedia of Molecular Cell Biology and Molecular Medicine: Epigenetic Regulation and Epigenomics*, 2nd edn. Wiley-VCH, Weinheim, Germany, pp. 1–45.

Xia, K., Wang, R., Ou, X., Fang, Z., Tian, C. *et al.* (2012) OsTIR1 and OsAFB2 downregulation via OsmiR393 overexpression leads to more tillers, early flowering and less tolerance to salt and drought in rice. *PLoS ONE* 7(1), e30039.

Yaish, M.W., Al-Lawati, A., Al-Harrasi, I. and Patankar, H.V. (2018) Genome-wide DNA methylation analysis in response to salinity in the model plant caliph medic (*Medicago truncatula*). *BMC Genomics* 19(1), 78.

Yang, R., Hong, Y., Ren, Z., Tang, K., Zhang, H. *et al.* (2019) A role for PICKLE in the regulation of cold and salt stress tolerance in *Arabidopsis*. *Frontiers in Plant Science* 10, 900.

Yang, Y., Karthikeyan, A., Yin, J., Jin, T., Ren, R. *et al.* (2022) The E3 ligase GmPUB21 negatively regulates drought and salinity stress response in soybean. *International Journal of Molecular Sciences* 23(13), 6893.

Yolcu, S., Ozdemir, F., Güler, A. and Bor, M. (2016) Histone acetylation influences the transcriptional activation of POX in *Beta vulgaris* L. and *Beta maritima* L. under salt stress. *Plant Physiology and Biochemistry* 100, 37–46.

Yu, X., Meng, X., Liu, Y., Li, N., Zhang, A. *et al.* (2018) The chromatin remodeler ZmCHB101 impacts expression of osmotic stress-responsive genes in maize. *Plant Molecular Biology* 97, 451–465.

Yu, Y., Wang, P., Bai, Y., Wang, Y., Wan, H. *et al.* (2020) The soybean F-box protein GmFBX176 regulates ABA-mediated responses to drought and salt stress. *Environmental and Experimental Botany* 176, 104056.

Yuan, S., Zhao, J., Li, Z., Hu, Q., Yuan, N. *et al.* (2019) MicroRNA396-mediated alteration in plant development and salinity stress response in creeping bentgrass. *Horticulture Research* 6, 48.

Yung, W.S., Li, M.W., Sze, C.C., Wang, Q. and Lam, H.M. (2021) Histone modifications and chromatin remodelling in plants in response to salt stress. *Physiologia Plantarum* 173(4), 1495–1513.

Zander, M., Willige, B.C., He, Y., Nguyen, T.A., Langford, A.E. *et al.* (2019) Epigenetic silencing of a multifunctional plant stress regulator. *Elife* 8, e47835.

Zhang, H., Cui, F., Wu, Y., Lou, L., Liu, L. *et al.* (2015) The RING finger ubiquitin E3 ligase SDIR1 targets SDIR1-INTERACTING PROTEIN1 for degradation to modulate the salt stress response and ABA signaling in *Arabidopsis*. *The Plant Cell* 27(1), 214–227.

Zhang, H., Lang, Z. and Zhu, J.K. (2018) Dynamics and function of DNA methylation in plants. *Nature Reviews Molecular Cell Biology* 19(8), 489–506.

Zhang, T., Zhang, W. and Jiang, J. (2015) Genome-wide nucleosome occupancy and positioning and their impact on gene expression and evolution in plants. *Plant Physiology* 168(4), 1406–1416.

Zhang, X., Yazaki, J., Sundaresan, A., Cokus, S., Chan, S.W.L. *et al.* (2006) Genome-wide high-resolution mapping and functional analysis of DNA methylation in *Arabidopsis*. *Cell* 126(6), 1189–1201.

Zhang, Z., Zhang, S., Zhang, Y., Wang, X., Li, D. *et al.* (2011) *Arabidopsis* floral initiator SKB1 confers high salt tolerance by regulating transcription and pre-mRNA splicing through altering histone H4R3 and small nuclear ribonucleoprotein LSM4 methylation. *The Plant Cell* 23(1), 396–411.

Zhao, J., Zhang, J., Zhang, W., Wu, K., Zheng, F. *et al.* (2015) Expression and functional analysis of the plant-specific histone deacetylase HDT701 in rice. *Frontiers in Plant Science* 5, 764.

Zhao, J., Li, M., Gu, D., Liu, X., Zhang, J. *et al.* (2016) Involvement of rice histone deacetylase HDA705 in seed germination and in response to ABA and abiotic stresses. *Biochemical and Biophysical Research Communications* 470(2), 439–444.

Zhao, J., Yuan, S., Zhou, M., Yuan, N., Li, Z. *et al.* (2019) Transgenic creeping bentgrass overexpressing Osa-miR393a exhibits altered plant development and improved multiple stress tolerance. *Plant Biotechnology Journal* 17(1), 233–251.

Zheng, M., Lin, J., Liu, X., Chu, W., Li, J. *et al.* (2021) Histone acetyltransferase TaHAG1 acts as a crucial regulator to strengthen salt tolerance of hexaploid wheat. *Plant Physiology* 186(4), 1951–1969.

Zheng, Y., Ding, Y., Sun, X., Xie, S., Wang, D. *et al.* (2016) Histone deacetylase HDA9 negatively regulates salt and drought stress responsiveness in *Arabidopsis*. *Journal of Experimental Botany* 67(6), 1703–1713.

Zhong, L., Xu, Y.H. and Wang, J.B. (2009) DNA-methylation changes induced by salt stress in wheat *Triticum aestivum*. *African Journal of Biotechnology* 8(22), 6201–6207.

Zhou, C., Zhang, L., Duan, J., Miki, B. and Wu, K. (2005) Histone deacetylase19 is involved in jasmonic acid and ethylene signaling of pathogen response in *Arabidopsis*. *The Plant Cell* 17(4), 1196–1204.

Zhou, C., Sczepanski, J.T. and Greenberg, M.M. (2013) Histone modification via rapid cleavage of C4′-oxidized abasic sites in nucleosome core particles. *Journal of the American Chemical Society* 135(14), 5274–5277.

Zhu, N., Cheng, S., Liu, X., Du, H., Dai, M. *et al.* (2015) The R2R3-type MYB gene OsMYB91 has a function in coordinating plant growth and salt stress tolerance in rice. *Plant Science* 236, 146–156.

5 Functional Role of RNA-Binding Proteins in Plant Salt Tolerance

Fergie Ann Quilloy[1] **and Vincent Pamugas Reyes**[2*]

[1]*Rice Breeding Innovations Platform, International Rice Research Institute, Metro Manila, Philippines;* [2]*Graduate School of Bioagricultural Sciences, Nagoya University, Nagoya, Japan*

Abstract

Salt stress is one of the major abiotic stress factors that limit plant growth and productivity. RNA-binding proteins (RBPs) are a class of proteins that play crucial roles in the regulation of gene expression and post-transcriptional processing of RNA molecules. In recent years, it has become increasingly clear that RBPs play important functional roles in plant salt tolerance. RBPs are involved in several key processes related to plant salt tolerance, including the regulation of alternative splicing of pre-mRNA, mRNA stability, translation, and RNA editing. These processes are important for the production of functional proteins that help plants to cope with salt stress. RBPs have been shown to be involved in regulation of the expression of key salt-stress response genes, such as those encoding ion transporters and enzymes involved in osmolyte synthesis. This chapter explores the diverse regulatory functions of RBPs in plant stress adaptation, with a focus on salt stress. Advances in the detection of RBPs in plants, including the use of high-throughput sequencing technologies (HSTs), are also discussed. Additionally, various studies on the roles of RBPs in salt-stress tolerance across different plant models are presented. Finally, this chapter offers insights into the potential applications of RBPs in the development of salt stress-tolerant plants.

5.1 Introduction

The advent of climate change has been projected to have a significant impact on the yield loss of crops. Wing *et al.* (2021) estimated a reduction in global crop yields by 3–12% by the middle of this century, and 11–25% by the end of the century. Among the various extremities brought by climate change, soil salinity is one of the major problems, especially for agricultural areas. The Food and Agriculture Organization of the United Nations (FAO) has estimated that around 1 billion ha of land in more than 100 countries suffer from salinity (FAO, 2015). This problem is expected to increase by 1.5–3 million ha per year and eventually decrease crop production by more than 20% (Porcel *et al.*, 2012; FAO, 2015).

Salt stress is a serious limiting factor for plant growth and production. In general, plants show growth inhibition in salt-stress conditions due to reduced ability to take up water (osmotic-deficit effect) and injury to the cells of transpiring leaves (ion-excess effect) (Greenway and Munns, 1980; ElSayed *et al.*, 2022). To date, these effects have been identified to damage various physio-biochemical processes in plants such as *Oryza*

*Email: reyesvp@nagoya-u.jp

 The Genetics of Salt Tolerance in Plants: A Central Dogma Perspective and Strategies for Enhancement (S. Ganie and S.H. Wani eds)
DOI: 10.1079/9781800623033.0005

sativa (Hussain *et al.*, 2017; Ganie, 2020), *Zea mays* (Hu *et al.*, 2022), *Brassica juncea* L. (Singh *et al.*, 2022), and *Arabidopsis thaliana* (Zhang *et al.*, 2023). In addition, several studies have demonstrated that salinity alters the water uptake by seeds due to a lower osmotic potential (Ganie, 2021). This results in a cascade of changes in the activities of nucleic acid metabolism, protein metabolism, hormonal balance, and reduction in seed reserves (Khan and Weber, 2006). Similarly, negative effects of salinity on the yield productivity of some plants have been observed (Cui *et al.*, 1995; Ganie *et al.*, 2017; Cucci *et al.*, 2019). To mitigate such deleterious effects and safeguard yield productivity, understanding the physiological and molecular networks governing salinity tolerance in plants is necessary.

Several gene expression controls are known to date, one of which is the RNA–protein interaction (Prall *et al.*, 2019). The interaction between RNAs and proteins serves as a critical point for fine-tuning gene expression at co- and post-transcriptional levels. Generally, the RNA-binding proteins (RBPs) connect with *cis*-acting regulatory areas of RNAs to generate dynamic ribonucleoprotein (RNP) complexes that control the function of RNA. RBPs regulate RNA synthesis, editing, processing (including capping, splicing, and polyadenylation), transport and localization, storage, translation, and turnover. Classical RNA-binding domains (RBDs) are typically utilized to annotate RBPs *in silico* using knowledge of conserved domain structure and function (Pancaldi and Bähler, 2011). Recently, knowledge of RBPs in plants was gained primarily from studying specific proteins and *in silico* approaches based on sequence homology reported in other species (Marondedze *et al.*, 2016; Reichel *et al.*, 2016). And to date, several studies in model plants have demonstrated that RBPs play a major role in abiotic stress adaptation (Chen *et al.*, 2010; Li *et al.*, 2014; Wang *et al.*, 2014; Sun *et al.*, 2020).

This chapter discusses the various regulatory roles of RBPs in plant stress adaptation, specifically salt stress. The current advances in methods for detecting RBPs in plants, including the utilization of high-throughput sequencing technologies (HSTs), are also highlighted. Several studies across different plant models on RBPs and salt-stress tolerance are also presented. Lastly, this chapter presents an outlook on the utilization of RBPs and how RBPs could aid in developing salt stress-tolerant plants.

5.2 Regulatory Roles of RNA-Binding Proteins in Plant Adaptation

Plants can respond to both internal and external changes by regulating the plant transcriptome through RBPs. One known mechanism is the binding to mRNAs using either a single protein or an assembly of several proteins (Corley *et al.*, 2020). The interplay between RNA and RBP is dictated by one or more well-conserved RBDs at the protein's N-terminus end while the C-terminus end contains various auxiliary motifs (Muthusamy *et al.*, 2021).

RBDs dictate the specificity of RBPs to bind to particular RNAs and the functionality of ribonucleoprotein (RNA + RBP) complexes during plant growth and adaptation (Lee and Kang, 2016). The characteristics and roles of RBDs in plant growth and abiotic stress response have been discussed in several papers (Lorković, 2009; Kim *et al.*, 2010; Jiang *et al.*, 2013; Lee and Kang, 2016, 2020; Marondedze, 2020; Huh, 2021; Muthusamy *et al.*, 2021; Yan *et al.*, 2022). A recent review emphasized that RBPs also carry several disordered regions that lack stable tertiary structure, which improves their ability to bind with RNAs but complicates the analysis of RBPs (Järvelin *et al.*, 2016).

During crop growth, plants rely on RBP-dependent stabilization of mRNAs that are needed to maintain cellular activities and alleviate cellular stress damage (Muthusamy *et al.*, 2021). The interactions between RBPs and RNAs and their effects on RNA metabolism are discussed below.

5.2.1 RNA splicing

Splicing involves the removal of introns from RNA for cells to generate various forms of functional transcripts. In plants, around 60% of mRNAs can produce multiple protein products from a single RNA transcript due to alternative splicing events (Prall *et al.*, 2019). This increases a plant's phenotypic plasticity which enables it to respond to environmental changes.

Plants utilize alternative splicing to generate multiple functional transcripts from a single mRNA molecule, allowing for phenotypic plasticity in response to environmental changes. The splicing process involves small nucleolar ribonucleoprotein (snRNP) complexes, or spliceosomes, which recognize the splice sites in the precursor mRNA. The U1 snRNP binds to the 5′ splice site, and the U2 auxiliary factor binds to the 3′ splice site. The U2 snRNP then forms a pre-spliceosome or complex A, followed by the assembly of U4/U6 and U5 snRNPs to form a pre-catalytic complex B. The activated Bact complex then catalyses cleavage at the 5′ splice site, ligation of exons, and release of the mRNA. The resulting mRNA is further processed in response to environmental conditions. Lariats, or discarded by-products, are formed and subsequently degraded, while snRNPs are recycled (Will and Luhrmann, 2011; Cordin and Beggs, 2013; Meyer *et al.*, 2015; Prall *et al.*, 2019).

5.2.2 RNA folding

Folding allows RNA to exist in a stable and functional form within the cell. RNA molecules can have a variety of configurations that are prone to faulty secondary structures and poor tertiary binding, resulting in non-functional structures (Woodson, 2010). Misfolding is caused by two major factors: (i) kinetic folding, in which the cell must generate enough kinetic energy to accomplish its native structure; and (ii) thermodynamic folding, in which the native conformation competes with other forms with the same energy stabilities (Herschlag, 1995).

RNA chaperones are RBPs with low sequence specificity that rescue various misfolded RNAs and improve RNA nucleation specificity to ensure proper structural rearrangements (Ivanyi-Nagy *et al.*, 2005; Woodson, 2010). During stress, many plant mRNAs become misfolded, requiring the action of RNA chaperones to achieve their functional state and accurately respond to the plant's environment (Ganie, 2020). Semrad (2010) notes that there are two major activities facilitated by RNA chaperones: annealing and unwinding.

5.2.3 RNA editing

Editing is a post-transcription process where: cytidine (C) is converted to uridine (U), adenosine (A) is converted to inosine (I), insertions and deletions of U are made, or insertions of guanosine (G) are made (Small *et al.*, 2020). In plants, C-to-U conversion is the predominant form of RNA editing, occurring in the nucleus, cytosol, mitochondria, and plastids (Hao *et al.*, 2021). Many of these editing events occur in the first or second codon position which could provide an evolutionary advantage since this serves as a checkpoint in translation initiation that contributes to plant development, flowering, organelle biogenesis, environmental stimuli response, signal transduction, RNA structure, and indirect repair to correct possible DNA mutations at the RNA level (Gualberto *et al.*, 1989; Li *et al.*, 2019; Small *et al.*, 2020).

The RNA editosome is an assembly of several proteins located in the mitochondria and chloroplasts (Takenaka, 2014) that influence RNA editing. Li *et al.* (2014) enumerated some components of the plant editosome identified through the years, namely pentatricopeptide repeat (PPR) proteins, RNA editing factor interacting proteins (RIPs)/multiple organellar RNA editing factors (MORFs), organelle RNA recognition motif (ORRM) proteins, organelle zinc-finger (OZ) proteins, and protoporphyrinogen oxidase 1 (PPO1). Small *et al.* (2020) succinctly explained that PPR proteins primarily recognize specific cytidines and bind four nucleotides upstream of the editing site. The DYW (aspartate (D), tyrosine (Y), and tryptophan (W)) domain carries cytidine deaminase activity of the complex resulting in C-to-U conversion.

5.2.4 RNA modification, transport and localization

RNA modification and RNA transport and localization are two essential processes that affect gene expression in plants. RNA modification refers to the chemical modification of RNA nucleotides, which can alter RNA configuration and ultimately lead to changes in the phenotype. On the other hand, RNA transport and localization involve the proper delivery of mRNAs to the appropriate cellular regions, which is essential for efficient stress response and growth.

One of the most well-understood RNA modification mechanisms is N^6-methyladenosine (m6A), which is initiated by mRNA adenosine methylase (MTA) and affects the RNA's secondary structure, RBP interactions, and mRNA stability (Vandivier and Gregory, 2018). Similarly, 5-methylcytosine (m5C) marks rely on transfer RNA (tRNA)-specific methyltransferase 4 (TRM4) and can affect plant development and oxidative response (David *et al.*, 2017). Other modification mechanisms, 8-hydroxyguanosine (8-OHG) and 8-nitroguanosine (8-NO_2-G), are governed by the action of reactive oxygen species (ROS), but their molecular mechanisms are not well understood (Chmielowska-Bąk *et al.*, 2019).

In contrast, RNA transport and localization require the proper delivery of mRNAs to the appropriate cellular regions. This process helps to conserve cellular energy, concentrate transcripts at intracellular locations where they are needed, prevent unwanted protein–protein complexes, and facilitate multiprotein complex assemblies (Tian *et al.*, 2022).

In summary, RNA modification and RNA transport and localization are two important epitranscriptomic processes that play a vital role in regulating gene expression in plants. While RNA modification alters RNA configuration through chemical modifications, RNA transport and localization ensure the proper delivery of mRNAs to the right cellular regions for efficient stress response and growth.

5.2.5 RNA stability and decay

RNA stability and decay are opposing forces that ensure accurate plant phenotypic responses to internal or external stimuli during crop growth. Stability is achieved by polyadenylation, which adds a 3′ poly(A) tail to mRNA, protecting it from degradation by nucleases and acting as a binding site for poly(A)-binding proteins (PABPs). PABPs provide a platform for factors responsible for polyadenylation, export, and translation, while also acting as antagonists to mRNA degradation complexes. Decay is governed by deadenylation, which involves the gradual shortening of the 3′ poly(A) tail or the cap, followed by decapping and further decay by the RNA exosome. Three deadenylase complexes are involved in the process: the CCR4–NOT complex, PARN, and PAN. These mechanisms ensure precise regulation of gene expression and response to stress in plants (Mangus *et al.*, 2003; Abbasi *et al.*, 2013; Sieburth and Vincent, 2018; Prall *et al.*, 2019).

5.2.6 RNA translation

Translation is an essential process in the central dogma where proteins are produced from the information carried by mRNAs. A genetic code guides this process which defines how an mRNA sequence can be translated into a protein composed of a series of amino acids.

Translation is a well-elucidated process as described by Staley and Woolford (2009) that requires the activity of ribosomes. Ribosomes contain two RNP subunits: (i) a large subunit with 47 proteins and three ribosomal RNAs (rRNAs); and (ii) a small subunit with 32 proteins and one rRNA. Ribosome assembly starts with rRNA synthesis and folding in the nucleolus, followed by exonuclease and endonuclease cleavage, and ribosomal protein binding, assisted by 180 assembly factors and 100 snRNPs. The generated complex is called a preribosome which is then transported to the cytoplasm. During transport, preribosomes still undergo assembly by releasing most of the assembly factors to make sure that they are small enough to fit in the nuclear pores. The final ribosome configuration carries the rRNA and the integral ribosomal proteins for translation.

5.3 Salt Tolerance and RNA-Binding Protein Studies in Plants

Significant progress has been made in studying transcriptional regulation, perception, and signalling cascades of stress-induced changes in plants. The RNA regulatory mechanisms of RBPs in synthesis, translation, editing, and degradation of RNA molecules are some of the promising key processes in the modulation of stress response to salinity (Ambrosone *et al.*, 2012; Nakaminami *et al.*, 2012). In this section, studies on the roles of RBPs in the salt-stress adaptation of selected plants are presented.

5.3.1 *Arabidopsis thaliana*

In the model plant *A. thaliana*, several studies have demonstrated that RBPs regulate tolerance to salt stress. In a study conducted by Kim *et al.* (2008a), the glycine-rich RNA-binding protein 7 (GRP7) was observed to regulate the stomatal aperture. As a result, under salt-stress conditions, transgenic plants overexpressing GRP7 showed an increase in stomatal aperture. Similarly, using transgenic and mutant plants, Kim *et al.* (2008b) observed that overexpression of *AtRH9* and *AtRH25* negatively affects seed germination in *A. thaliana* under salt-stress conditions. The regulatory role of cold shock-domain proteins (CSDPs), CSDP1 and CSDP2, in *A. thaliana* under salt stress was examined in the study by Park *et al.* (2009). On one hand, the overexpression of CSDP1, which harbours a long C-terminal glycine-rich region with seven CCHC-type zinc fingers, delayed the seed germination under salt-stress conditions. On the other hand, the overexpression of CSDP2, which has a shorter glycine-rich region with two CCHC-type zinc fingers, accelerated seed germination. Nevertheless, these two RBPs were able to rescue mutant plants from freezing damage.

Serine/arginine-rich (SR) and SR-like proteins, a conserved family of RBPs, are essential for pre- and post-mRNA transcriptional processes. Albaqami *et al.* (2019) showed that the Arabidopsis *SR45* and its splice variant play an integral role in salt-stress tolerance. Their study demonstrated that the loss of *SR45* positively regulates salt tolerance, and only the long isoform (*SR45.1*) is necessary for salt tolerance. In addition, this RBP altered the gene expression and splicing of *Arabidopsis* mutant lines. Recently, Li *et al.* (2021) elucidated the role of *SR45a* in salt-stress tolerance. Their study demonstrated that the *SR45a* mediates salt-stress response through its interaction with the cap-binding complex (CBC) subunit cap-binding protein (CBP20). This RBP was also observed to be involved in post-transcriptional regulation such as alternative splicing and mRNA maturation.

Another RBP, which has been studied in *A. thaliana*, is the S1 domain-containing protein (SDP), a nucleus-encoded chloroplast protein. In a study by Dinh *et al.* (2019), the *A. thaliana* chloroplast SDP was identified as an organellar component response to different abiotic stresses. The *sdp* complementation and overexpression lines were observed to have similar survival rates to wild-type plants under salt-stress conditions. Interestingly, the transcript levels of the genes associated with salt-stress tolerance were not altered. These findings suggest that SDP dysfunction directly affects chloroplast transcript translation and biogenesis, which in turn modulates the expression of a subset of nuclear genes throughout salt-stress tolerance.

The RBP Tudor-SN was found to be essential for stress tolerance in *Arabidopsis*, stabilizing levels of stress-responsive mRNAs encoding secreted proteins (Dit Frey *et al.*, 2010). Another study showed that the RALF1–FERONIA complex affects splicing dynamics, modulating stress responses and growth in plants (Wang *et al.*, 2020). The MUG13.4 RBP was found to interact with AtAGO2 to modulate salinity tolerance in *Arabidopsis*, and the SCL30a SR protein was found to regulate abscisic acid (ABA)-dependent seed traits and salt-stress tolerance during germination (Wang *et al.*, 2019). These findings provide valuable insights into the molecular mechanisms underlying stress tolerance in plants and suggest that RBPs could be promising targets for improving crop productivity under adverse environmental conditions.

5.3.2 *Oryza sativa*

Rice (*O. sativa*) is one of the major crops which has been widely affected by salinity stress. To date, several genes and quantitative trait loci (QTLs) have been associated with salinity tolerance (Bonilla *et al.*, 2002; Tiwari *et al.*, 2016; Rahman *et al.*, 2017; Amoah *et al.*, 2020; Goto *et al.*, 2022). However, only a few of them have been successfully incorporated into breeding programmes. Other than these genes and QTLs, other molecular networks are involved in regulating salinity tolerance. For example, Macovei *et al.* (2012) identified a new DEAD-box helicase, ATP-binding protein (OsABP), which interacts with proteins involved in RNA metabolism, signal transduction, and stress response. Under salinity stress, gene transcripts associated with salt, dehydration, and ABA are upregulated. RNA chaperones were identified as potential candidates for engineering salt tolerance in rice, highlighting the importance of RNA processing in salt-stress tolerance (Ganie, 2020). Tuteja *et al.* (2013) elucidated the role of SUV3 helicase, an RBP that exhibits DNA and RNA helicase and ATPase activities. The study showed that SUV3 was induced in rice seedlings under salinity stress. In addition, this RBP facilitates salt-stress adaptation by improving photosynthesis and antioxidant production. Cui *et al.* (2015) described the role of a specific RBP, DCA1, in regulating salt tolerance in rice. The study found that DCA1 acts as a transcriptional co-activator of *DST*, a gene that plays a crucial role in regulating stress responses in rice. Specifically, DCA1 physically interacts with the promoter region of *DST* and enhances its expression by recruiting other transcriptional factors. Chen *et al.* (2018) conducted a comprehensive genome-wide analysis of genes encoding PPR proteins in rice. Interestingly, PPR genes, *LOC_Os02g46980* and *LOC_Os07g09370*, were differentially expressed indicating that they are possibly involved in salinity-stress response in rice. In a recent study by Zhang *et al.* (2020), the chloroplast RNA splicing and ribosome maturation (CRM) domain proteins were associated with salinity tolerance. The study showed 14 *CRM* genes were significantly downregulated. Generally, these proteins play a significant role in plant growth and development through RNA-binding activity (Barkan *et al.*, 2007).

5.3.3 *Triticum aestivum* L.

In wheat, only a few studies have been conducted to identify the roles and mechanisms of RBPs in salt-stress regulation. For example, Xu *et al.* (2014) conducted a functional analysis of *Triticum aestivum* L. zinc finger-containing glycine-rich RNA-binding proteins (TaRZs) namely TaRZ1, TaRZ2, and TaRZ3. The study demonstrated that two of three TaRZs (TaRZ1 and TaRZ3) are localized to the nucleus whereas the TaRZ2 was localized on the nucleus, endoplasmic reticulum, and cytoplasm. In addition, transgenic plants expressing *TaRZ1* showed suppressed growth under salt-stress conditions, while seedling growth of transgenic plants expressing *TaRZ2* was suppressed under salt-stress conditions. Using HST, Guo *et al.* (2020) conducted genome-wide analysis and global profiling of alternative splicing in wheat under salt-stress conditions. The study showed strong enrichment of transcriptional regulation, specifically RNA binding, at 12, 24, and 48 h after salt stress. Similarly, Sun *et al.* (2020) used the same approach and identified genes that encode YTH domain-containing proteins that are responsive under salt-stress conditions. As compared to the model plants, research on the roles of RBPs in wheat has remained limited. Therefore, future research on RBPs and their role in salinity stress in wheat is necessary to further identify and understand the molecular networks involved.

5.3.4 Other plants

To date, RBPs on other non-model plants were also identified to be involved in salinity stress regulation. In a review conducted by Kawa and Testerink (2017), the RBPs such as AtGRP8, AtC3H14, and AtCAF1B were found to promote mRNA decay during salinity-stress response, whereas RBPs such as AtRZ1A, AtRZ1B, and AtGRP7 were found to stabilize mRNAs under salinity-stress conditions. Moreover, the RBPs such as AtBUD13 and AtCCR4 were identified as crucial factors in the deadenylation of mRNAs, whereas AtXRN4, which interacts with RBPs such as AtC3H14 and AtCAF1B, was found to promote mRNA decay through 5′-to-3′ exonucleolytic activity. The study emphasizes the importance of these RBPs in regulating mRNA decay during salinity-stress response in plants, which provides insights into improving plant salinity tolerance. In a study conducted by Tada *et al.* (2019), the glycine-rich RNA-binding protein (SvGR-RBP1) was found to play a crucial role in enhancing salt tolerance in plants. The study showed that *SvGR-RBP1* was upregulated under salt stress, and transgenic plants expressing the protein exhibited increased salt tolerance. Transcriptomic and metabolomic analysis revealed that SvGR-RBP1 regulated the expression of genes involved in stress response and metabolism, leading to improved salt tolerance. These findings suggest that RBPs such as SvGR-RBP1 could serve as promising targets for improving crop productivity under adverse environmental conditions. In *Brassica rapa*, Yoon *et al.* (2018) demonstrated that SR proteins form a family of conserved RBPs and are involved in the regulation of salinity. Gu *et al.* (2020) conducted a genome-wide identification of SR proteins that are involved in salinity tolerance in cassava (*Manihot esculenta* Crantz). The overexpression of *MeSR34* in their study was associated with salinity tolerance as observed in seedlings grown in soil with 300 mM NaCl for 2 weeks. Interestingly, the MeSR34 functions by relieving the inhibiting effect of salinity in plants through the maintenance of K^+/Na^+ homeostasis. In *Solanum lycopersicum*, Zhu *et al.* (2015) showed that the DEAD-box RNA helicase gene, *SlDEAD31*, plays a regulatory role in salinity tolerance where it functions by enhancing photosynthetic activity and water balance. Interestingly, lower malondialdehyde (MDA) was also observed in transgenic plants which could be attributed to enhanced capacity for scavenging ROS. Yeap *et al.* (2012) investigated the biological role of RBP in oil palm (*Elaeis guineensis* Jacq.) and identified EgRBP42, a heterogeneous nuclear ribonucleoprotein (hnRNP)-like RNA-binding protein, to be induced under abiotic stresses, including salinity. This RBP co-regulates several transcripts through various post-transcriptional processes such as mRNA splicing, export, stability, and translation. The PPR proteins belong to the α-solenoid RBP superfamily and were previously reported as regulators of mRNA (Kramer *et al.*, 2018; Müller-McNicoll *et al.*, 2019). In *Populus trichocarpa*, Xing *et al.* (2018) explored the expression patterns of *P. trichocarpa* pentatricopeptide repeat (*PtrPPR*) genes and proteins that are modulated under salt-stress conditions. However, the study still needs detailed characterization of the identified genes through additional physiological and biochemical experiments.

In conclusion, RBPs play a crucial role in plant response to salinity stress, and their function varies between different plant species. The above examples illustrate the diversity of RBPs and their role in the response of different plant species to salinity stress. Further research in this area will help us to understand the mechanisms underlying plant response to salinity stress and develop strategies to enhance salt tolerance in crops.

5.4 Advances in Methods of Detecting RNA-Binding Proteins in Plants

Burjoski and Reddy (2021) recently provided an exhaustive list of methods used in detecting RBPs based on three categories: (i) protein-to-RNA detection; (ii) RNA-to-protein detection; and (iii) global detection approaches. RNA-to-protein approaches where RNA is used as bait to identify the protein targets have not been used in plant studies. Here we summarize the methods used in plants and their salient characteristics (Table 5.1).

Other protein-to-RNA detection methods are yet to be adapted for the evaluation of plant RBPs. Several of these methods also show promise as they can overcome some of the limitations of the currently used methods. Details of these techniques can be further examined in the detailed review of Burjoski and Reddy (2021).

5.5 Future Outlook and Conclusion

This chapter presented a comprehensive overview of plant RBPs and established studies that identified useful RBP–RNA interactions. Several RBP-encoding genes regulated under salinity-stress conditions were collected as resources that may provide information to conduct future studies for this abiotic stress. Among the latest tools, the application of HST and RNA interactome capture (RIC) has great potential for the discovery of novel RBPs across multiple plant genomes. Although advances have been made in terms of technology and functions of RBPs, a window for future research on the target specificity, motif structures, regulation of stress response, and functional roles of individual RBPs is still necessary.

Considering the worsening climate scenarios and their adverse effects on crop production, future work focusing on RBPs and their role in salinity-stress tolerance is needed. The integration of the latest biotechnological tools such as clustered regularly interspaced short palindromic repeats/CRISPR-associated protein 9 (CRISPR/Cas9) and the expansion of the use of HTS and RIC to non-model plants will benefit researchers in developing salinity-tolerant plants.

Table 5.1. RBP detection methods used in plants.

Detection category	Method	Features
Protein-to-RNA	RNA immunoprecipitation (RIP)	Applies an immunoprecipitation technique where a lysate of cells expressing the target protein is incubated with antibodies immobilized on beads. The proteins are digested and washed. Only the pool of RNA remaining on the beads is used to identify the RNA targets of the protein of interest. This method is combined with reversible crosslinking using formaldehyde to stabilize the protein–RNA interaction
	RNA immunoprecipitation sequencing (RIP-Seq)	RIP method which has been adapted for high-throughput sequencing platforms
	Crosslinking and immunoprecipitation sequencing (CLIP-Seq)	Uses crosslinking by ultraviolet (UV) light instead of formaldehyde to address possible DNA–protein and protein–protein crosslinking. Immunoprecipitation steps are followed as in RIP followed by partial digestion of RNA to roughly detect the binding site of the protein. SDS-PAGE separates the RNA–protein complex and it is transferred to a nitrocellulose membrane that is then treated with proteinase K. Linkers are ligated to the RNA fragments which can be used to create the fragment library for sequencing
	High-throughput sequencing of RNA isolated by crosslinking immunoprecipitation (HITS-CLIP)	Collects 1000-fold more data from the extracted CLIP fragment libraries. However, the volume of data generated can be confused with noise signals
	Individual-nucleotide resolution by crosslinking immunoprecipitation (iCLIP)	Applies a complementary DNA (cDNA) self-circularization step instead of an RNA ligation step during library preparation to overcome the premature termination of reverse transcriptase whenever it encounters a nucleotide sequence linked to residual amino acids
Global RNA	RNA interactome capture (RIC)	Begins with UV-crosslinking as in the CLIP method. Poly(A) RNAs and denatured proteins are then extracted using oligo(dT)-magnetic beads under conditions that favour denaturation. Stringent washing separates the denatured proteins and RNAs. The RNA is digested and the protein is analysed using mass spectrometry. However, this method is limited to poly(A) RNAs
	Plant RNA interactome capture (ptRIC)	Optimized RIC method for leaf tissues where: (i) UV conditions are adjusted to irradiate the leaf surfaces; (ii) stringency of washing conditions is increased; and (iii) genomic DNA is sheared by passing the RNA-loaded beads through a narrow needle
	Orthogonal organic phase separation (OOPS)	Uses UV-crosslinking and acidic guanidinium thiocyanate–phenol–chloroform (AGPC) phase separation to extract the RBPs. This does not require an mRNA pulldown which can capture RBP interactions with all types of RNAs instead of only mRNAs

References

Abbasi, N., Park, Y.I. and Choi, S.B. (2013) RNA deadenylation and decay in plants. *Journal of Plant Biology* 56, 198–207.

Albaqami, M., Laluk, K. and Reddy, A.S.N. (2019) The *Arabidopsis* splicing regulator SR45 confers salt tolerance in a splice isoform-dependent manner. *Plant Molecular Biology* 100(4–5), 379–390.

Ambrosone, A., Costa, A., Leone, A. and Grillo, S. (2012) Beyond transcription: RNA-binding proteins as emerging regulators of plant response to environmental constraints. *Plant Science* 182, 12–18.

Amoah, N.K.A., Akromah, R., Kena, A.W., Manneh, B., Dieng, I. and Bimpong, I.K. (2020) Mapping QTLs for tolerance to salt stress at the early seedling stage in rice (*Oryza sativa* L.) using a newly identified donor 'Madina Koyo'. *Euphytica* 216(10), 156.

Barkan, A., Klipcan, L., Ostersetzer, O., Kawamura, T., Asakura, Y. and Watkins, K.P. (2007) The CRM domain: an RNA binding module derived from an ancient ribosome-associated protein. *RNA* 13(1), 55–64.

Bonilla, P., Dvorak, J., Mackill, D., Deal, K. and Gregorio, G. (2002) RFLP and SSLP mapping of salinity tolerance genes in chromosome 1 of rice (*Oryza sativa* L.) using recombinant inbred lines. *Philippine Agricultural Scientist* 85, 68–76.

Burjoski, V. and Reddy, A.S.N. (2021) The landscape of RNA–protein interactions in plants: approaches and current status. *International Journal of Molecular Sciences* 22, 2845.

Chen, G., Zou, Y., Hu, J. and Ding, Y. (2018) Genome-wide analysis of the rice PPR gene family and their expression profiles under different stress treatments. *BMC Genomics* 19(1), 720.

Chen, X., Zeng, Q., Lu, X. Yu, D. and Li, W. (2010) Characterization and expression analysis of four glycine-rich RNA-binding proteins involved in osmotic response in tobacco (*Nicotiana tabacum* cv. Xanthi). *Agricultural Sciences in China* 9(11), 1577–1587.

Chmielowska-Bąk, J., Arasimowicz-Jelonek, M. and Deckert, J. (2019) In search of the mRNA modification landscape in plants. *BMC Plant Biology* 19, 421.

Cordin, O. and Beggs, J.D. (2013) RNA helicases in splicing. *RNA Biology* 10(1), 83–95.

Corley, M., Burns, M.C. and Yeo, G.W. (2020) How RNA-binding proteins interact with RNA: molecules and mechanisms. *Molecular Cell* 78, 9–29.

Cucci, G., Lacolla, G., Boari, F., Mastro, M.A. and Cantore, V. (2019) Effect of water salinity and irrigation regime on maize (*Zea mays* L.) cultivated on clay loam soil and irrigated by furrow in Southern Italy. *Agricultural Water Management* 222, 118–124.

Cui, H., Takeoka, Y. and Wada, T. (1995) Effect of sodium chloride on the panicle and spikelet morphogenesis in rice. II. Developmental morphology of the panicle. *Japanese Journal of Crop Science* 64(3), 593–600.

Cui, L.-G., Shan, J.-X., Shi, M., Gao, J.-P. and Lin, H.-X. (2015) DCA1 acts as a transcriptional co-activator of DST and contributes to drought and salt tolerance in rice. *PLoS Genetics* 11(10), e1005617. DOI: 10.1371/journal.pgen.1005617.

David, R., Burgess, A., Parker, B., Li, J., Plusford, K. *et al.* (2017) Transcriptome-wide mapping of RNA 5-methylcytosine in *Arabidopsis* mRNAs and noncoding RNAs. *The Plant Cell* 29, 445–460.

Dinh, S.N., Park, S.J., Han, J.H. and Kang, H. (2019) A chloroplast-targeted S1 RNA-binding domain protein plays a role in *Arabidopsis* response to diverse abiotic stresses. *Journal of Plant Biology* 62(1), 74–81. DOI: 10.1007/s12374-018-0325-y.

Dit Frey, N.F., Muller, P., Jammes, F., Kizis, D., Leung, J. *et al.* (2010) The RNA binding protein Tudor-SN is essential for stress tolerance and stabilizes levels of stress-responsive mRNAs encoding secreted proteins in *Arabidopsis*. *The Plant Cell* 22(5), 1575–1591. DOI: 10.1105/tpc.109.070680.

ElSayed, A.I., Rafudeen, M.S., Ganie, S.A., Hossain, M.S. and Gomaa, A.M. (2022) Seed priming with cypress leaf extract enhances photosynthesis and antioxidative defense in zucchini seedlings under salt stress. *Scientia Horticulturae* 293, 110707.

FAO (2015) *Status of the World's Soil Resources (SWSR) – Main Report*. Food and Agriculture Organization of the United Nations, Rome.

Ganie, S.A. (2020) RNA chaperones: potential candidates for engineering salt tolerance in rice. *Crop Science* 60(2), 530–540. DOI: 10.1002/csc2.20134.

Ganie, S.A. (2021) Amino acids other than proline and their participation in abiotic stress tolerance. In: Wani, S.H., Gangola, M.P. and Ramadoss, B.R. (eds) *Compatible Solutes Engineering for Crop Plants Facing Climate Change*. Springer, Cham, Switzerland, pp. 47–96.

Ganie, S.A., Debnath, A.B., Gumi, A.M. and Mondal, T.K. (2017) Comprehensive survey and evolutionary analysis of genome-wide miRNA genes from ten diploid *Oryza* species. *BMC Genomics* 18(1), 711.

Goto, I., Neang, S., Kuroki, R., Reyes, V.P., Doi, K. *et al.* (2022) QTL analysis for sodium removal ability in rice leaf sheaths under salinity using an IR-44595/318 F2 population. *Frontiers in Plant Science* 13, 1002605. DOI: 10.3389/fpls.2022.1002605.

Greenway, H. and Munns, R. (1980) Mechanisms of salt tolerance in nonhalophytes. *Annual Review of Plant Physiology* 31(1), 149–190.

Gualberto, J.M., Lamattina, L., Bonnard, G., Weil, J.-H. and Grienenberger, J.-M. (1989) RNA editing in wheat mitochondria results in the conservation of protein sequences. *Nature* 341, 660–662.

Gu, J., Ma, S., Zhang, Y., Wang, D., Cao, S. and Wang, Z.-Y. (2020) Genome-wide identification of cassava serine/arginine-rich proteins: insights into alternative splicing of pre-mRNAs and response to abiotic stress. *Plant & Cell Physiology* 61(1), 178–191.

Guo, W., Yu, K., Han, L., Li, X., Wang, H. *et al.* (2020) Global profiling of alternative splicing landscape responsive to salt stress in wheat (*Triticum aestivum* L.). *Plant Growth Regulation* 92(1), 107–116.

Hao, W., Liu, G., Wang, W., Shen, W., Zhao, Y. *et al.* (2021) RNA editing and its roles in plant organelles. *Frontiers in Genetics* 12, 757109.

Herschlag, D. (1995) RNA chaperones and the RNA folding problem. *Journal of Biochemical Chemistry* 270(36), 20871–20874.

Hu, D., Li, R., Dong, S., Zhang, J., Zhao, B. *et al.* (2022) Maize (*Zea mays* L.) responses to salt stress in terms of root anatomy, respiration and antioxidative enzyme activity. *BMC Plant Biology* 22(1), 602. DOI: 10.1186/s12870-022-03972-4.

Huh, S.U. (2021) The role of pumilio RNA binding protein in plants. *Biomolecules* 11, 1851.

Hussain, S., Zhang, J., Zhong, C., Zhu, L., Cao, X. *et al.* (2017) Effects of salt stress on rice growth, development characteristics, and the regulating ways: a review. *Journal of Integrative Agriculture* 16(11), 2357–2374. DOI: 10.1016/S2095-3119(16)61608-8.

Ivanyi-Nagy, R., Khandjian, E. and Davidovic, L. (2005) Disordered RNA chaperone proteins: from functions to disease. *Cellular and Molecular Life Sciences* 62, 1409–1417.

Järvelin, A.I., Noerenberg, M., Davis, I. and Castello, A. (2016) The new (dis)order in RNA regulation. *Cell Communication and Signaling* 14, 9.

Jiang, J., Wang, B., Shen, Y., Wang, H., Feng, Q. and Shi, H. (2013) The *Arabidopsis* RNA binding protein with K homology motifs, SHINY1, interacts with the C-terminal domain phosphatase-like 1 (CPL1) to repress stress-inducible gene expression. *PLoS Genetics* 9, 7.

Kawa, D. and Testerink, C. (2017) Regulation of mRNA decay in plant responses to salt and osmotic stress. *Cellular and Molecular Life Sciences* 74(7), 1165–1176. DOI: 10.1007/s00018-016-2376-x.

Khan, M.A. and Weber, D.J. (eds) (2006) *Ecophysiology of High Salinity Tolerant Plants*. Tasks for Vegetation Science, Vol. 40. Springer, Dordrecht, The Netherlands.

Kim, J.S., Jung, H.J., Lee, H.J., Kim, K.A., Goh, C.-H. *et al.* (2008a) Glycine-rich RNA-binding protein7 affects abiotic stress responses by regulating stomata opening and closing in *Arabidopsis thaliana*. *The Plant Journal* 55(3), 455–466.

Kim, J.S., Kim, K.A., Oh, T.R., Park, C.M. and Kang, H. (2008b) Functional characterization of DEAD-box RNA helicases in *Arabidopsis thaliana* under abiotic stress conditions. *Plant & Cell Physiology* 49(10), 1563–1571.

Kim, W.Y., Kim, J.Y., Jung, H.J., Oh, S.H., Han, Y.S. and Kang, H. (2010) Comparative analysis of *Arabidopsis* zinc finger-containing glycine-rich RNA-binding proteins during cold adaptation. *Plant Physiology and Biochemistry* 48(10), 866–872.

Kramer, M.C., Anderson, S.J. and Gregory, B.D. (2018) The nucleotides they are a-changin': function of RNA binding proteins in post-transcriptional messenger RNA editing and modification in *Arabidopsis*. *Current Opinion in Plant Biology* 45, 88–95.

Lee, K. and Kang, H. (2016) Emerging roles of RNA-binding proteins in plant growth, development, and stress responses. *Molecules and Cells* 39(3), 179–185.

Lee, K. and Kang, H. (2020) Roles of organellar RNA-binding proteins in plant growth, development, and abiotic stress responses. *International Journal of Molecular Sciences* 21, 4548.

Li, D., Zhang, H., Hong, Y., Huang, L., Li, X. *et al.* (2014) Genome-wide identification, biochemical characterization, and expression analyses of the YTH domain-containing RNA-binding protein family in *Arabidopsis* and rice. *Plant Molecular Biology Reporter* 32(6), 1169–1186.

Li, M., Xia, L., Zhang, Y., Niu, G., Li, M. *et al.* (2019) Plant editosome database: a curated database of RNA editosome in plants. *Nucleic Acids Research* 47, 170–174.

Li, Y., Guo, Q., Liu, P., Huang, J., Zhang, S. *et al.* (2021) Dual roles of the serine/arginine-rich splicing factor SR45a in promoting and interacting with nuclear cap-binding complex to modulate the salt-stress response in *Arabidopsis*. *New Phytologist* 230(2), 641–655.

Lorković, Z.J. (2009) Role of plant RNA-binding proteins in development, stress response and genome organization. *Trends in Plant Science* 14(4), 229–236.

Macovei, A., Vaid, N., Tula, S. and Tuteja, N. (2012) A new DEAD-box helicase ATP-binding protein (OsABP) from rice is responsive to abiotic stress. *Plant Signaling & Behavior* 7(9), 1138–1143.

Mangus, D.A., Evans, M.C. and Jacobson, A. (2003) Poly(A)-binding proteins: multifunctional scaffolds for the post- transcriptional control of gene expression. *Genome Biology* 4, 223.

Marondedze, C. (2020) The increasing diversity and complexity of the RNA-binding protein repertoire in plants. *Proceedings of the Royal Society B: Biological Sciences* 287, 20201397.

Marondedze, C., Thomas, L., Serrano, N.L., Lilley, K.S. and Gehring, C. (2016) The RNA-binding protein repertoire of *Arabidopsis thaliana*. *Scientific Reports* 6(1), 29766.

Meyer, K., Koester, T. and Staiger, D. (2015) Pre-mRNA splicing in plants: *in vivo* functions of RNA-binding proteins implicated in the splicing process. *Biomolecules* 5, 1717–1740.

Müller-McNicoll, M., Rossbach, O., Hui, J. and Medenbach, J. (2019) Auto-regulatory feedback by RNA-binding proteins. *Journal of Molecular Cell Biology* 11(10), 930–939.

Muthusamy, M., Kim, J.-H., Kim, J.A. and Lee, S.-I. (2021) Plant RNA binding proteins as critical modulators in drought, high salinity, heat, and cold stress responses: an updated overview. *International Journal of Molecular Sciences* 22, 6731.

Nakaminami, K., Matsui, A., Shinozaki, K. and Seki, M. (2012) RNA regulation in plant abiotic stress responses. *Biochimica et Biophysica Acta (BBA) – Gene Regulatory Mechanisms* 1819(2), 149–153.

Pancaldi, V. and Bähler, J. (2011) *In silico* characterization and prediction of global protein–mRNA interactions in yeast. *Nucleic Acids Research* 39(14), 5826–5836.

Park, S.J., Kwak, K.J., Oh, T.R., Kim, Y.O. and Kang, H. (2009) Cold shock domain proteins affect seed germination and growth of *Arabidopsis thaliana* under abiotic stress conditions. *Plant & Cell Physiology* 50(4), 869–878.

Porcel, R., Aroca, R. and Ruiz-Lozano, J.M. (2012) Salinity stress alleviation using arbuscular mycorrhizal fungi. A review. *Agronomy for Sustainable Development* 32(1), 181–200.

Prall, W., Sharma, B. and Gregory, B.D. (2019) Transcription is just the beginning of gene expression regulation: the functional significance of RNA-binding proteins to post-transcriptional processes in plants. *Plant & Cell Physiology* 60(9), 1939–1952.

Rahman, M.A., Bimpong, I.K., Bizimana, J.B., Pascual, E.D., Arceta, M. *et al.* (2017) Mapping QTLs using a novel source of salinity tolerance from Hasawi and their interaction with environments in rice. *Rice* 10(1), 47.

Reichel, M., Liao, Y., Rettel, M., Ragan, C., Evers, M. *et al.* (2016) *In planta* determination of the mRNA-binding proteome of *Arabidopsis* etiolated seedlings. *The Plant Cell* 28(10), 2435–2452.

Semrad, K. (2010) Proteins with RNA chaperone activity: a world of diverse proteins with a common task – impediment of RNA misfolding. *Biochemistry Research International* 2011, 532908.

Sieburth, L.E. and Vincent, J.N. (2018) Beyond transcription factors: roles of mRNA decay in regulating gene expression in plants. *F1000Research* 7, 1940.

Singh, J., Singh, V., Dutt, V., Walia, N., Kumawat, G. *et al.* (2022) Insights into salt tolerance of mustard (*Brassica juncea* L. Czern & Coss): a metabolomics perspective. *Environmental and Experimental Botany* 194, 104760. DOI: 10.1016/j.envexpbot.2021.104760.

Small, I.D., Schallenberg-Rudinger, M., Takenaka, M., Mireau, H. and Ostersetzer-Biran, O. (2020) Plant organellar RNA editing: what 30 years of research has revealed. *The Plant Journal* 101, 1040–1056.

Staley, J.P. and Woolford, J.L.J. (2009) Assembly of ribosomes and spliceosomes: complex ribonucleoprotein machines. *Current Opinion in Cell Biology* 21(1), 109–118.

Sun, J., Bie, X.M., Wang, N., Zhang, X.S. and Gao, X.-Q. (2020) Genome-wide identification and expression analysis of YTH domain-containing RNA-binding protein family in common wheat. *BMC Plant Biology* 20(1), 351.

Tada, Y., Kawano, R., Komatsubara, S., Nishimura, H., Katsuhara, M. *et al.* (2019) Functional screening of salt tolerance genes from a halophyte *Sporobolus virginicus* and transcriptomic and metabolomic analysis of salt tolerant plants expressing glycine-rich RNA-binding protein. *Plant Science* 278, 54–63. DOI: 10.1016/j.plantsci.2018.10.019.

Takenaka, M. (2014) How complex are the editosomes in plant organelles? *Molecular Plant* 7(4), 582–585.

Tian, L., Chou, H.-L., Fukuda, M., Kumamaru, T. and Okita, T.W. (2022) mRNA localization in plant cells. *Plant Physiology* 182, 97–109.

Tiwari, S., Krishnamurthy, S.I., Kumar, V., Singh, B., Rao, A. *et al.* (2016) Mapping QTLs for salt tolerance in rice (*Oryza sativa* L.) by bulked segregant analysis of recombinant inbred lines using 50K SNP chip. *PLoS ONE* 11(4), e0153610.

Tuteja, N., Sahoo, R.K., Garg, B. and Tuteja, R. (2013) OsSUV3 dual helicase functions in salinity stress tolerance by maintaining photosynthesis and antioxidant machinery in rice (*Oryza sativa* L. cv. IR64). *The Plant Journal* 76(1), 115–127.

Vandivier, L.E. and Gregory, B.D. (2018) New insights into the plant epitranscriptome. *Journal of Experimental Botany* 69(20), 4659–4665.

Wang, H., Liu, C., Ren, Y., Wu, M., Wu, Z. *et al.* (2019) An RNA-binding protein MUG13.4 interacts with AtAGO2 to modulate salinity tolerance in *Arabidopsis*. *Plant Science* 288, 110218. DOI: 10.1016/j.plantsci.2019.110218.

Wang, L., Yang, T., Wang, B., Lin, Q., Zhu, S. *et al.* (2020) RALF1–FERONIA complex affects splicing dynamics to modulate stress responses and growth in plants. *Science Advances* 6(21), eaaz1622. DOI: 10.1126/sciadv.aaz1622.

Wang, N., Yue, Z., Liang, D. and Ma, F. (2014) Genome-wide identification of members in the YTH domain-containing RNA-binding protein family in apple and expression analysis of their responsiveness to senescence and abiotic stresses. *Gene* 538(2), 292–305.

Will, C.L. and Luhrmann, R. (2011) Spliceosome structure and function. *Cold Spring Harbor Perspectives in Biology* 3, 7.

Wing, I.S., De Cian, E. and Mistry, M.N. (2021) Global vulnerability of crop yields to climate change. *Journal of Environmental Economics and Management* 109, 102462.

Woodson, S.A. (2010) Taming free energy landscapes with RNA chaperones. *RNA Biology* 7(6), 677–686.

Xing, H., Fu, X., Yang, C., Tang, X., Guo, L. *et al.* (2018) Genome-wide investigation of pentatricopeptide repeat gene family in poplar and their expression analysis in response to biotic and abiotic stresses. *Scientific Reports* 8(1), 2817.

Xu, T., Gu, L., Choi, M.J., Kim, R.J., Suh, M.C. and Kang, H. (2014) Comparative functional analysis of wheat (*Triticum aestivum*) zinc finger-containing glycine-rich RNA-binding proteins in response to abiotic stresses. *PLoS ONE* 9(5), e96877.

Yan, Y., Gan, J., Tao, Y., Okita, T.W. and Tian, L. (2022) RNA-binding proteins: the key modulator in stress granule formation and abiotic stress response. *Frontiers in Plant Science* 13, 882596.

Yeap, W.-C., Ooi, T.E.K., Namasivayam, P., Kulaveerasingam, H. and Ho, C.-L. (2012) *EgRBP42* encoding an hnRNP-like RNA-binding protein from *Elaeis guineensis* Jacq. is responsive to abiotic stresses. *Plant Cell Reports* 31(10), 1829–1843.

Yoon, E.K., Krishnamurthy, P., Kim, J.A., Jeong, M.-J. and Lee, S.I. (2018) Genome-wide characterization of *Brassica rapa* genes encoding serine/arginine-rich proteins: expression and alternative splicing events by abiotic stresses. *Journal of Plant Biology* 61(4), 198–209.

Zhang, Q., Shen, L., Ren, D., Hu, J., Zhu, L. *et al.* (2020) Characterization of the CRM gene family and elucidating the function of OsCFM2 in rice. *Biomolecules* 10(2), 327.

Zhang, X., Zhang, J., Tuluhong, G. and Zhang, F. (2023) A nuclear-located glyceraldehyde-3-phosphate dehydrogenase affects salt stress response processes in *Arabidopsis thaliana* as a senescence component. *Journal of Plant Biochemistry and Biotechnology* (2023). DOI: 10.1007/s13562-022-00815-z.

Zhu, M., Chen, G., Dong, T., Wang, L., Zhang, J. *et al.* (2015) *SlDEAD31*, a putative DEAD-box RNA helicase gene, regulates salt and drought tolerance and stress-related genes in tomato. *PLoS ONE* 10(8), e0133849.

6 Sequestration of mRNAs: Role of Stress Granules and Processing Bodies in Plant Salt Tolerance

Nancy McMulkin*

Plant Molecular Science and Centre of Systems and Synthetic Biology, Department of Biological Sciences, Royal Holloway University of London, Egham, UK

Abstract

Salinity stress is a global issue, affecting crop yields drastically year upon year. Understanding how naturally salt-tolerant species mediate growth and development in the presence of high salinity is crucial to genetically engineer more salt-tolerant crops for the future. In response to salinity stress, dramatic alterations in the transcriptome and proteome take place, whereby genes involved in growth and development are silenced and translation halted, while rapid expression of stress-response genes begins. Emerging evidence shows that cytoplasmic foci, stress granules (SGs) and processing bodies (PBs), take part in these cellular processes. SGs form in response to salinity stress and stably silence transcripts, sequestering their mRNAs via interaction with RNA-interacting proteins, heat-shock proteins, and RNA helicases. mRNA transcripts within SGs are stored and protected, ready for the stress-recovery phase during which transcripts are released and rapidly resume translation. PBs may play a differing role in response to salinity stress. PBs are the site of 5′–3′ mRNA degradation and emerging evidence reveals the degradation of transcripts involved in abscisic acid biosynthesis to attenuate salt-stress responses. This chapter highlights the role of plant SGs and PBs in sequestering mRNAs in response to salinity stress and discusses how SG core components regulate salt-stress tolerance. Despite some progress, research on this aspect of mRNA metabolism in plants in response to salt stress is in its infancy with large questions remaining.

6.1 Introduction

A critical problem of the 21st century is global food security because the demand for agricultural crops is drastically increasing due to accelerated population growth and limited resources. Two solutions are available to tackle the problem of limited crop production: the area of land used to grow crops could be increased or the productivity of the already available land could be enhanced, with the latter option being preferable because it does not rely on the further destruction of environmentally important land (Edgerton, 2009). Crop productivity could be drastically increased by improving the growth and development of the most important grain-crop species such as *Triticum aestivum* (wheat), *Oryza sativa* (rice), *Zea mays* (maize), and *Hordeum vulgare* (barley), especially under environmental stress conditions. Several crucial abiotic stresses threaten crop plants across the globe such as drought, cold, heat, flooding, salinity, and lack of nutrients in the soil (Ganie, 2021). Salinity in the soil affects 6% of the available land for crop

*Email: Nancy.McMulkin.2019@live.rhul.ac.uk

 The Genetics of Salt Tolerance in Plants: A Central Dogma Perspective and Strategies for Enhancement (S. Ganie and S.H. Wani eds)
DOI: 10.1079/9781800623033.0006

production, and by 2050, this will significantly increase to 50%, highlighting salinity as a major abiotic stressor that affects crop production globally (Ludwig *et al.*, 2018). There are two possible solutions to tackling soil salinity; one option is to treat the soil with chemicals to remove the NaCl and the other is to genetically engineer salt-tolerant crops that have a better capability to grow and develop within a high-salinity environment (recently reviewed in Wani *et al.*, 2020). Salinity negatively impacts all biological functions, leading to less investment in growth and development and more in protection against the disturbance of ion homeostasis and inhibition of photosynthesis (reviewed in Zhu, 2002). These effects, in turn, decrease germination, growth, and biomass, cause shortening of the vegetative phase and withering of leaves, ultimately leading to a decrease in crop yield (reviewed in Zhu, 2002). There are several naturally occurring salt-tolerant plant species known as halophytes which have evolved a variety of protective mechanisms such as osmolyte biosynthesis, ion homeostasis, water transport, scavenging of reactive oxygen species (ROS), and translational arrest (Hasegawa *et al.*, 2000). Orchestration of such protective mechanisms in response to salinity involves rapid large-scale changes in the plant transcriptome, metabolome, proteome, and degradome, which are brought about by Ca^{2+} signals, ion transporters, and hormones (Kollist *et al.*, 2019).

Salt stress imposes osmotic stress on a plant by limiting its water-uptake ability, since an increase in salt concentration within the soil drives water from the plant into the soil via osmosis. In this initial phase of salt stress, water deficiency instead of toxic levels of Na^+ and Cl^- ions is the main factor leading to reduced plant growth, further supported by the fact that initially Na^+ and Cl^- are below toxic concentrations in both the leaves and the root (Jeschke, 1984; Jeschke *et al.*, 1986; Hu and Schmidhalter, 1998; Fricke, 2004). The next phase of salinity stress is the ionic stress caused by an influx of Na^+ entering the root via non-selective cation channels (NSCCs) and there is limited evidence for the potential role of high-affinity K^+ transporters during this Na^+ influx from both the KUP/HAK/KT and AKT families, that requires further investigation (reviewed by Kronzucker and Britto, 2011). In turn, this Na^+ influx triggers a rapid rise (within 10 s of salt treatment) in the levels of the second messenger Ca^{2+} in the root which then travels rapidly to the leaves (within 30 s of salt treatment), and this calcium wave was found to be specific to salt treatment (Choi *et al.*, 2014). A calcium wave is regulated by an array of Ca^{2+} channels, Ca^{2+} pumps, and H^+/Ca^{2+} antiporters located on cell and organelle membranes (White and Broadley, 2003). This Ca^{2+} burst is the initial signal that activates Na^+ transporters to discharge Na^+ out into two locations, the cytoplasm and the vacuoles (Seifikalhor *et al.*, 2019). The well-characterized Na^+ transporter salt overly sensitive 1 (SOS1) is a plasma membrane-localized H^+/cation antiporter which functions in the CBL–CIPK pathway, whereby calcineurin B-like proteins (CBLs) bind Ca^{2+} and cause protein phosphorylation through their interaction with CBL-interacting protein kinases (CIPKs). The Ca^{2+} signal is initially sensed by SOS3/CBL4 (Liu and Zhu, 1998); once SOS3 is in a complex with Ca^{2+}, it interacts with SOS2/CIPK24, activating the SOS2 kinase activity, and together the SOS2/SOS3 complex phosphorylates and activates SOS1 leading to Na^+ efflux across the plasma membrane (Halfter *et al.*, 2000; Ishitani *et al.*, 2000; Liu *et al.*, 2000). Other transporters, for example Na^+/H^+ exchanger 1 and 2 (NHX1 and -2), are involved in discharging Na^+ from the cytoplasm. They are H^+/cation antiporters located on intracellular membranes and are hypothesized to transport Na^+ into vacuoles in response to salt stress. These NHXs also transport K^+ with equal affinity, playing an important role in potassium homeostasis, while also maintaining vacuole pH (Sottosanto *et al.*, 2004; Barragán *et al.*, 2012). The calcium wave followed by the CBL–CIPK cascade that activates SOS1, mediating efflux of Na^+ from the cell, aided by the activity of NHX1 and NHX2 storing Na^+ in vacuoles, together lead to the reduction of toxic levels of Na^+ within the cytoplasm.

Quick robust changes in gene expression must take place for a cell to respond to environmental stresses by dynamically altering the transcription and translation of genes involved in growth and stress responses. One way in which gene expression is regulated is via the shuttling of mRNAs into stress granules (SGs) (reviewed in Chodasiewicz *et al.*, 2022), discrete cytoplasmic foci composed of protein and RNA complexes which self-assemble in response to external stress signals via liquid–liquid phase separation (LLPS), using by-products of mRNA metabolism. In response to an external stress, translation of non-stress-response genes is halted, and translation of crucial stress-response genes begins (Haak *et al.*, 2017). Once the stress has subsided, translation of non-stress-response genes is rapidly resumed, suggesting that the untranslated mRNAs are readily available from the SGs. These untranslated mRNAs are stably stored in SGs (Chodasiewicz *et al.*, 2022), which were first identified in tomato (*Lycopersicon peruvianum*), where they were found to form complexes with heat-shock proteins (HSPs) and were thus originally named as heat-shock granules (HSGs) (Nover *et al.*, 1983, 1989; Scharf *et al.*, 1998). SGs have recently been found to a play a clear role in mediating tolerance to other abiotic stresses including salt stress in plants (recently reviewed in Yan *et al.*, 2022).

SGs are a particular subset of the cytoplasmic foci that form in plants. There are three groups of such foci characterized by their formation and composition: (i) HSGs; (ii) SGs; and (iii) processing bodies (PBs). HSGs form after prolonged heat stress and do not contain poly(A) mRNAs, while SGs form in response to short-term stress and they uniquely contain poly(A) mRNAs. Contrastingly, PBs are cytoplasmic foci that are constitutively present within plant cells and play a role in translational repression and mRNA decay. Characterization of these three main groups of cytoplasmic foci in mammalian systems has been extensively investigated and recently reviewed (Ivanov *et al.*, 2019; Riggs *et al.*, 2020). PBs contain mRNA degradation factors, namely mRNA decapping factors, decapping 1 (DCP1) and decapping 2 (DCP2), that form a complex together with varicose (VCS) and the cytoplasmic 5′-to-3′ exoribonuclease, exoribonuclease 4 (XRN4) (Xu *et al.*, 2006; Goeres *et al.*, 2007; Iwasaki *et al.*, 2007; Weber *et al.*, 2008; Chantarachot and Bailey-Serres, 2018). Our understanding of SGs is much greater in mammals than in plants. To begin, the core composition of mammalian SGs and their homologues that have been identified in plants will be described.

Nunes *et al.* (2019) identified approximately 500 protein components localized within mammalian SGs and provided the first open-access database of the mammalian proteome, describing molecular function and the cell types in which SG proteins were detected. A large proportion of the mammalian proteome contains intrinsically disordered proteins (IDPs) which have an unstable native structure and interact with or function as hubs in protein interaction networks; many IDPs are RNA-binding proteins and have ATPase activity required for ATP-dependent remodelling complexes (Ward *et al.*, 2004; van der Lee *et al.*, 2014; Deiana *et al.*, 2019; Zhou *et al.*, 2021). Together, this reveals that mammalian SGs play key roles in protein–protein interactions and RNA processing (Kedersha and Anderson, 2002; Kedersha *et al.*, 2005; Banani *et al.*, 2017; Markmiller *et al.*, 2018; Youn *et al.*, 2018, 2019). The core components of yeast and mammalian SGs were identified, with these core SG proteins being fundamental in SG formation and function regardless of the stress stimulus and mainly possessing: the 40S ribosomal subunit, two eukaryotic translation initiation factors eIF3 and 4G, subunit poly(A)-binding protein cytoplasmic 1 (PAB1, which interacts with poly(A) tails of mRNA and facilitates mRNA–protein interactions), and prion-like domain (PrLD)-containing proteins (Kedersha *et al.*, 2002, 2005). The SG core proteins, Ras-GAP SH3 domain-binding proteins 1 and 2 (G3BP1 and G3BP2) (Tourrière *et al.*, 2003) and T-cell-restricted intracellular antigen-1 and TIA1-related protein (TIA1 and TIAL1, respectively), are found to play a role in the initial nucleation stage of SG formation. G3BP1 and G3BP2 unwind partial RNA/DNA and RNA/RNA duplexes in an ATP-dependent fashion while TIA1 and TIAL1 possess nucleolytic activity (Tourrière *et al.*, 2003; Gilks *et al.*, 2004; Jain *et al.*, 2016; Kedersha *et al.*, 2016; Huang *et al.*, 2020). The two core components of plant SGs, RNA-binding protein 47b (RBP47b) and oligouridylate-binding protein 1B (UBP1B), have been identified based on their homology with mammalian and yeast core SG proteins. For example, both RBP47b and UBP1B are homologues of the TIA1 involved in the initial nucleation stage of SG formation (Kedersha *et al.*, 1999; Gilks *et al.*, 2004). Under normal growth conditions, RBP47b and UBP1B localize to the nucleus and/or cytoplasm, and upon stress, are transported to SGs (Weber *et al.*, 2008; Nguyen *et al.*, 2016, 2017). In plants, the

SG core protein angustifolia (AN) was identified by its conserved localization to SGs in response to heat, salt, osmotic, and low-oxygen stresses, and has been found to a play role in Golgi-associated membrane trafficking (Bhasin and Hülskamp, 2017). AN is a homologue of the animal C-terminal-binding protein/brefeldin A-ADP ribosylated substrate (CtBP/BARS) which functions as a transcriptional co-repressor within the nucleus. An interactome study of AN, using an N-terminal yellow fluorescent protein (YFP)-tagged AN with a Strep tag at the C-terminus, revealed a plethora of AN interactors including the SG core protein RBP47 and its close relative RBP45, while also identifying potentially novel plant SG core proteins, i.e. eukaryotic initiation factor (eIF4E1), tandem zinc finger 3 (TZF3), and poly(A)-binding protein 2 (PAB2), which have animal homologues already characterized as core SG proteins (Bhasin and Hülskamp, 2017). Affinity purification mass spectrometry (AP-MS) and green fluorescent protein (GFP)-tagged RBP47 were utilized to reveal the interactome of RBP47, and proteins were identified with functions in RNA processing, protein scaffolds (disordered region-containing proteins), post-translational modification (PTM) activities, ROS-related enzymes, and ethylene biosynthetic enzymes (Kosmacz *et al.*, 2019). RBP47 was found to interact with polyadenylate binding proteins (PABPs), which are a group of RNA-binding proteins that enable binding of eukaryotic initiation factor 4 complex (eIF4G) directly to poly(A) tails of mRNA (200–250 nts long), such as PABP2, PABP4, PABP5, and PABP8 (Kosmacz *et al.*, 2019). PABP2 was found to be essential for SG aggregation and is used as a marker for SG formation (Kosmacz *et al.*, 2019). Tudor staphylococcal nuclease protein (TSN) is highly conserved in eukaryotes and was first identified as a transcriptional activator and part of the RNA-induced silencing complex (RISC) with nuclease activity; it is also a target of caspases and metacaspases during apoptosis (Tong *et al.*, 1995; Leverson *et al.*, 1998; Caudy *et al.*, 2003; Sundström *et al.*, 2009). TSN2, the homologue of TSN in plants, was first identified as a core component of plant SGs and found to localize to SGs under salt stress using confocal microscopy (Yan *et al.*, 2014). TSN2 interacts with a whole range of SG-localized proteins and changes its interactome dependent on the type of stress that induces SG formation, and therefore is hypothesized to be a key docking platform for SG formation (Gutierrez-Beltran *et al.*, 2021). RNA helicases (RHs) play a key role in RNA metabolism by rearranging RNA secondary structure utilizing ATP, thus they have been hypothesized to function as chaperones mediating RNA stability (Richardson *et al.*, 1998; Tanner and Linder, 2001; Carlotto *et al.*, 2016; Ganie, 2020). One newly identified component of SGs that is a TSN1-binding protein, RH31, a DEAD-box RNA helicase in the model organism *Arabidopsis thaliana* (Arabidopsis), was found to localize to salt stress-induced SGs (Liu *et al.*, 2021). A review by Bailey-Serres *et al.* (2009) identified proteins predicted to localize to SGs that function in cytoplasmic mRNA management: nucleolin-like 1 (AtNUC-L1), nuclear transcription factor–RNA recognition motif (NTF-RRM) (nine family members), RHAU-like (nine family members), and PUM (26 family members). Arabidopsis genome encodes 26 Pumilio or Puf domain (PUM) proteins that contain an RNA-binding motif and are candidates for guiding mRNAs with a specific sequence or structural features; in yeast, they function in intracellular transport of specific mRNAs (Hogan *et al.*, 2008; Huh, 2021).

Recently, SGs have emerged as key regulators in the inducement of stress response, gene transcription, and silencing of non-stress genes in plants, resulting in quicker and more efficient stress-response signalling. This chapter describes the role of plant SGs formed in response to stress and discusses how these SGs regulate stress response via hormones such as gibberellic acid (GA) and abscisic acid (ABA) to mediate salt-stress signalling.

6.2 Brief Overview of Heat-Shock Stress Granules in Plants

SGs form in response to a range of abiotic stresses, appearing in response to hypoxia, high-salt conditions, and exposure to methyl jasmonate, arsenite, potassium cyanide, and myxothiazol (Nover *et al.*, 1983, 1989; Weber *et al.*, 2008; Pomeranz *et al.*, 2010b; Sorenson and Bailey-Serres, 2014; Yan *et al.*, 2014; Gutierrez-Beltran *et al.*, 2015). The most well-characterized abiotic stress triggering SG formation is heat shock. In response to heat shock in Arabidopsis, the core components assemble to form SGs; one core component such as UBP1B assembles into cytoplasmic foci via LLPS, binds to the 3′ untranslated region (UTR) and protects mRNA transcripts from degradation in SGs (Nguyen *et al.*, 2016; Muleya and Marondedze, 2020). The poly(A)-binding protein PAB1 forms the same cytoplasmic foci and is crucial for survival during heat shock in yeast (reviewed in Omkar and Truman, 2022). From yeast and mammalian studies, it is understood that mRNAs are shuttled in an unspecific manner into SGs in response to stress for storage and temporary silencing; then, after the stress is over, stored mRNAs are returned to polysomes to resume translation (Jain *et al.*, 2016; Markmiller *et al.*, 2018; Youn *et al.*, 2018). Arabidopsis heat-shock protein 101 (HSP101) knock-out mutant (*hsp101*) displayed impaired translation recovery and SG disaggregation post heat-stress treatment, highlighting the importance of HSPs in returning stored mRNAs to translation during recovery (Merret *et al.*, 2017). Interestingly, one subset of mRNAs stored in the SGs encode for ribosomal proteins, highlighting that SGs protect mRNAs important for translation and polysome assembly. Once the stress is over, the stably stored ribosomal mRNAs can once again begin translation to provide the ribosomal proteins to quickly reform polysomes and have a positive feedback loop on the activity of translation and polysome assembly (Merret *et al.*, 2017). The yeast homologue of the Arabidopsis HSP101, HSP104, together with HSP70, is involved in SG disaggregation and restoration of translation activity during heat-stress recovery, suggesting a conserved mechanism across organisms of SG disassembly after the stress (Cherkasov *et al.*, 2013).

A cell-cycle regulator, cyclin-dependent kinase 1 (CDKA1) which regulates plant growth and is important in the G1/S phase transition (Dissmeyer *et al.*, 2007; Nowack *et al.*, 2012; Zhao *et al.*, 2012), has been found to remain in an inactive form when localized to SGs in response to heat stress (Kosmacz *et al.*, 2019). CDKA1 activity is drastically reduced in response to abiotic stress and causes cell-cycle arrest and inhibition of plant growth (West *et al.*, 2004; Jang *et al.*, 2005; Skirycz *et al.*, 2011). A mechanism for how CDKA1 activity is inhibited in response to abiotic stress has now been proposed by Kosmacz *et al.* (2019): aggregation of CDKA1 into SGs may repress CDKA1 activity but the protein is stabilized in an inactive state, so during stress recovery the SGs disassemble and the CDKA1 is released and active once again, allowing CDKA1 to reinitiate plant growth. The core SG component RBP45a, which interacts with other core SG proteins such as AN (Bhasin and Hülskamp, 2017), has been reported as a putative interaction partner for CDKA1, further supporting the role of CDKA1 at the centre of SG function in response to heat stress (Van Leene *et al.*, 2007). A transcriptional co-activator, multiprotein bridging factor 1 (MBF1), binds to transcription factors and mediates their activation and has been found to play crucial roles in heat stress in Arabidopsis (Suzuki *et al.*, 2005, 2008). Transcriptome studies highlighted that AtMBF1 regulates a plethora of heat stress-response genes (e.g. *DREB2A*, *HSFB2A*, and *HSFB2B*) (Suzuki *et al.*, 2011), while expression analysis of knock-out *AtMBF1c* plants revealed MBF1 regulated two more heat-shock proteins: HSA23 and HSP70T-2 (Kim *et al.*, 2015). The *T. aestivum* (wheat) homologue was first identified and characterized to be a regulator of thermotolerance by showing that overexpression of *TaMBF1c* improved heat tolerance in rice (Qin *et al.*, 2015). Expression of *TaMBF1c* in response to heat stress is upregulated in wheat and the protein found to localize to SGs. A yeast two-hybrid screen and immunoprecipitation of an MYC-tagged TaMBF1c followed by liquid chromatography–tandem mass spectrometry (LC-MS/MS) analysis in Arabidopsis revealed TaMBF1c interacts with the SG component RNA-binding Ras-GAP SH3-binding protein (TaG3BP). Polysome profiling technology revealed TaG3BP regulates the translation efficiency of heat stress-response genes and heat-shock proteins, together suggesting TaMBF1c plays a role in response to heat stress in wheat, regulating translation of specific heat-shock proteins within SGs (Tian *et al.*, 2022). The recently characterized acetylation lowers binding affinity (ALBA) proteins have been found to play crucial roles in heat tolerance and SGs; in Arabidopsis there are six ALBA proteins, ALBA1 to ALBA6 (Aravind *et al.*, 2003). ALBA4, ALBA5, and ALBA6 belong to the Mdp2-like clade and possess additional RGG (Arg-Gly-Gly) motifs, which mediate nucleic acid binding and protein–protein interactions

(Thandapani *et al.*, 2013). ALBA4, ALBA5, and ALBA6 localize to SGs in response to heat stress and were found to directly bind to heat stress transcription factor (HSF) mRNAs, suggesting ALBA proteins function in SG formation and binding of HSF mRNAs to localize specific mRNAs to SGs in response to heat shock. Only the *alba456* triple mutant displayed hypersensitivity to heat stress, suggesting these three ALBA proteins play redundant roles in binding to HSF mRNAs and stabilizing targets inside heat-shock SGs. To test whether the degradation of HSF mRNAs in *alba456* is dependent on the 5′-to-3′ exoribonuclease activity of XRN4, a core component of PBs, Tong *et al.* (2022) generated two *xrn4* mutations, *xrn4-8* and *xrn4-9*, in the *alba456* background; knock-out of *xrn4* rescued the developmental defects seen in *alba456*, showing that XRN4 is responsible for the degradation of HSF mRNAs in the absence of ALBA4, ALBA5, and ALBA6 (Tong *et al.*, 2022). This once again fits the hypothesis that PBs reduce the levels of transcripts stabilized in SGs to ensure an appropriate response, in the form of negative regulation.

Identification of two RNA-binding glycine-rich group D (RBGD) genes (*rbgd2* and *rbgd4*) in Arabidopsis aided characterization of their roles in the heat-shock response, wherein their expression levels were upregulated and the RBGD2 and -4 proteins localized to SGs in response to heat shock. The role of RBGD2 and RBGD4 in heat-shock tolerance was further analysed using loss-of-function mutants (*rbgd2-1* and *rbgd4-1*). The mutants showed hypersensitivity to heat stress and this phenotype was rescued by complementation with pRBGD2::RBGD2-YFP transgene, showing that the heat shock susceptibility in *rbgd2-1* and *rbgd4-1* is directly due to the lack of function of RBGD2. AP-MS of RBGD2-containing SGs was carried out to identify the novel protein components of heat-shock SGs and the RBGD2 interactome. SG marker proteins including PAB2, PAB4, and PAB8 are associated with RBGD2 and RBGD4 under both normal and heat-shock growth conditions, while other SG components such as TSN1, TSN2, and EIF2 gamma only interact with RBGD2 and RBDG4 after heat shock. RBGD2 and RBGD4 contain prion-like low-complexity domains (LCDs) that are required for protein–protein interactions and phase separation into heat stress-induced SGs (Zhu *et al.*, 2022). Altogether, this study shows how RNA-binding proteins are crucial for phase separation and SG formation in response to heat shock and thermotolerance. In Arabidopsis, a new player in heat shock-responsive SGs was identified, metacaspase 1 (MC1), which is part of a family of cysteine proteases already characterized in stress responses and known to play a role in regulating senescence and protein stabilization (Coll *et al.*, 2014; Lema Asqui *et al.*, 2018). MC1 was found to be recruited to SGs upon proteotoxic stress (a characteristic of heat shock) and heat stress itself via two disordered regions, the prodomain and the 360 loop (Ruiz-Solaní *et al.*, 2023). The question of whether plant metacaspases play a role in clearing protein aggregates via recruitment to SGs in response to salt stress is yet to be answered.

6.3 Chloroplast Stress Granules and Abiotic Stress

SG-like foci inside the unicellular green alga *Chlamydomonas reinhardtii* (cpSGs) were identified in response to oxidative stress treatment and were found to contain a similar subset of protein–RNA complexes already identified within SGs; for example, polysomes, poly(A)-binding proteins, and small ribosomal subunits. Interestingly, cpSGs contained photosynthetic enzymes such as ribulosebisphosphate carboxylase/oxygenase (RuBisCO) (Uniacke and Zerges, 2008). These cpSGs form in response to oxidative stress and disassemble during the recovery phase, suggesting photosynthetic machinery such as RuBisCO is potentially inhibited within cpSGs or RuBisCO itself plays a role in cpSG formation during stress (Uniacke and Zerges, 2008). Arabidopsis chloroplast-localized SGs (chSGs) have been characterized in response to heat stress and found to contain two PrLD-containing proteins and several RNA recognition motif (RRM) proteins, the core components of cytosolic SGs which play a role in SG assembly, showing that chSGs and cytosolic SGs have the same core components (Chodasiewicz *et al.*, 2020). Similarly to oxidative stress-induced cpSGs in *C. reinhardtii*, Arabidopsis chSGs were found to contain RuBisCO activase and RuBisCO accumulation factors and the magnesium chelatase complex, highlighting the conserved role of chSGs to contain critical enzymes for photosynthesis. It remains to be investigated how recruitment of photosynthetic machinery within chSGs alters the rate of photosynthesis and if the enzymes are inactivated within the chSGs. Additionally, the question remains of whether chloroplast-located SGs play a role in response to salt stress and could be a new avenue of exploration for understanding how chloroplast SGs could potentially protect key photosynthetic enzymes from high salinity.

6.4 Processing Bodies and Their Role in Abiotic Stress

SGs interact with other cytoplasmic foci such as PBs, with which they have been found to interact physically and share many core components and functions, such as mediating mRNA decay (Buchan and Parker, 2009; Chantarachot and Bailey-Serres, 2018). Time-lapse analyses revealed SGs can be generated in a PB-independent manner although SGs are often closely associated with PBs but cannot perfectly overlap with SGs under the microscope, suggesting SG–PB fusion does not take place (Hamada *et al.*, 2018). Recently, in human cells, single-molecule imaging of mRNA and use of mRNA reporter lines have revealed there is little mRNA trafficking between SGs and PBs, instead mRNA in polysomes moves directly to either SGs or PBs in response to stress (Wilbertz *et al.*, 2019). Contrastingly, in yeast, mRNAs must pass through PBs before they can contribute to assembly of SGs (Buchan *et al.*, 2008). It is currently unclear which mechanism of SG formation and mRNA trafficking takes place in plants. The roles of PBs were first characterized in humans (Kedersha *et al.*, 1999) and yeast whereby they were found to contain mRNA decapping abilities and 5′-to-3′ mRNA degradation (Sheth and Parker, 2003). As previously mentioned, PBs contain the mRNA degradation factors DCP1 and DCP2, the 5′-to-3′ exoribonuclease XRN4, and the eukaryotic translation initiation factor eIF4E, confirming the role of mRNA degradation in PBs (Xu *et al.*, 2006; Goeres *et al.*, 2007; Iwasaki *et al.*, 2007; Weber *et al.*, 2008). Surprisingly, PBs also can stabilize mRNAs instead of mediating degradation, which has led to confusion and a debate on their function (Chantarachot and Bailey-Serres, 2018).

In plants, PBs were first identified and characterized in Arabidopsis (Xu *et al.*, 2006; Goeres *et al.*, 2007; Iwasaki *et al.*, 2007). Unlike SGs which form in response to stress, PBs are constitutively present in the cytoplasm, but the dynamics and localization of PBs may still be affected by environmental signals. For example, in humans, PB formation is increased specifically in response to oxidative stress (Kedersha *et al.*, 2005) and a similar increase in PBs is seen in yeast in response to glucose starvation (Teixeira and Parker, 2007) and osmotic stress (Huch and Nissan, 2017). PB dynamics are regulated by plant immune responses, triggered by microbe-associated molecular patterns (MAMPs) and mRNA degradation factors DCP1 and DCP2, previously mentioned to be core components of PBs, whereby they positively regulate plant MAMP-triggered responses and immunity against pathogenic bacteria via degradation of target genes. MAMP-activated MAPKs directly phosphorylate DCP1 at the Ser^{237} residue, which further stimulates its interaction with XRN4, an exonuclease executing 5′-to-3′ degradation of decapped mRNA, and promotes mRNA degradation of MAMP-downregulated genes that play crucial roles in plant immunity (Yu *et al.*, 2019). Recently, a study on the role of PBs during Cauliflower mosaic virus (CaMV) infection in Arabidopsis discovered several PB components form viral factories that support virus multiplication aiding viral mRNA translation, instead of their characteristic role in mRNA degradation (Hoffmann *et al.*, 2022).

PBs also respond to different abiotic stresses such as cold, drought, and salt stress, and while their role in salt-stress signalling remains elusive, recent work has begun to show their importance in salt-stress tolerance. SM-like proteins (LSMs) are RNA-binding proteins involved in RNA metabolism in eukaryotes and are part of the decapping machinery; there are 11 LSM proteins in Arabidopsis (LSM1–LSM8) whereby they form two heteroheptameric ring complexes, LSM1-7 and LSM2-8, and the LSM1-7 complex were found to be required for PB formation (Perea-Resa *et al.*, 2012). Alternative splicing analysis in *lsm4* and *sad1/lsm5* mutants revealed the LSM complex acts as a positive regulator of salt tolerance in Arabidopsis (Zhang *et al.*, 2011;

Cui *et al.*, 2014). Arabidopsis LSM1 proteins (LSM1A and LSM1B) were found to localize to PBs in response to the different stresses and bind to stress-inducible transcripts, targeting transcripts for decapping and degradation. Two target transcripts NCED3 and NCED5, two key enzymes in ABA biosynthesis, were identified and revealed that the LSM1-7 complex negatively regulates ABA biosynthesis, ensuring appropriate levels of ABA in Arabidopsis plants exposed to salt stress. This was further confirmed for plants lacking a functional LSM1-7 complex (*lsm1a* and *lsm1b* double mutants), which displayed salt hypersensitivity (Perea-Resa *et al.*, 2015). Members of the highly conserved class of BEACH domain-containing proteins (BDCPs) play a role in protein–protein interactions and membrane dynamics. The Arabidopsis BDCP SPIRRIG (SPI) is important for membrane integrity, as *spi* mutants exhibit split vacuoles (Saedler *et al.*, 2009). Studies showed that SPI interacts with the PB core component DCP1 and localizes to PBs in response to salt stress. *spi* mutant plants exhibit salt hypersensitivity and show a reduced number of PBs in response to salt stress. Two mRNA transcripts, TZF3 (previously introduced as a core component of plant SGs in Section 6.1 above and discussed in greater detail in Section 6.5.4 below) and ABA insensitive 1 (ABI1) which catalyses the first step of ABA biosynthesis, were found to be degraded by SPI via their recruitment to PBs because these transcripts were stabilized in the *spi* mutants (Steffens *et al.*, 2015). Together, this suggests SPI is crucial for PB formation and PB localization of salt-responsive mRNAs in response to salt stress (Steffens *et al.*, 2015). DCP1 and DCP2 form a complex together with VCS, the core components of PBs, and DCP1 is phosphorylated by mitogen-activated protein kinase 3 (MAPK3/MPK3). Dehydration stress activates MPK3 activity and MPK3 phosphorylates Ser^{237} on DCP1, promoting mRNA decapping (Xu and Chua, 2012). It is currently unknown if this MPK3 activity and activation of DCP1 within PBs takes place in response to salt stress.

6.5 Salt Hypersensitivity of Stress Granule Component Mutants

Salt stress induces the formation of SGs containing all the core components previously described in Section 6.1. There are several similarities between heat shock-induced SGs (described in Section 6.2) and salt stress-induced SGs. This section discusses in detail how different SG components play a role in salt-stress responses across a variety of model and crop species.

6.5.1 Heat-shock proteins

HSPs function as molecular chaperones, preventing their target protein from denaturation, misfolding, and aggregation, and a subgroup of HSPs, small HSPs (sHSPs) (12–40 kDa), play key roles in responding to abiotic stress and mediating stress tolerance (Sun *et al.*, 2002; Sun and MacRae, 2005; Song and Ahn, 2011; Yang *et al.*, 2014; Sun *et al.*, 2016). One sHSP identified in *O. sativa* (rice), OsHSP20, was found to localize to SGs and interact with a viral RNA-dependent RNA polymerase (RdRp); expression of *OsHSP20* mRNA and protein levels increased in response to salt treatment, showing clear salt inducibility (Li *et al.*, 2015; Guo *et al.*, 2020). Moreover, transgenic overexpression of *OsHSP20* conferred heat and salt tolerance in rice, reducing growth inhibition in response to both abiotic stresses, revealing that OsHSP20 acts as a molecular chaperone that localizes to SGs mediating salt tolerance (Guo *et al.*, 2020). sHSP LimHSP16.45, a novel sHSP characterized in David lily (*Lilium davidii* var. *willmottiae* (E.H. Wilson) Raffill), was found to be expressed in response to high salt and oxidative stress, recruited to SGs in response to high temperature and high-salt stress and enhanced the activity of ROS-scavenging enzymes (Mu *et al.*, 2013). A novel class II sHSP, PpHsp16.4, was identified and characterized in the moss *Physcomitrella patens* (Physcomitrella), a model for primitive land plants, that showed inducibility in response to ABA, osmotic, and salt stresses. Knock-out of *PpHsp16.4* resulted in sensitivity to heat, salt, and osmotic stress, and hindered recovery. Transient expression analysis of PpHsp16.4:GFP in tobacco protoplasts revealed PpHsp16.4 localized to cytosolic granules close to chloroplasts under non-stress conditions and became more abundant in response to heat and osmotic stress, suggesting a role for PpHsp16.4 in SG formation in response to abiotic stress (Ruibal *et al.*, 2013). Further studies to identify if ABA treatment and high-salt stress induce the accumulation of PpHSP16.4 into cytosolic granules are required to fully confirm if PpHSP16.4 plays a role in salt stress-induced SGs. Together these studies of plant HSPs highlight that HSPs play a key role not just in heat shock but also other abiotic stresses such as salt stress and regulate the stress response via stabilizing target proteins in SGs.

6.5.2 RNA helicases

In Arabidopsis, the RNA helicase RH31 (previously introduced as a core component of plant SGs in Section 6.1) was found to localize in salt stress-induced SGs; its knock-out lines showed a salt-stress susceptibility phenotype alongside reduced expression of salt-inducible genes (*PROAtCAPE-3*, *-7*, *-8* and *-9*, *P5CS1*, *AREB1*, and *RD29B*) (Liu *et al.*, 2021). In response to salt-stress treatment, *PROAtCAPE3* showed the lowest expression in the *rh31* mutant and highest expression in the *rh31*-overexpressing plant and was therefore selected as the main candidate target of RH31 (Liu *et al.*, 2021). Together these data suggest that RH31 may act as an RNA chaperone required for efficient translation of salt stress-response genes such as *PROAtCAPE3* within SGs. Other plant RHs may play roles in RNA stability and SG formation; for example, the *Arachis hypogaea* (peanut) RNA helicase 47 (AhRH47) has been associated with salt-stress tolerance but not yet identified within SGs (Mahesh and Udayakumar, 2018). This hypothesis is further supported by the fact that in mammals SG formation and RHs are necessary for efficient translation of stress-response genes (Buchan and Parker, 2009; Vanderweyde *et al.*, 2013). Another role for RHs within SGs was identified from the finding that the DHH1/DDX6-like RHs degrade stress-response genes under normal growth conditions and in knock-out lines of three particular RH-encoding genes (*rh-6*, *-8*, and *-12*) there is reduced SG formation (Chantarachot *et al.*, 2020). This role of RHs could be characteristic of salt stress-induced SGs, since mutation in three redundant *rh6*, *-8*, and *-12* causes reduction in SG formation and increases the stability of stress-response genes, highlighting that some RHs within SGs degrade target mRNA transcripts. Together these studies show that different types of RHs can stabilize and degrade salt-inducible genes under salt-stress conditions when they are shuttled into SGs, potentially suggesting a shift in the function of RHs dependent on the protein and RNA complex they are part of within SGs.

6.5.3 DUF proteins

Domains of unknown function (DUF) are a large set of uncharacterized protein families. In rice, SIDP361 that contains a DUF1644 motif is induced by high salt, drought, and ABA (Li *et al.*, 2016). Transgenic rice overexpressing *SIDP361* displayed improved tolerance to salt stress and higher levels of several proline synthetases and products of stress-response genes were identified: four late embryogenesis abundant (LEA) proteins (OsRab21/OsLEA29, OsRab16b/OsLEA26, OsLEA3, and OsLEA3-1), ABA biosynthesis enzymes (9-*cis*-epoxycarotenoid dioxygenase (NCED) and β-carotene hydroxylase DSM2), and ABA-dependent NAC-type transcription factors SNAC1 and SNAC2 (Li *et al.*, 2016). Another member of the DUF family, DUF1644/OsSIDP366 (stress-induced DUF1644 protein), was found to a play role in the salt-stress response in rice; ABA-deficient mutants (*psh3*-1 and *psh3*-2) displayed the same expression pattern of SIDP366 as the wild type, suggesting ABA is not required for SIDP366 inducement in response to salt stress unlike the previously mentioned SIDP361 (Guo *et al.*, 2016). RNA interference (RNAi) knock-down of *SIDP366* revealed plants were more sensitive to salinity and drought, and confocal microscopy of rice protoplasts overexpressing OsSIDP366:GFP revealed SIDP366 localized to PBs and SGs, alongside their corresponding markers AtDCP2, an mRNA decapping protein, and AtPABP8, a poly(A)-binding protein, respectively (Guo *et al.*, 2016).

6.5.4 Tandem CCCH zinc finger (TZF) proteins

CCCH zinc finger proteins contain one or more CCCH-type zinc finger motifs (three cysteines followed by a histidine); the subgroup of tandem CCCH zinc finger (TZF) proteins contain two identical CCH motifs and were first introduced in Section 6.1 as core components of SGs (Wang *et al.*, 2008; Bogamuwa and Jang, 2013). TZF proteins are RNA-binding proteins, possessing a direct interaction to adenine- and uridine-rich (AU-rich) elements (AREs) within the 3′ UTR of mRNAs. This interaction between TZF and mRNAs can recruit the carbon catabolite repression–negative on TATA-less (CCR4–NOT) complex which has well-characterized roles in regulation of gene expression through ubiquitination; when TZF interacts with CCR4–NOT together they mediate mRNA decay (Otsuka *et al.*, 2019). TZF proteins have been found to translocate and localize to PBs and SGs in Arabidopsis and rice, and several TZFs, i.e. AtTZF1, AtTZF9, and OsTZF1, were found to directly bind RNA while AtTZF1 was found to mediate mRNA decay (Pomeranz *et al.*, 2010a,b; Bogamuwa and Jang 2013; Jan *et al.*, 2013; Maldonado-Bonilla *et al.*, 2014; Qu *et al.*, 2014). TZFs translocate between the nucleus, cytoplasm, polysomes, SGs, and PBs (reviewed in Bogamuwa and Jang, 2014). All members of the Arabidopsis TZF (subfamily IX) show induced gene expression in response to abiotic stress and hormone treatments including salt, cold, mannitol, ABA, hypoxia, and osmotic stress (Wang *et al.*, 2008). Two Arabidopsis TZF proteins were found to play a role in the salt-stress response: AtSZF1/AtTZF11 and AtSZF2/AtTZF10. They were induced by high salt treatment and loss-of-function mutants displayed hypersensitivity to salt stress, while gain-of-function plants showed a salt-tolerant phenotype (Sun *et al.*, 2007). Another TZF family member, AtTZF, was found to activate ABA hormone responses, whereby overexpression of *AtTZF1* displayed a characteristic ABA-sensitivity phenotype while transgenic RNAi knock-down of *AtTZF1* resulted in abiotic stress sensitivity. Overexpression of *AtTZF1* delayed plant growth and development but enhanced stress tolerance, characteristic of GA deficiency and ABA over-accumulation (Lin *et al.*, 2011). AtTZF4, AtTZF5, and AtTZF6 have been found to play a role in ABA- and GA-mediated seed germination, since all three TZFs are positive regulators of ABA signalling via regulation of ABA and GA biosynthetic and signalling gene expression and have been found to localize to PBs and SGs (Bogamuwa and Jang, 2013). AtTZF4, AtTZF5, and AtTZF6 interact via their TZF motif with several ABA/stress-related proteins, such as mediator of ABA and regulated dormancy 1 (MARD1) and responsive to dehydration 21A (RD21A). Together these TZF target proteins, MARD1 and RD21A, co-localize with each other and with PB and SG markers in plant cells (Bogamuwa and Jang, 2016). A yeast two-hybrid screen revealed the 35 interactors of AtTZF5: 19 out of 35 proteins were found to be involved in stress responses, further supporting the role of TZFs in stress responses and SGs (Bogamuwa and Jang, 2016). A rice homologue of the TZF family, OsTZF7, positively regulates both drought and high salt stress because *OsTZF7* expression was upregulated by drought, salt stress, and ABA (Guo *et al.*, 2022). Overexpression of *OsTZF7* in rice seedlings resulted in hypersensitivity to exogenous ABA, whereas RNAi knock-down lines were insensitive to ABA and displayed drought sensitivity. Salt and polyethylene glycol (PEG) treatment promoted the accumulation of OsTZF7 in PB-like foci (Guo *et al.*, 2022). Together these studies across Arabidopsis and rice highlight that TZFs play crucial roles in regulating the ABA signalling pathways in response to both drought and salt stresses, functioning within SGs and PBs.

6.5.5 Core components of stress granules are required for salt-stress responses

AN (previously mentioned in Section 6.1), a core component of plant SGs, was found to play a role in the salt-stress response because salt treatment induced the localization of AN to SGs and transgenic *an* mutant plants displayed salt tolerance with a hyposensitivity phenotype in response to high salt (Bhasin and Hülskamp, 2017). UBP1B (also previously introduced in Section 6.1), a core component of plant SGs and involved in the initial nucleation stage of SG formation, positively regulates salt-stress signalling since *ubp1b* mutants displayed hypersensitivity to ABA, salt, and osmotic stress, suggesting core SG components are required for salt-stress responses and tolerance (McCue *et al.*, 2012; Sorenson and Bailey-Serres, 2014; Nguyen *et al.*, 2017). In *ubp1b*-overexpressing Arabidopsis, mRNA of transcription factors such as WRKY33 was stabilized suggesting UBP1B-containing SGs protected WRKY33 mRNA during ABA-mediated stress responses (Nguyen *et al.*, 2017). The contrasting salt-susceptibility phenotypes of the *an* and *ubp1b* mutants highlights that different core components of SGs have opposing roles in the salt-stress response, one component may be a positive regulator, for example UBP1B, while the other may be a negative regulator such as AN.

TSN1 and TSN2 (previously mentioned in Section 6.1), components of plant SGs were found to regulate growth in response to salt stress, whereby transgenic lines with knock-down of *TSN1* and *TSN2* via RNAi displayed slower growth in response to high salt stress while contrastingly a *TSN1* overexpression transgenic line displayed greater tolerance to high salt stress (dit Frey *et al.*, 2010). Expression levels of GA20ox3 mRNA transcripts, a crucial enzyme in GA biosynthesis, decreased significantly in the *TSN1/TSN2* transgenic RNAi lines, yet expression increased upon high salt stress. In response to high salt TSN1 localized to SGs and binds to GA20ox3 mRNA *in vivo*, highlighting TSN1 is a regulator of GA20ox3 mRNA stability during salt stress (Yan *et al.*, 2014). T-DNA mutant lines of *GA20ox3-1* and *-2* also showed slower growth than the wild type (Yan *et al.*, 2014). GA has an antagonist interaction with the growth inhibition hormone jasmonic acid (JA); therefore GA biosynthesis mutants show a reduced growth phenotype characteristic of biotic/abiotic stress and of JA signalling, suggesting TSN1 in response to salt stress sequesters GA20ox3 in SGs, silencing this mRNA transcript and therefore activating stress signalling and potentially JA responses.

Previously introduced in Section 6.1, a review by Bailey-Serres *et al.* (2009) highlighted that Arabidopsis has 26 PUM proteins that are candidates for guiding mRNAs with specific sequence or structural features. In mammals, there are two PUM proteins, PUM1 and PUM2, which have been well-characterized in their ability to repress protein expression by inhibiting translation and promoting mRNA decay and that localize to SGs in response to oxidative stress (Morris *et al.*, 2008; van Etten *et al.*, 2012; Khong *et al.*, 2017). In plants, the first characterized PUM protein was the Arabidopsis APUM23 which regulates the salt-stress response. For example, an *sahy9/apum23* mutant lacking a functional APUM23 protein displayed hypersensitivity to high salt stress, and transcriptome and proteome analysis revealed differential expression of genes encoding proteins involved in ABA biosynthesis and signalling, abiotic stress responses, and ribosome biogenesis-related functions, such as NCED3, ABI2, PP2CA, RD20, and RD29B. Additionally, *sahy9/apum23* mutants displayed reduced accumulation of ABA and expression of genes encoding LEA proteins, which are small molecules that function in protecting larger molecules or cellular compartmental membranes from deleterious environmental effects in response to salt stress (Hundertmark and Hincha, 2008; Saha *et al.*, 2016). Together these experiments highlight that APUM23 is associated with ABA-mediated stress-response signalling (Huang *et al.*, 2018). It remains to be determined whether PUM proteins predicted to play a role in SGs, such as APUM23, function within plant SGs in response to abiotic stresses including salt stress.

6.6 Conclusion

In response to salinity stress across model and crop species, a dramatic change in the transcriptome and proteome is required to mediate the physiological responses needed to survive in high-salt growth conditions, initiating protective mechanisms such as osmolyte biosynthesis, water transport, ROS scavenging, and translational arrest (Hasegawa *et al.*, 2000). SGs form and the number of PBs increases in the cytoplasm in response to salinity stress. Therefore it can be postulated that mRNAs alongside RNA-interacting proteins, a variety of HSPs and RHs, transcription factors, and core SG components are shuttled into SGs for silencing, preventing translation, and promoting polysome disassembly to inhibit the translation of mRNAs involved in growth and development.

HSPs play key roles as molecular chaperones, stabilizing target proteins in response to heat shock and salt stress. Across several different species (rice, David lily, and Physcomitrella), HSPs have been found to localize to SGs in response to salt stress and knock-out of these HSPs leads to salt hypersensitivity (Mu *et al.*, 2013; Ruibal *et al.*, 2013; Guo *et al.*, 2020). RHs have a well-characterized role in mammalian systems, being required for SG formation and necessary for efficient translation of stress-response genes (Buchan and Parker, 2009; Vanderweyde *et al.*, 2013). Fitting this role of RHs in mammals, the Arabidopsis RH31 acts as an RNA chaperone required for efficient translation of salt-stress response genes such as that encoding PROAtCAPE3 within SGs (Liu *et al.*, 2021). Another group required for SG formation is the DHH1/DDX6-like RHs such as RH6, -8, and -12, but these RHs are associated with degradation of stress-response genes because in their knock-out lines there is stabilization of stress-response mRNAs (Chantarachot *et al.*, 2020). Together these studies of different plant RHs highlight the contrasting roles that RHs can have within SGs, for some RHs can stabilize transcripts leading to translation of stress-response genes while other RHs promote degradation, highlighting the need to unpick how specific RHs within SGs have differing roles in stabilization and degradation of mRNA in response to stress. The DUF group of plant proteins are a large set of uncharacterized protein families found in rice which localize to SGs and are responsible for salt-stress tolerance, since two knock-out lines of DUF proteins (SIDP361 and SIDP366) are hypersensitive to salt stress (Guo *et al.*, 2016; Li *et al.*, 2016). SIDP361 activated the expression of several proline synthetases and ABA biosynthesis genes (Li *et al.*, 2016). Together, these two papers show the DUF group of plant proteins localize to SGs and activate the expression of stress-response genes and ABA biosynthesis, highlighting the importance of SGs in activating salt-stress signalling. Another well-characterized group of RNA-interacting proteins within plants are the TZFs; in both Arabidopsis and rice these proteins are induced in response to salt and drought stress and localize to SGs and PBs. Studies using overexpression, knock-out, and knock-down mutant lines revealed TZFs activate ABA signalling and biosynthesis and that functional TZFs are required for salt tolerance (Sun *et al.*, 2007; Wang *et al.*, 2008; Lin *et al.*, 2011; Bogamuwa and Jang, 2013; Guo *et al.*, 2022). Core components of SGs, such as AN and UBP1B, contrastingly regulate salt-stress signalling, with AN being a repressor and UBP1B a positive regulator of salt-stress responses, suggesting different protein complexes within SGs have differing roles in response to salt stress (Bhasin and Hülskamp, 2017; Nguyen *et al.*, 2017). Another core component of SGs, TSN1 is involved in salt-stress signalling whereby salt treatment increases the levels of TSN1 and directs TSN1 into SGs. TSN1 binds to mRNA of the GA biosynthesis gene *GA20ox3* and therefore GA signalling is inhibited in response to salt stress. This fits the hypothesis that SGs stably store and silence mRNAs involved in growth and development during salt stress. Currently, these studies reveal that ABA and GA are the predominant signalling pathways found to be regulated within SG and PBs in response to salt stress. However, further exploration of the role of SGs and PBs in regulating other hormones involved in abiotic stress responses, such as JA and ethylene, will be key to understanding how universal the role of SGs and PBs is in regulating hormone signalling pathways in response to salt stress.

Questions of how SGs and PBs orchestrate salt tolerance in plants remain. For example, what is the mechanism by which mRNAs are shuttled between SGs, PBs, and polysomes (it is currently unknown in plants) and could PUM proteins play a role? It will be critical to understand how SGs can have dual functions of silencing and degrading transcripts via DHH1/DDX6-like RHs but activating others, for example the activation of PROAtCAPE3 translation via RH31 in response to heat stress, and whether it involves differing multiprotein complexes (Chantarachot *et al.*, 2020; Liu *et al.*, 2021). It will also be critical to understand what marks stress-inducible transcripts for rapid translation in the cytoplasm and prevents their trafficking and silencing in SGs. It can be hypothesized that shuttling of mRNAs via a combination of interactions with specific HSPs and RHs could ensure mRNA transcripts involved in growth and development are either stably stored in SGs or degraded in PBs, while opposingly mRNA transcripts involved in stress responses could potentially become more concentrated within the cytoplasm or other types of SGs and have greater availability at polysomes for more efficient and quicker translation. Together these cytoplasmic foci are at the centre of regulating abiotic stress signalling and play a pivotal role in efficiently changing the state of a plant cell from growth and development to protect and survive, and many avenues of exploration remain. It will be critical to understand how naturally salt-tolerant species mediate growth and development in the presence of high salinity and to discover if SGs and PBs are responsible for this salt-tolerant phenotype. Therefore it would be interesting to genetically engineer crops with enhanced or more efficient SG and PB formation with the aim to create more salt-tolerant crops for the future.

References

Aravind, L., Iyer, L.M. and Anantharaman, V. (2003) The two faces of Alba: the evolutionary connection between proteins participating in chromatin structure and RNA metabolism. *Genome Biology* 4(10), R64.

Bailey-Serres, J., Sorenson, R. and Juntawong, P. (2009) Getting the message across: cytoplasmic ribonucleoprotein complexes. *Trends in Plant Science* 14(8), 443–453.

Banani, S.F., Lee, H.O., Hyman, A.A. and Rosen, M.K. (2017) Biomolecular condensates: organizers of cellular biochemistry. *Nature Reviews Molecular Cell Biology* 18(5), 285–298.

Barragán, V., Leidi, E.O., Andrés, Z., Rubio, L., De Luca, A. *et al.* (2012) Ion exchangers NHX1 and NHX2 mediate active potassium uptake into vacuoles to regulate cell turgor and stomatal function in *Arabidopsis*. *The Plant Cell* 24(3), 1127–1142.

Bhasin, H. and Hülskamp, M. (2017) ANGUSTIFOLIA, a plant homolog of CtBP/BARS localizes to stress granules and regulates their formation. *Frontiers in Plant Science* 8, 1004.

Bogamuwa, S. and Jang, J.C. (2013) The *Arabidopsis* tandem CCCH zinc finger proteins AtTZF4, 5 and 6 are involved in light-, abscisic acid- and gibberellic acid-mediated regulation of seed germination. *Plant, Cell & Environment* 36(8), 1507–1519.

Bogamuwa, S.P. and Jang, J.C. (2014) Tandem CCCH zinc finger proteins in plant growth, development and stress response. *Plant & Cell Physiology* 55(8), 1367–1375.

Bogamuwa, S. and Jang, J.C. (2016) Plant tandem CCCH zinc finger proteins interact with ABA, drought, and stress response regulators in processing-bodies and stress granules. *PLoS ONE* 11(3), e0151574.

Buchan, J.R. and Parker, R. (2009) Eukaryotic stress granules: the ins and outs of translation. *Molecular Cell* 36(6), 932–941.

Buchan, J.R., Muhlrad, D. and Parker, R. (2008) P bodies promote stress granule assembly in *Saccharomyces cerevisiae*. *The Journal of Cell Biology* 183(3), 441–455.

Carlotto, N., Wirth, S., Furman, N., Ferreyra Solari, N., Ariel, F. *et al.* (2016) The chloroplastic DEVH-box RNA helicase increased size exclusion limit 2 involved in plasmodesmata regulation is required for group II intron splicing. *Plant, Cell & Environment* 39(1), 165–173.

Caucy, A.A., Ketting, R.F., Hammond, S.M., Denli, A M., Bathoorn, A.M.P. *et al.* (2003) A micrococcal nuclease homologue in RNAi effector complexes. *Nature* 425, 411–414.

Chantarachot, T. and Bailey-Serres, J. (2018) Polysomes, stress granules, and processing bodies: a dynamic triumvirate controlling cytoplasmic mRNA fate and function. *Plant Physiology* 176(1), 254–269.

Chantarachot, T., Sorenson, R.S., Hummel, M., Ke, H., Kettenburg, A.T. *et al.* (2020) DHH1/DDX6-like RNA helicases maintain ephemeral half-lives of stress-response mRNAs. *Nature Plants* 6(6), 675–685.

Cherkasov, V., Hofmann, S., Druffel-Augustin, S., Mogk, A., Tyedmers, J. *et al.* (2013) Coordination of translational control and protein homeostasis during severe heat stress. *Current Biology* 23(24), 2452–2462.

Chodasiewicz, M., Sokolowska, E.M., Nelson-Dittrich, A.C., Masiuk, A., Beltran, J.C.M. *et al.* (2020) Identification and characterization of the heat-induced plastidial stress granules reveal new insight into *Arabidopsis* stress response. *Frontiers in Plant Science* 11, 595792.

Chodasiewicz, M., Jang, J.C. and Gutierrez-Beltran, E. (2022) Editorial: Biology of stress granules in plants. *Frontiers in Plant Science* 13, 938654.

Choi, W.G., Toyota, M., Kim, S.H., Hilleary, R. and Gilroy, S. (2014) Salt stress-induced Ca^{2+} waves are associated with rapid, long-distance root-to-shoot signaling in plants. *Proceedings of the National Academy of Sciences USA* 111(17), 6497–6502.

Coll, N.S., Smidler, A., Puigvert, M., Popa, C., Valls, M. and Dangl, J.L. (2014) The plant metacaspase AtMC1 in pathogen-triggered programmed cell death and aging: functional linkage with autophagy. *Cell Death & Differentiation* 21(9), 1399–1408.

Cui, P., Zhang, S., Ding, F., Ali, S. and Xiong, L. (2014) Dynamic regulation of genome-wide pre-mRNA splicing and stress tolerance by the Sm-like protein LSm5 in *Arabidopsis*. *Genome Biology* 15(1), R1.

Deiana, A., Forcelloni, S., Porrello, A. and Giansanti, A. (2019) Intrinsically disordered proteins and structured proteins with intrinsically disordered regions have different functional roles in the cell. *PLoS ONE* 14(8), e0217889.

Dissmeyer, N., Nowack, M.K., Pusch, S., Stals, H., Inzé, D. *et al.* (2007) T-loop phosphorylation of *Arabidopsis* CDKA;1 is required for its function and can be partially substituted by an aspartate residue. *The Plant Cell* 19(3), 972–985.

dit Frey, N.F., Muller, P., Jammes, F., Kizis, D., Leung, J. *et al.* (2010) The RNA binding protein Tudor-SN is essential for stress tolerance and stabilizes levels of stress-responsive mRNAs encoding secreted proteins in *Arabidopsis*. *The Plant Cell* 22(5), 1575–1591.

Edgerton, M.D. (2009) Increasing crop productivity to meet global needs for feed, food, and fuel. *Plant Physiology* 149(1), 7–13.

Fricke, W. (2004) Rapid and tissue-specific accumulation of solutes in the growth zone of barley leaves in response to salinity. *Planta* 219(3), 515–525.

Ganie, S.A. (2020) RNA chaperones: potential candidates for engineering salt tolerance in rice. *Crop Science* 60(2), 530-540.

Ganie, S.A. (2021) Amino acids other than proline and their participation in abiotic stress tolerance. In: Wani, S.H., Gangola, M.P. and Ramadoss, B.R. (eds) *Compatible Solutes Engineering for Crop Plants Facing Climate Change*. Springer, Cham, Switzerland, pp. 47-96.

Gilks, N., Kedersha, N., Ayodele, M., Shen, L., Stoecklin, G. *et al.* (2004) Stress granule assembly is mediated by prion-like aggregation of TIA-1. *Molecular Biology of the Cell* 15(12), 5383–5398.

Goeres, D.C., Van Norman, J.M., Zhang, W., Fauver, N.A., Spencer, M.L. and Sieburth, L.E. (2007) Components of the *Arabidopsis* mRNA decapping complex are required for early seedling development. *The Plant Cell* 19(5), 1549–1564.

Guo, C., Luo, C., Guo, L., Li, M., Guo, X. *et al.* (2016) OsSIDP366, a DUF1644 gene, positively regulates responses to drought and salt stresses in rice. *Journal of Integrative Plant Biology* 58(5), 492–502.

Guo, C., Chen, L., Cui, Y., Tang, M., Guo, Y. *et al.* (2022) RNA binding protein OsTZF7 traffics between the nucleus and processing bodies/stress granules and positively regulates drought stress in rice. *Frontiers in Plant Science* 13, 802337.

Guo, L.M., Li, J., He, J., Liu, H. and Zhang, H.M. (2020) A class I cytosolic HSP20 of rice enhances heat and salt tolerance in different organisms. *Scientific Reports* 10(1), 1383.

Gutierrez-Beltran, E., Moschou, P.N., Smertenko, A.P. and Bozhkov, P.V. (2015) Tudor staphylococcal nuclease links formation of stress granules and processing bodies with mRNA catabolism in *Arabidopsis*. *The Plant Cell* 27, 926-943.

Gutierrez-Beltran, E., Elander, P.H., Dalman, K., Dayhoff, G.W. II, Moschou, P.N. *et al.* (2021) Tudor staphylococcal nuclease is a docking platform for stress granule components and is essential for SnRK1 activation in *Arabidopsis*. *EMBO Journal* 40(17), e105043.

Haak, D.C., Fukao, T., Grene, R., Hua, Z., Ivanov, R. *et al.* (2017) Multilevel regulation of abiotic stress responses in plants. *Frontiers in Plant Science* 8, 1564.

Halfter, U., Ishitani, M. and Zhu, J. K. (2000) The *Arabidopsis* SOS2 protein kinase physically interacts with and is activated by the calcium-binding protein SOS3. *Proceedings of the National Academy of Sciences USA* 97(7), 3735–3740.

Hamada, T., Yako, M., Minegishi, M., Sato, M., Kamei, Y. *et al.* (2018) Stress granule formation is induced by a threshold temperature rather than a temperature difference in *Arabidopsis*. *Journal of Cell Science* 131(16), jcs216051.

Hasegawa, P.M., Bressan, R.A., Zhu, J.K. and Bohnert, H.J. (2000) Plant cellular and molecular responses to high salinity. *Annual Review of Plant Physiology and Plant Molecular Biology* 51, 463–499.

Hoffmann, G., Mahboubi, A., Bente, H., Garcia, D., Hanson, J. and Hafrén, A. (2022) *Arabidopsis* RNA processing body components LSM1 and DCP5 aid in the evasion of translational repression during Cauliflower mosaic virus infection. *The Plant Cell* 34(8), 3128–3147.

Hogan, D.J., Riordan, D.P., Gerber, A.P., Herschlag, D. and Brown, P.O. (2008) Diverse RNA-binding proteins interact with functionally related sets of RNAs, suggesting an extensive regulatory system. *PLoS Biology* 6(10), e255.

Hu, Y. and Schmidhalter, U. (1998) Spatial distributions and net deposition rates of mineral elements in the elongating wheat (*Triticum aestivum* L.) leaf under saline soil conditions. *Planta* 204, 212–219.

Huang, C., Chen, Y., Dai, H., Zhang, H., Xie, M. *et al.* (2020) UBAP2L arginine methylation by PRMT1 modulates stress granule assembly. *Cell Death & Differentiation* 27(1), 227–241.

Huang, K.C., Lin, W.C. and Cheng, W.H. (2018) Salt hypersensitive mutant 9, a nucleolar APUM23 protein, is essential for salt sensitivity in association with the ABA signaling pathway in *Arabidopsis*. *BMC Plant Biology* 18(1), 40.

Huch, S. and Nissan, T. (2017) An mRNA decapping mutant deficient in P body assembly limits mRNA stabilization in response to osmotic stress. *Scientific Reports* 7, 44395.

Huh, S.U. (2021) The role of pumilio RNA binding protein in plants. *Biomolecules* 11(12), 1851.

Hundertmark, M. and Hincha, D.K. (2008) LEA (late embryogenesis abundant) proteins and their encoding genes in *Arabidopsis thaliana*. *BMC Genomics* 9, 118–140.

Ishitani, M., Liu, J., Halfter, U., Kim, C.S., Shi, W. and Zhu, J.K. (2000) SOS3 function in plant salt tolerance requires N-myristoylation and calcium binding. *The Plant Cell* 12(9), 1667–1678.

Ivanov, P., Kedersha, N. and Anderson, P. (2019) Stress granules and processing bodies in translational control. *Cold Spring Harbor Perspectives in Biology* 11(5), a032813.

Iwasaki, S., Takeda, A., Motose, H. and Watanabe, Y. (2007) Characterization of *Arabidopsis* decapping proteins AtDCP1 and AtDCP2, which are essential for post-embryonic development. *FEBS Letters* 581(13), 2455–2459.

Jain, S., Wheeler, J.R., Walters, R.W., Agrawal, A., Barsic, A. and Parker, R. (2016) ATPase-modulated stress granules contain a diverse proteome and substructure. *Cell* 164(3), 487–498.

Jan, A., Maruyama, K., Todaka, D., Kidokoro, S., Abo, M. *et al.* (2013) OsTZF1, a CCCH-tandem zinc finger protein, confers delayed senescence and stress tolerance in rice by regulating stress-related genes. *Plant Physiology* 161(3), 1202–1216.

Jang, S.J., Shin, S.H., Yee, S.T., Hwang, B., Im, K.H. and Park, K.Y. (2005) Effects of abiotic stresses on cell cycle progression in tobacco BY-2 cells. *Molecules and Cells* 20(1), 136–141.

Jeschke, W.D. (1984) K^+–Na^+ exchange at cellular membranes, intercellular compartmentation of cations and salt tolerance. In: Staples, R.C. and Toenniessen, G.H. (eds) *Salinity Tolerance in Plants: Strategies for Crop Improvement*. Wiley, New York, pp. 37–66.

Jeschke, W.D., Aslam, Z. and Greenway, H. (1986) Effects of NaCl on ion relations and carbohydrate status of roots and on osmotic regulation of roots and shoots of *Atriplex amnicola*. *Plant, Cell & Environment* 9, 559–569.

Kedersha, N. and Anderson, P. (2002) Stress granules: sites of mRNA triage that regulate mRNA stability and translatability. *Biochemical Society Transactions* 30(6), 963–969.

Kedersha, N., Gupta, M., Li, W., Miller, I. and Anderson, P. (1999) RNA-binding proteins TIA-1 and TIAR link the phosphorylation of eIF-2α to the assembly of mammalian stress granules. *The Journal of Cell Biology* 147(7), 1431–1441.

Kedersha, N., Chen, S., Gilks, N., Li, W., Miller, I.J. *et al.* (2002) Evidence that ternary complex (eIF2-GTP-$tRNA_i^{Met}$)-deficient preinitiation complexes are core constituents of mammalian stress granules. *Molecular Biology of the Cell* 13, 195–210.

Kedersha, N., Stoecklin, G., Ayodele, M., Yacono, P., Lykke-Andersen, J. *et al.* (2005) Stress granules and processing bodies are dynamically linked sites of mRNP remodeling. *The Journal of Cell Biology* 169(6), 871–884.

Kedersha, N., Panas, M.D., Achorn, C.A., Lyons, S., Tisdale, S. *et al.* (2016) G3BP–Caprin1–USP10 complexes mediate stress granule condensation and associate with 40S subunits. *The Journal of Cell Biology* 212(7), 845–860.

Khong, A., Matheny, T., Jain, S., Mitchell, S.F., Wheeler, J.R. and Parker, R. (2017) The stress granule transcriptome reveals principles of mRNA accumulation in stress granules. *Molecular Cell* 68(4), 808-820.

Kim, G.D., Cho, Y.H. and Yoo, S.D. (2015) Regulatory functions of evolutionarily conserved AN1/A20-like zinc finger family proteins in *Arabidopsis* stress responses under high temperature. *Biochemical and Biophysical Research Communications* 457(2), 213–220.

Kollist, H., Zandalinas, S.I., Sengupta, S., Nuhkat, M., Kangasjärvi, J. and Mittler, R. (2019) Rapid responses to abiotic stress: priming the landscape for the signal transduction network. *Trends in Plant Science* 24(1), 25–37.

Kosmacz, M., Gorka, M., Schmidt, S., Luzarowski, M., Moreno, J.C. *et al.* (2019) Protein and metabolite composition of *Arabidopsis* stress granules. *New Phytologist* 222(3), 1420–1433.

Kronzucker, H.J. and Britto, D.T. (2011) Sodium transport in plants: a critical review. *New Phytologist* 189(1), 54–81.

Lema Asqui, S., Vercammen, D., Serrano, I., Valls, M., Rivas, S. *et al.* (2018) AtSERPIN1 is an inhibitor of the metacaspase AtMC1-mediated cell death and autocatalytic processing in planta. *New Phytologist* 218(3), 1156–1166.

Leverson, J.D., Koskinen, P.J., Orrico, F.C., Rainio, E.-M., Jalkanen, K.J. *et al.* (1998) Pim-1 kinase and p100 cooperate to enhance c-Myb activity. *Molecular Cell* 2(4), 417–425.

Li, J., Xiang, C.Y., Yang, J., Chen, J.P. and Zhang, H.M. (2015) Interaction of HSP20 with a viral RdRp changes its sub-cellular localization and distribution pattern in plants. *Scientific Reports* 5, 14016.

Li, M., Guo, L., Guo, C., Wang, L. and Chen, L. (2016) Over-expression of a DUF1644 protein gene, SIDP361, enhances tolerance to salt stress in transgenic rice. *Journal of Plant Biology* 59(1), 62–73.

Lin, P.C., Pomeranz, M.C., Jikumaru, Y., Kang, S.G., Hah, C. *et al.* (2011) The *Arabidopsis* tandem zinc finger protein AtTZF1 affects ABA- and GA-mediated growth, stress and gene expression responses. *The Plant Journal* 65(2), 253–268.

Liu, J. and Zhu, J.K. (1998) A calcium sensor homolog required for plant salt tolerance. *Science* 280(5371), 1943–1945.

Liu, J., Ishitani, M., Halfter, U., Kim, C.S. and Zhu, J.K. (2000) The *Arabidopsis thaliana* SOS2 gene encodes a protein kinase that is required for salt tolerance. *Proceedings of the National Academy of Sciences USA* 97(7), 3730–3734.

Liu, Y., Liu, S., Shi, H., Ma, J., Jing, M. and Han, Y. (2021) The TSN1 binding protein RH31 is a component of stress granules and participates in regulation of salt-stress tolerance in *Arabidopsis*. *Frontiers in Plant Science* 12, 804356.

Ludwig, M., Wilmes, P. and Schrader, S. (2018) Measuring soil sustainability via soil resilience. *Science of the Total Environment* 626, 1484–1493.

Mahesh, S. and Udayakumar, M. (2018) Peanut RNA helicase AhRH47 sustains protein synthesis under stress and improves stress adaptation in *Arabidopsis*. *Plant Molecular Biology Reporter* 36(1), 58–70.

Maldonado-Bonilla, L.D., Eschen-Lippold, L., Gago-Zachert, S., Tabassum, N., Bauer, N. *et al.* (2014) The *Arabidopsis* tandem zinc finger 9 protein binds RNA and mediates pathogen-associated molecular pattern-triggered immune responses. *Plant & Cell Physiology* 55(2), 412–425.

Markmiller, S., Soltanieh, S., Server, K.L., Mak, R., Jin, W. *et al.* (2018) Context-dependent and disease-specific diversity in protein interactions within stress granules. *Cell* 172(3), 590–604.

McCue, A.D., Nuthikattu, S., Reeder, S.H. and Slotkin, R.K. (2012) Gene expression and stress response mediated by the epigenetic regulation of a transposable element small RNA. *PLoS Genetics* 8(2), e1002474.

Merret, R., Carpentier, M.C., Favory, J.J., Picart, C., Descombin, J. *et al.* (2017) Heat shock protein HSP101 affects the release of ribosomal protein mRNAs for recovery after heat shock. *Plant Physiology* 174(2), 1216–1225.

Morris, A.R., Mukherjee, N. and Keene, J.D. (2008) Ribonomic analysis of human Pum1 reveals *cis–trans* conservation across species despite evolution of diverse mRNA target sets. *Molecular and Cellular Biology* 28(12), 4093–4103.

Mu, C., Zhang, S., Yu, G., Chen, N., Li, X. and Liu, H. (2013) Overexpression of small heat shock protein LimHSP16.45 in *Arabidopsis* enhances tolerance to abiotic stresses. *PLoS ONE* 8(12), e82264.

Muleya, V. and Marondedze, C. (2020) Functional roles of RNA-binding proteins in plant signaling. *Life* 10(11), 288.

Nguyen, C.C., Nakaminami, K., Matsui, A., Kobayashi, S., Kurihara, Y. *et al.* (2016) Oligouridylate binding protein 1b plays an integral role in plant heat stress tolerance. *Frontiers in Plant Science* 7, 853.

Nguyen, C.C., Nakaminami, K., Matsui, A., Watanabe, S., Kanno, Y. *et al.* (2017) Overexpression of oligouridylate binding protein 1b results in ABA hypersensitivity. *Plant Signaling & Behaviour* 12(2), e1282591.

Nover, L., Scharf, K.D. and Neumann, D. (1983) Formation of cytoplasmic heat shock granules in tomato cell cultures and leaves. *Molecular and Cellular Biology* 3(9), 1648–1655.

Nover, L., Scharf, K.D. and Neumann, D. (1989) Cytoplasmic heat shock granules are formed from precursor particles and are associated with a specific set of mRNAs. *Molecular and Cellular Biology* 9(3), 1298–1308.

Nowack, M.K., Harashima, H., Dissmeyer, N., Zhao, X., Bouyer, D. *et al.* (2012) Genetic framework of cyclin-dependent kinase function in *Arabidopsis*. *Developmental Cell* 22(5), 1030–1040.

Nunes, C., Mestre, I., Marcelo, A., Koppenol, R., Matos, C.A. and Nóbrega, C. (2019) MSGP: the first database of the protein components of the mammalian stress granules. *Database: The Journal of Biological Databases and Curation* 2019, baz031.

Omkar, S. and Truman, A.W. (2022) Feeling the heat: how chaperones deal with biomolecular condensates. *Trends in Biochemical Sciences* 47(9), 728–729.

Otsuka, H., Fukao, A., Funakami, Y., Duncan, K.E. and Fujiwara, T. (2019) Emerging evidence of translational control by AU-rich element-binding proteins. *Frontiers in Genetics* 10, 332.

Perea-Resa, C., Hernández-Verdeja, T., López-Cobollo, R., del Mar Castellano, M. and Salinas, J. (2012) LSM proteins provide accurate splicing and decay of selected transcripts to ensure normal *Arabidopsis* development. *The Plant Cell* 24(12), 4930–4947.

Perea-Resa, C., Carrasco-López, C., Catalá, R., Turečková, V., Novak, O. *et al.* (2015) The LSM1-7 complex differentially regulates *Arabidopsis* tolerance to abiotic stress conditions by promoting selective mRNA decapping. *The Plant Cell* 28(2), 505–520.

Pomeranz, M.C., Hah, C., Lin, P.C., Kang, S.G., Finer, J.J. *et al.* (2010a) The *Arabidopsis* tandem zinc finger protein AtTZF1 traffics between the nucleus and cytoplasmic foci and binds both DNA and RNA. *Plant Physiology* 152(1), 151–165.

Pomeranz, M., Lin, P.C., Finer, J. and Jang, J.C. (2010b) AtTZF gene family localizes to cytoplasmic foci. *Plant Signaling & Behavior* 5(2), 190–192.

Qin, D., Wang, F., Geng, X., Zhang, L., Yao, Y. *et al.* (2015) Overexpression of heat stress-responsive TaMBF1c, a wheat (*Triticum aestivum* L.) multiprotein bridging factor, confers heat tolerance in both yeast and rice. *Plant Molecular Biology* 87(1–2), 31–45.

Qu, J., Kang, S.G., Wang, W., Musier-Forsyth, K. and Jang, J.C. (2014) The *Arabidopsis thaliana* tandem zinc finger 1 (AtTZF1) protein in RNA binding and decay. *The Plant Journal* 78(3), 452–467.

Richardson, A., Landry, S.J. and Georgopoulos, C. (1998) The ins and outs of a molecular chaperone machine. *Trends in Biochemical Sciences* 23(4), 138–143.

Riggs, C.L., Kedersha, N., Ivanov, P. and Anderson, P. (2020) Mammalian stress granules and P bodies at a glance. *Journal of Cell Science* 133(16), jcs242487.

Ruibal, C., Castro, A., Carballo, V., Szabados, L. and Vidal, S. (2013) Recovery from heat, salt and osmotic stress in *Physcomitrella patens* requires a functional small heat shock protein PpHsp16.4. *BMC Plant Biology* 13, 174.

Ruiz-Solaní, N., Salguero-Linares, J., Armengot, L., Santos, J., Pallarès, I. *et al.* (2023) *Arabidopsis* metacaspase MC1 localizes in stress granules, clears protein aggregates, and delays senescence. *The Plant Cell* 35(9), 3325–3344.

Saedler, R., Jakoby, M., Marin, B., Galiana-Jaime, E. and Hülskamp, M. (2009) The cell morphogenesis gene SPIRRIG in *Arabidopsis* encodes a WD/BEACH domain protein. *The Plant Journal* 59(4), 612–621.

Saha, B., Mishra, S., Awasthi, J.P., Sahoo, L. and Panda, S.K. (2016) Enhanced drought and salinity tolerance in transgenic mustard [*Brassica juncea* (L.) Czern & Coss.] overexpressing *Arabidopsis* group 4 late embryogenesis abundant gene (AtLEA4-1). *Environmental and Experimental Botany* 128, 99–111.

Scharf, K., Heider, H., Höhfeld, I., Lyck, R., Schmidt, E. and Nover, L. (1998) The tomato Hsf system: HsfA2 needs interaction with HsfA1 for efficient nuclear import and may be localized in cytoplasmic heat stress granules. *Molecular and Cellular Biology* 18(4), 2240–2251.

Seifikalhor, M., Aliniaeifard, S., Shomali, A., Azad, N., Hassani, B. *et al.* (2019) Calcium signaling and salt tolerance are diversely entwined in plants. *Plant Signaling & Behavior* 14(11), 1665455.

Sheth, U. and Parker, R. (2003) Decapping and decay of messenger RNA occur in cytoplasmic processing bodies. *Science* 300(5620), 805–808.

Skirycz, A., Claeys, H., De Bodt, S., Oikawa, A., Shinoda, S. *et al.* (2011) Pause-and-stop: the effects of osmotic stress on cell proliferation during early leaf development in *Arabidopsis* and a role for ethylene signaling in cell cycle arrest. *The Plant Cell* 23(5), 1876–1888.

Song, N.H. and Ahn, Y.J. (2011) DcHsp17.7, a small heat shock protein in carrot, is tissue-specifically expressed under salt stress and confers tolerance to salinity. *New Biotechnology* 28(6), 698–704.

Sorenson, R. and Bailey-Serres, J. (2014) Selective mRNA sequestration by oligouridylate-binding protein 1 contributes to translational control during hypoxia in *Arabidopsis*. *Proceedings of the National Academy of Sciences USA* 111(6), 2373–2378.

Sottosanto, J.B., Gelli, A. and Blumwald, E. (2004) DNA array analyses of *Arabidopsis thaliana* lacking a vacuolar Na^+/H^+ antiporter: impact of AtNHX1 on gene expression. *The Plant Journal* 40(5), 752–771.

Steffens, A., Bräutigam, A., Jakoby, M. and Hülskamp, M. (2015) The BEACH domain protein SPIRRIG is essential for *Arabidopsis* salt stress tolerance and functions as a regulator of transcript stabilization and localization. *PLoS Biology* 13(7), e1002188.

Sun, J., Jiang, H., Xu, Y., Li, H., Wu, X. *et al.* (2007) The CCCH-type zinc finger proteins AtSZF1 and AtSZF2 regulate salt stress responses in *Arabidopsis*. *Plant & Cell Physiology* 48(8), 1148–1158.

Sun, W., Van Montagu, M. and Verbruggen, N. (2002) Small heat shock proteins and stress tolerance in plants. *Biochimica et Biophysica Acta* 1577(1), 1–9.

Sun, X., Sun, C., Li, Z., Hu, Q., Han, L. and Luo, H. (2016) AsHSP17, a creeping bentgrass small heat shock protein modulates plant photosynthesis and ABA-dependent and independent signalling to attenuate plant response to abiotic stress. *Plant, Cell & Environment* 39(6), 1320–1337.

Sun, Y. and MacRae, T.H. (2005) Small heat shock proteins: molecular structure and chaperone function. *Cellular and Molecular Life Sciences* 62(21), 2460–2476.

Sundström, J.F., Vaculova, A., Smertenko, A.P., Savenkov, E.I., Golovko, A. *et al.* (2009) Tudor staphylococcal nuclease is an evolutionarily conserved component of the programmed cell death degradome. *Nature Cell Biology* 11(11), 1347–1354.

Suzuki, N., Rizhsky, L., Liang, H., Shuman, J., Shulaev, V. and Mittler, R. (2005) Enhanced tolerance to environmental stress in transgenic plants expressing the transcriptional coactivator multiprotein bridging factor 1c. *Plant Physiology* 139(3), 1313–1322.

Suzuki, N., Bajad, S., Shuman, J., Shulaev, V. and Mittler, R. (2008) The transcriptional co-activator MBF1c is a key regulator of thermotolerance in *Arabidopsis thaliana*. *The Journal of Biological Chemistry* 283(14), 9269–9275.

Suzuki, N., Sejima, H., Tam, R., Schlauch, K. and Mittler, R. (2011) Identification of the MBF1 heat-response regulon of *Arabidopsis thaliana*. *The Plant Journal* 66(5), 844–851.

Tanner, N.K. and Linder, P. (2001) DExD/H box RNA helicases: from generic motors to specific dissociation functions. *Molecular Cell* 8(2), 251–262.

Teixeira, D. and Parker, R. (2007) Analysis of P-body assembly in *Saccharomyces cerevisiae*. *Molecular Biology of the Cell* 18(6), 2274–2287.

Thandapani, P., O'Connor, T.R., Bailey, T.L. and Richard, S. (2013) Defining the RGG/RG motif. *Molecular Cell* 50(5), 613–623.

Tian, X., Qin, Z., Zhao, Y., Wen, J., Lan, T. *et al.* (2022) Stress granule-associated TaMBF1c confers thermotolerance through regulating specific mRNA translation in wheat (*Triticum aestivum*). *New Phytologist* 233(4), 1719–1731.

Tong, J., Ren, Z., Sun, L., Zhou, S., Yuan, W. *et al.* (2022) ALBA proteins confer thermotolerance through stabilizing HSF messenger RNAs in cytoplasmic granules. *Nature Plants* 8(7), 778–791.

Tong, X., Drapkin, R., Yalamanchili, R., Mosialos, G. and Kieff, E. (1995) The Epstein-Barr virus nuclear protein 2 acidic domain forms a complex with a novel cellular coactivator that can interact with TFIIE. *Molecular and Cellular Biology* 15, 4735–4744.

Tourrière, H., Chebli, K., Zekri, L., Courselaud, B., Blanchard, J.M. *et al.* (2003) The RasGAP-associated endoribonuclease G3BP assembles stress granules. *The Journal of Cell Biology* 160(6), 823–831.

Uniacke, J. and Zerges, W. (2008) Stress induces the assembly of RNA granules in the chloroplast of *Chlamydomonas reinhardtii*. *The Journal of Cell Biology* 182(4), 641–646.

van der Lee, R., Buljan, M., Lang, B., Weatheritt, R.J., Daughdrill, G.W. *et al.* (2014) Classification of intrinsically disordered regions and proteins. *Chemical Reviews* 114(13), 6589–6631.

Vanderweyde, T., Youmans, K., Liu-Yesucevitz, L. and Wolozin, B. (2013) Role of stress granules and RNA-binding proteins in neurodegeneration: a mini-review. *Gerontology* 59(6), 524–533.

van Etten, J., Schagat, T.L., Hrit, J., Weidmann, C.A., Brumbaugh, J. *et al.* (2012) Human pumilio proteins recruit multiple deadenylases to efficiently repress messenger RNAs. *Journal of Biological Chemistry* 287(43), 36370–36383.

Van Leene, J., Stals, H., Eeckhout, D., Persiau, G., Van De Slijke, E. *et al.* (2007) A tandem affinity purification-based technology platform to study the cell cycle interactome in *Arabidopsis thaliana*. *Molecular & Cellular Proteomics* 6(7), 1226–1238.

Wang, D., Guo, Y., Wu, C., Yang, G., Li, Y. and Zheng, C. (2008) Genome-wide analysis of CCCH zinc finger family in *Arabidopsis* and rice. *BMC Genomics* 9, 44.

Wani, S.H., Kumar, V., Khare, T., Guddimalli, R., Parveda, M. *et al.* (2020) Engineering salinity tolerance in plants: progress and prospects. *Planta* 251(4), 76.

Ward, J.J., Sodhi, J.S., McGuffin, L.J., Buxton, B.F. and Jones, D.T. (2004) Prediction and functional analysis of native disorder in proteins from the three kingdoms of life. *Journal of Molecular Biology* 337(3), 635–645.

Weber, C., Nover, L. and Fauth, M. (2008) Plant stress granules and mRNA processing bodies are distinct from heat stress granules. *The Plant Journal* 56(4), 517–530.

West, G., Inzé, D. and Beemster, G.T. (2004) Cell cycle modulation in the response of the primary root of *Arabidopsis* to salt stress. *Plant Physiology* 135(2), 1050–1058.

White, P.J. and Broadley, M.R. (2003) Calcium in plants. *Annals of Botany* 92(4), 487–511.

Wilbertz, J.H., Voigt, F., Horvathova, I., Roth, G., Zhan, Y. and Chao, J.A. (2019) Single-molecule imaging of mRNA localization and regulation during the integrated stress response. *Molecular Cell* 73(5), 946–958.

Xu, J. and Chua, N.H. (2012) Dehydration stress activates *Arabidopsis* MPK6 to signal DCP1 phosphorylation. *EMBO Journal* 31(8), 1975–1984.

Xu, J., Yang, J.Y., Niu, Q.W. and Chua, N.H. (2006) *Arabidopsis* DCP2, DCP1, and VARICOSE form a decapping complex required for postembryonic development. *The Plant Cell* 18(12), 3386–3398.

Yan, C., Yan, Z., Wang, Y., Yan, X. and Han, Y. (2014) Tudor-SN, a component of stress granules, regulates growth under salt stress by modulating GA20ox3 mRNA levels in *Arabidopsis*. *Journal of Experimental Botany* 65(20), 5933–5944.

Yan, Y., Gan, J., Tao, Y., Okita, T.W. and Tian, L. (2022) RNA-binding proteins: the key modulator in stress granule formation and abiotic stress response. *Frontiers in Plant Science* 13, 882596.

Yang, G., Wang, Y., Zhang, K. and Gao, C. (2014) Expression analysis of nine small heat shock protein genes from *Tamarix hispida* in response to different abiotic stresses and abscisic acid treatment. *Molecular Biology Reports* 41(3), 1279–1289.

Youn, J.Y., Dunham, W.H., Hong, S.J., Knight, J.D.R., Bashkurov, M. *et al.* (2018) High-density proximity mapping reveals the subcellular organization of mRNA-associated granules and bodies. *Molecular Cell* 69(3), 517–532.

Youn, J.Y., Dyakov, B.J.A., Zhang, J., Knight, J.D.R., Vernon, R.M. *et al.* (2019) Properties of stress granule and P-body proteomes. *Molecular Cell* 76(2), 286–294.

Yu, X., Li, B., Jang, G.J., Jiang, S., Jiang, D. *et al.* (2019) Orchestration of processing body dynamics and mRNA decay in *Arabidopsis* immunity. *Cell Reports* 28(8), 2194–2205.

Zhang, Z., Zhang, S., Zhang, Y., Wang, X., Li, D. *et al.* (2011) *Arabidopsis* floral initiator SKB1 confers high salt tolerance by regulating transcription and pre-mRNA splicing through altering histone H4R3 and small nuclear ribonucleoprotein LSM4 methylation. *The Plant Cell* 23(1), 396–411.

Zhao, X., Harashima, H., Dissmeyer, N., Pusch, S., Weimer, A.K. *et al.* (2012) A general G1/S-phase cell-cycle control module in the flowering plant *Arabidopsis thaliana*. *PLoS Genetics* 8(8), e1002847.

Zhou, Q., Usluer, S., Zhang, F., Lenard, A.J. and Madl, T. (2021) ATP regulates RNA-driven cold inducible RNA binding protein phase separation. *Protein Science* 30(7), 1438–1453.

Zhu, J.K. (2002) Salt and drought stress signal transduction in plants. *Annual Review of Plant Biology* 53, 247–273.

Zhu, S., Gu, J., Yao, J., Li, Y., Zhang, Z. *et al.* (2022) Liquid–liquid phase separation of RBGD2/4 is required for heat stress resistance in *Arabidopsis*. *Developmental Cell* 57(5), 583–597.

7 Updates on Protein Post-Translational Modifications for Modulating Response to Salinity

Mushtaq Ahmad Najar*
Department of Biological Sciences, Unified Academic Campus, Kolkata, West Bengal, India

Abstract

Post-translational modifications (PTMs) precisely regulate the location, activity, and interactions of proteins. PTMs of proteins are crucial for plant development and stress adaptation. They facilitate the coordination of diverse stress signalling pathways, enabling complex regulation of plant responses to diverse environmental cues. Soil salinity has become a severe problem, risking global crop production and food security, and unravelling the molecular mechanisms regulating plant salt-stress responses could circumvent this problem. Protein PTMs are among the quickest responses to stressful conditions and thus understanding the PTMs of salt-responsive proteins could be crucial for getting deeper insights into the regulation of salt tolerance in plants. However, research on the role of PTMs in plant salt tolerance is in a very early stage, which warrants more efforts on how these modifications regulate protein functions under salt stress. This chapter provides a detailed report on the role of PTMs, particularly phosphorylation, ubiquitinoylation, lipidation, and SUMOylation, in regulating plant responses to salt stress.

7.1 Introduction

Agriculture feeds over 7 billion people globally and this number is expected to increase in the coming decades. By 2050, the demand for food production is expected to increase by 70–100% due to rapid population growth (Bortesi and Fischer, 2015; Zhang *et al.*, 2019; Molotoks *et al.*, 2021; Nazir *et al.*, 2022). The rise in crop production has resulted in soil fertility depletion and salinization, making such soils unsuitable for the cultivation of crops (Bortesi and Fischer, 2015; Ogunniyi *et al.*, 2020). Being immobile organisms, plants experience fluctuations in environmental conditions. Plants respond to challenging biotic and abiotic stresses by modulating their developmental programmes and adaptive mechanisms across physiological, biochemical, cellular, and molecular levels (Hashiguchi and Komatsu, 2016; Kim *et al.*, 2017; Ganie, 2020). As per the Global Map of Salt-affected Soils (GSASmap) V1.0.0, approximately 424 million ha of topsoil (0–30 cm) and 833 million ha of subsoil (30–100 cm) are impacted by salinity (FAO, 2021).

To overcome the agricultural problems associated with soil salinity, it is imperative to comprehend the molecular mechanisms governing

*Email: mushtaqnajar149@gmail.com

 The Genetics of Salt Tolerance in Plants: A Central Dogma Perspective and Strategies for Enhancement (S. Ganie and S.H. Wani eds)
DOI: 10.1079/9781800623033.0007

plant responses to salt stress (Ganie, 2020). This knowledge could be exploited through modern breeding platforms, genome editing technologies, and other state-of-the-art techniques to develop salt-tolerant crop cultivars (Bortesi and Fischer, 2015; Zhang *et al.*, 2019; Nazir *et al.*, 2022). The proteome pool after translation undergoes several types of modifications, controlling the location, stability, and function of target proteins. This chapter provides progress updates on research into the role of post-translation modifications (PTMs) in regulating salinity-stress responses in different crop species (Kurepa *et al.*, 2003; Hwang *et al.*, 2016). Further work on protein PTMs is critically needed to better understand the pathways regulating the salinity-stress responses of plants.

7.2 Post-Translational Modifications: An Overview of Common Classes

After protein translation, PTM is necessary to ensure proper function of proteins by regulating their conformation, stability, activity, and subcellular localization (Zhang *et al.*, 2020; Xiang *et al.*, 2021). Various types of protein modifications in plants have been documented, encompassing glycosylation, phosphorylation, ubiquitination, acetylation, SUMOylation, and lipidation (Millar *et al.*, 2019). This section presents a thorough yet succinct overview of the key PTMs observed in plants that have been intensively studied under salinity stress during the last decade.

7.2.1 Phosphorylation

Phosphorylation of proteins is the most extensively researched reversible protein modification, particularly in terms of functional understanding, and accounts for up to 30% of total modifications. The regulation of proteins phosphorylation status is predominantly carried out by protein kinases and phosphatases, primarily targeting serine (80–85%), threonine (10–15%), and tyrosine (0–5%) (Wang *et al.*, 2014). Five per cent of the *Arabidopsis* genome encodes kinases, with 10% of these kinases associated with mitogen-activated protein kinase (MAPK) pathways. Approximately 41% of the expressed proteins in *Arabidopsis* undergo phosphorylation under various conditions. Phosphorylation of proteins takes place through ester bonds, primarily involving the hydroxyl groups of amino acids. In the *Arabidopsis* genome, protein kinases (1052) catalyse this process, and it is reversed by phosphatases (162) (Van Wijk *et al.*, 2014; Friso and Van Wijk, 2015; Yan *et al.*, 2021). Numerous comprehensive phosphorylation studies have been conducted for *Arabidopsis*, rice (*Oryza sativa*), *Medicago* spp., maize (*Zea mays*), and other plant species on proteins regulating various developmental and environmental perturbations (Zhao *et al.*, 2016). Section 7.3 below will cover reports exploring the involvement of plant protein phosphorylation which directly or indirectly is interconnected with salinity tolerance in plants.

7.2.2 N-terminal acetylation

Nα-acetylation, or N-terminal acetylation (NTA), stands out as one of the most prevalent protein modifications in eukaryotes. NTA transfers an acetyl group from acetyl coenzyme A (acetyl-CoA) to the α-amine of the N-terminal amino acid of an emerging protein chain. Nα-acetyltransferases (NAT family) catalyse the reaction, leading to the removal of a positive charge. The studies in plants started from the human Nats (N-acetyltransferases). In humans there are seven N-acetyltransferases (named NatA–F and NatH). Acetylation of the N-terminal α-amine typically takes place after co-translational N-terminal methionine excision (NME) in the cytosol. The presence of a plant cytosolic co-translational NTA machinery has been demonstrated through the characterization of N-termini in numerous photosynthetic organisms (Martinez *et al.*, 2008; Zybailov *et al.*, 2008; Liu *et al.*, 2013). Most proteins in plants and other eukaryotes undergo N-acetylation, leading to pleiotropic effects on development and growth. For instance, the loss of function of NAT-A in *Arabidopsis* results in developmental and growth defects (Ferrández-Ayela *et al.*, 2013). The regulation of NTA as a PTM is not well defined due to the prevalence of this modification In the majority of proteins (Starheim *et al.*, 2012). NTA has been demonstrated to play a pivotal role in governing essential processes, including protein–protein interactions, protein sorting, and subcellular targeting, as well as protein stability and turnover (Forte *et al.*, 2011; Scott *et al.*, 2011, 2017; Starheim *et al.*, 2012; Yang *et al.*, 2013; Holmes *et al.*, 2014). Surprisingly, the plant NTA machinery and its biological significance are not well comprehended, with only a limited number of publications delving into the crucial role of NATs in responding to salinity stress.

7.2.3 Ubiquitination

Ubiquitination of proteins stands as the third most common, reversible, and highly diverse type of PTM in plants. Even though ubiquitination modifications are a predominant PTM type, they constitute only 1% or less of the total proteome and notably in eukaryotes, plant genomes encode significant U-box protein-related genes (Lyzenga and Stone, 2012). Monoubiquitination entails the attachment of several lysine residues within the same protein, leading to multi-monoubiquitination.

Due to the nature of the linkages formed, ubiquitin chains can be categorized as either homotypic or heterotypic. In homotypic chains, all components of the chain are connected through the same lysine (Lys) or methionine (Met) residues, whereas heterotypic chains consist of mixed and branched polyubiquitin chains with different linkages within a single polymer. The process of ubiquitination typically involves the collaborative functions of ubiquitin-activating enzymes (E1s), ubiquitin-conjugating enzymes (E2s), and ubiquitin-ligase enzymes (E3s) (Song and Luo, 2019). Various forms of ubiquitination modifications initiate distinct downstream signalling pathways (Husnjak and Dikic, 2012; Kulathu and Komander, 2012; Wang, S. *et al.*, 2022). Significantly, ubiquitination is not limited to newly synthesized or misfolded proteins; it also plays a role in the molecular regulatory mechanisms through which organisms respond to diverse environmental stimuli.

Extensive crosstalk to other PTMs such as acetylation, phosphorylation, and ribosylation contributes to the broadening of the functional variations within the ubiquitin code, such as phosphorylation on of one of the sites of a protein increasing the tendency for addition of a ubiquitin tag (Filipčík *et al.*, 2017). Different ubiquitin modifications are detected by effector proteins equipped with ubiquitin-binding domains (UBDs) that link the modified substrate to subsequent events within the ubiquitination processing machinery. The ubiquitin–26S proteasome system (UPS) plays a vital role in the precise control of gene transcription, DNA replication and repair, the quality and abundance of short-lived regulatory proteins, misfolded proteins, and eukaryotic protein trafficking processes (Vierstra, 2009). Despite the inhibition of both proteasomes and deubiquitinating enzymes, the buildup of proteins modified by ubiquitin is constrained in cells. Consequently, investigating potential PTMs resistant to ubiquitination through proteomics offers a novel approach to understanding the molecular mechanisms governing organismal growth, development, and stress resilience.

7.2.4 SUMOylation

SUMOylation is an influential mechanism which adds a polypeptide tag (SUMO) to a target protein, causing modification and thus regulating the protein's activity. Small Ub-related protein modifiers (SUMOs) are compact polypeptides with a molecular weight of 10 kDa, featuring N-terminal extensions of 10–30 residues. They engage in polySUMOylation, similar to ubiquitin, by attaching to accessible lysine(s) on target proteins through an isopeptide bond. The minimum consensus needed for SUMO attachment to a lysine residue in the substrate is a lysine within the ψKXE/D motif (where ψ represents a bulky hydrophobic group and X can be any amino acid) (Geiss-Friedlander and Melchior, 2007; Srivastava *et al.*, 2016; Srivastava and Sadanandom, 2020).

SUMOylation involves a three-step process facilitated by enzymes: (i) SUMO-activating enzyme or E1 (SAE); (ii) SUMO-conjugating enzyme or E2 (SCE); and (iii) SUMO ligase (E3). These enzymes activate, transfer, and conjugate SUMO to specific targets. The activation of SUMO is dependent on ATP and results in the formation of SUMO adenylate, with the release of pyrophosphate. The activation of SUMO involves the E1 complex, a heterodimer consisting of SAE1 and SAE2, utilizing ATP to form SUMO–AMP. This process leads to the creation of a thioester bond, after which the activated SUMO is transferred to the cysteine residue of E2 (SCE1) and ultimately conjugated to a substrate protein with the assistance of E3 enzymes (Miura *et al.*, 2007, 2010). The regulatory function of SUMOylation relies on its reversibility, which is facilitated by deSUMOylating proteases (DSPs). The presence of a greater number of DSPs compared to E3s in plants implies the significance of selectivity not only in SUMOylation but also in deSUMOylation. The latter is essential for maintaining a dynamic pool of both SUMOylated and non-SUMOylated proteins, contributing to the diverse cellular responses (Augustine and Vierstra, 2018). The widespread occurrence and conservation of SUMOylation throughout eukaryotes underscore its essential role in maintaining proteome functions. The *Arabidopsis* proteome contains numerous SUMO-target proteins, indicating that SUMOylation is a significant PTM like phosphorylation and ubiquitination (Augustine and Vierstra, 2018). Increased SUMOylation of nuclear proteins represents one of the swift reactions to diverse stresses, yet the exact mechanism by which this PTM contributes to stress resilience remains unclear. Published findings have substantiated the pivotal involvement of SUMO proteins in plants experiencing salinity stress. A widespread approach in these studies involves *in vivo* experimentation, specifically the overexpression of particular SUMO proteins, and these findings will be summarized in Section 7.6.

7.2.5 Lipidation

Protein lipidation entails attaching hydrophobic moieties to proteins, typically to enhance their affinity for membranes or specific membrane microdomains. This modification occurs either co-translationally or post-translationally and is irreversible (Ku *et al.*, 2022).

Frequent lipidation modifications of proteins encompass myristoylation, palmitoylation (S-acylation), glycosylphosphatidylinositol (GPI) anchoring, prenylation, and oleation. Of these mechanisms, myristoylation stands out as the primary protein lipidation mechanism in eukaryotes. Prenylation, N-myristoylation, and S-acylation involve simple fatty acids or polyisoprene groups, while GPI and glycosylinositolphosphorylceramide (GIPC) anchors are extensive and intricate glycolipids.

The examination of lipidated proteins in plants has frequently trailed behind studies in animals and yeast, usually drawing inspiration from these systems. It is believed that lipidation in plants differs significantly, both in terms of the proteins modified and the physiological needs of lipidated proteins, warranting independent investigation distinct from findings in other organisms. Nevertheless, numerous publications focus on the impact of salinity stress in *Arabidopsis*, with particular emphasis on the extensive study of S-acylation and N-myristoylation.

7.2.6 S-acylation

S-acylation, formerly referred to as palmitoylation, differs from other protein lipid modifications due to its reversible nature. It is facilitated by a large family of integral membrane proteins (24 in *Arabidopsis*), known as protein S-acyltransferases (PATs), comprising a conserved DHHC motif cysteine-rich domain (Batistič *et al.*, 2012; Hemsley *et al.*, 2013; Hemsley, 2020). The abundance of PATs and their distribution across various cellular locations are likely due to the fact that PATs function as integral membrane proteins (Batistič *et al.*, 2012). S-acylation entails attaching mainly saturated fatty acids, typically palmitate or stearate, to cysteine residues of proteins through thioester bonds. Notably, S-acylation stands out among lipid modifications because it lacks a specific consensus sequence for modification by PAT enzymes. Palmitoylation serves to direct proteins to particular membrane compartments, governing their presence within and movement between membrane microdomains and managing protein conformation and activity. Consequently, there is limited existing literature on these emerging aspects of plant lipidation (Hemsley *et al.*, 2013).

7.2.7 N-myristoylation

Myristoylation involves linking a 14-carbon myristoyl chain to the N-terminal glycine of the protein through the creation of an amide bond facilitated by N-myristoyltransferase (NMT). Typically, N-myristoylation takes place co-translationally, with NMT binding directly to ribosomes. This process occurs right after the removal of the N-terminal methionine, but it can also occur if internal glycines become exposed through proteolytic processing. The least stringent consensus sequence for NMT is (M)GXXXS/TXX, where X represents any amino acid. However, in plants, this sequence has been further defined and differs from what is observed in metazoans and fungi (Yamauchi *et al.*, 2010). In *Arabidopsis*, there are two isoforms of NMT genes, namely *NMT1* and *NMT2*. The absence of NMT1, as observed in the *nmt1-1* mutant, leads to lethality, and plants with reduced NMT1 levels exhibit a stunted growth phenotype (Pierre *et al.*, 2007).

7.2.8 N-glycosylation

N-glycosylation of proteins and lipids occurs in the endoplasmic reticulum (ER) and is considered a ubiquitous PTM in eukaryotes. Addition of preassembled core oligosaccharides (Glc3Man9GlcNac2) to the conserved sequence (in NXT/S sequence) on a nascent polypeptide chain en bloc by enzymatic activity of oligosaccharyltransferase (OT complex) leads to the formation of glycosidic linkages, enabling proper folding of proteins, cellulose synthesis, interactions between proteins, and salt-stress tolerance in plants (Jae *et al.*, 2008). The composition of plant N-glycans is unique compared to animals and contains β1,2-xylose and core α1,3-fucose residues (Johnson and Chrispeels, 1987; Laurière *et al.*, 1989). Following the covalent attachment to the polypeptide chain, ER glucosidases I and II promptly eliminate the terminal glucose residues from the core oligosaccharides. Within the Golgi apparatus, glycosidases and glycosyltransferases then orchestrate the creation of intricate N-glycans, including both complex and hybrid forms.

In *Arabidopsis*, the mannose-1 phosphate guanylyl transferase (CYT1) gene encodes for the enzyme GDP-mannose. *Arabidopsis cyt1* mutants exhibited secondary abnormalities in cell-wall composition, N-glycosylation, and reduction in production of ascorbic acid (Johnson and Chrispeels, 1987; Laurière *et al.*, 1989). In addition, null mutations in the gene encoding ER α-glucosidase I (KNOPF/GCSI) lead to lethality and disrupt an early stage of the protein N-glycosylation pathway (Boisson *et al.*, 2001; Stewart Gillmor *et al.*, 2002). In plants, abnormalities in the ER N-glycosylation pathway manifest in various phenotypes. Section 7.8 below covers N-glycosylation in detail and how it influences salt stress through crosstalk with other pathways.

7.3 General Overview: Mechanism and Role of Phosphorylation in Regulating Salinity-Stress Responses

Results of studies on *Oryza sativa* have established that salt stress and drought stress induce abscisic acid (ABA)-activated protein kinase (OsSAPK10), which regulates rice basic leucine zipper (bZIP) family transcription factors (Lindemose *et al.*, 2013; Gao *et al.*, 2022). OsbZIP20 is phosphorylated by OsSAPK10, which promotes transcription of the vacuolar Na^{+}/H^{+} (NHX1) antiporter (Bassil *et al.*, 2011). Furthermore, *Arabidopsis* which expressed *AtNHX1* exhibited enhanced salt tolerance through sequestration of Na^{+} ions into vacuoles along with efficient scavenging of ROS by ascorbate and antioxidative enzymes under salt stress (Kumar *et al.*, 2017). Diverse signalling pathways are activated, and integration of these pathways helps the plant to resist and combat the stressful conditions. The salt stress-responsive calcium sensor system is central to the salt response, is highly conserved, and has been comprehensively investigated in *Arabidopsis*. The proteins involved in the signalling pathway include a calcium-binding protein SCaBP8 (which relays the signal), SOS3 (a calcineurin B-like (CBL) family calcium-binding protein), SOS2 (a CBL-interacting protein kinase (CIPK) family kinase), and the CBL8–CIPK24 complex including 14-3-3 proteins. When the plants grow without salt stress, PKS5 (protein kinase SOS2-like 5) phosphorylates SOS2 at position Ser^{294} inducing protein–protein interaction between SOS2 and the ubiquitin tag (Huang *et al.*, 2022) protein for degradation, which results in inhibition of SOS2 (kinase) and thus inhibits phosphorylation of plasma membrane (PM) Na^{+}/H^{+}-ATPase (Kim *et al.*, 2017; Huang *et al.*, 2022). During salt stress increased levels of Ca^{2+} bind to calcium-binding proteins SOS3 and SCaBP8 and cause repression of PKS5 whose function is to phosphorylate SOS2 at its Ser^{294} position and repress SOS2 activity (Kudla *et al.*, 2018; Jiang *et al.*, 2019; Steinhorst and Kudla, 2019; Chen, X. *et al.*, 2021). Due to low level of PKS5, SOS3 phosphorylates SOS2, resulting in activation of the downstream target SOS1 (a PM-localized member of the NHX family of Na^+/H^+ antiporters). PM Na^+/H^+ antiporters act to regulate Na^+ transport and pump out excess Na^+ across the membrane, thus reducing Na^+ concentration in the cell (Shi *et al.*, 2002; Yang and Guo, 2018; Huang *et al.*, 2022). Reduced protein level of PKS5 triggers activation of PM H^+-ATPase by binding to 14-3-3ω, which provides SOS1 with the proton gradient needed to drive Na^+ transportation. Overexpression studies in maize (*Z. mays*) revealed *ZmSTL2* encodes an Na^+/H^+ antiporter ZmSOS1, and 4 bp frameshift deletion in the reading frame was associated with hypersensitivity towards salt stress, thus validating the role of SOS1 (Yang and Guo, 2018). Recently studies using biochemical phosphorylation assays, protein–protein interactions, and quantitative methods of calcium signalling both *in planta* and *in vitro* have established that CBL4/CBL8–CIPK24 complexes act as calcium sensors. CBL proteins acquire function through association with CIPKs, making CBL–CIPK complexes. CIPK-mediated phosphorylation of the CBL subunit results in activation to carry out PTM of downstream targets such as SOS1, thus maintaining the osmotic balance. CBL8–CIPK24 carries out phosphorylation at high levels of Ca^{2+} concentration and CBL4 shows activity at low Ca^{2+} levels, these outcomes were established by *in vitro* enzymatic assay as well as *in vivo* assays, respectively (Kim *et al.*, 2014; Steinhorst *et al.*, 2022). It is noteworthy that flowering time regulator GIGANTEA (GI) and 14-3-3λ proteins cause repression of CIPK24 in the absence of salt stress, adjusting and integrating the SOS pathway into development programmes like flowering timing (Zhou *et al.*, 2014). The maintenance of proper ion homeostasis is very important for proper functioning of cell metabolism. Salt stress leads to accumulation of Na^+ ions especially in the aerial parts of plants and Na^+ excess disturbs the Na^+/K^+ ion balance in the cytosol of cells; thus the activity of transporters/channels of Na^+ and K^+ is required to maintain homeostasis (Assaha *et al.*, 2017). PTM of these transporters/channels (mainly phosphorylation) causes activation and maintains ion homeostasis and stable potential at the PM by the activity of Na^+/H^+ antiporters like SOS1 under high salt level (Shi *et al.*, 2000; Millar *et al.*, 2019) (Fig. 7.1A). During salt stress, the depolarization due to inward movement of Na^+ ions carried by high-affinity K^+ transporters (HKTs) and non-selective cation channels causes depletion of the K^+ ion pool and impairs cell metabolism, causing cells to undergo programmed cell death (PCD) (Su *et al.*, 2015). Most of the H^+-ATPase activity is upregulated by phosphorylation, which results in repolarization of the membrane and thus allows better retention of K^+ ions. Haem-oxygenase (HO) regulates H^+-ATPase activity, resulting in efficient K^+ repolarization of membrane potential which maintains optimum concentration of K^+ ions and extrusion of Na^+ ions under salt stress in plants (Su *et al.*, 2015). Under salt-stress conditions, PM H^+ pumps (ATPases) are upregulated to provide the proton motive force to electrogenic pumps, driven by the chemical energy of ATP; this constitutes a driving force for the movement of solutes and metabolites across the PM. These increased activities of ATPases are vital for driving transporters to adapt during salt stress (Shi *et al.*, 2000; Yang *et al.*, 2019; Khandelwal *et al.*, 2022) (Fig. 7.1A).

It is worthwhile to mention that through mass spectroscopy experiments various phosphorylation sites have been shown: ten phosphorylation sites were distributed at the N-termini, P-domain, and the C-terminus in the H^+-ATPases AHA1, AHA2, AHA3, and AHA4/11 of *Arabidopsis* (Rudashevskaya *et al.*, 2012).

7.3.1 Protein phosphorylation orchestrating ion homeostasis under salt stress

Under high salt concentration, a series of biochemical and physiological changes occurs in plants which ultimately result in changes in their physiological, metabolic, and genomic expression patterns (Chourasia *et al.*, 2021, 2022).

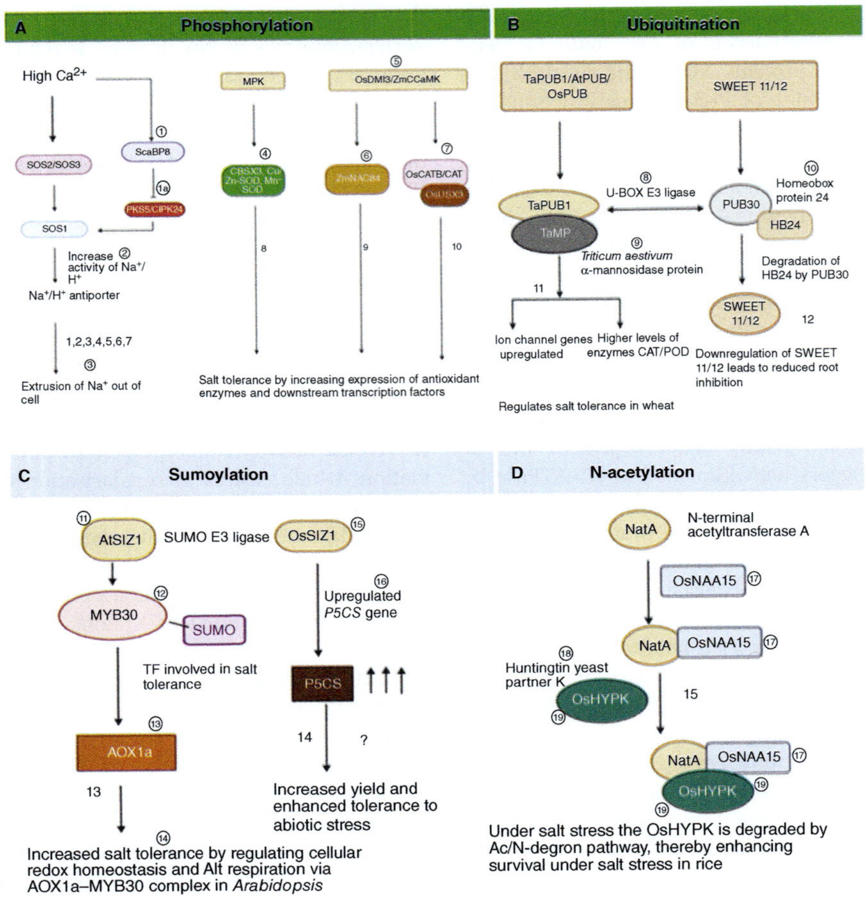

Fig. 7.1. Various PTMs and how PTMs regulate the interaction of other interacting protein partners and downstream transcription factors to regulate salinity stress. (A) In response to elevated levels of salt stress, the activation of Na^+/H^+-ATPase by SOS2 kinase induces the generation of a proton gradient. This gradient, in turn, facilitates the action of the SOS1 Na^+/H^+ antiporter, allowing the effective movement of sodium ions out of the cell (1 = Chen, X. *et al.*, 2021; 2 = Khandelwal *et al.*, 2022; 3 = Kim *et al.*, 2014; 4 = Millar *et al.*, 2019; 5 = Shi *et al.*, 2002; 6 = Steinhorst *et al.*, 2022; 7 = Zhu *et al.*, 2016). CCaMK (calcium- and calmodulin-dependent protein kinase) plays a beneficial role in abiotic stress signalling in plants. In maize (ZmCCaMK) and rice (OsDMI3), CCaMK phosphorylates ZmNAC84 and OsUSX3, respectively. This phosphorylation event serves to modulate the expression levels of genes associated with antioxidant activity, specifically those encoding superoxide dismutases (SODs) and catalase (CAT). Additionally, CAT, in turn, regulates the expression levels of *NDK1*, *CPPK8*, *NCA1*, *RLK1*, and *BAK1* (8 = Chen, X. *et al.*, 2021b; 9 = Gong *et al.*, 2004; 10 = Ni *et al.*, 2022). (B) In wheat, TaPUB1 degrades TaMP which encodes α-mannosidase protein, regulating ion channel genes and levels of antioxidant enzymes CAT and peroxidase (POD) under salt stress (11 = Wang, Q. *et al.*, 2022). The inhibition of root growth under salt stress in *Arabidopsis* involves the downregulation of the HB24–SWEET11 module by U-box type E3 ubiquitin ligase 30 (PUB30), leading to a reduction in sucrose supply (12 = Wang *et al.*, 2023). (C) The activation of AOX1 (alternative oxidase 1A) to regulate salinity stress in *Arabidopsis* involves the SUMOylation of the R2R3-type transcription factor MYB30 by SIZ1. Plants with elevated levels of SIZ1 demonstrated increased expression of the *P5CS* gene, which plays a role in the synthesis of proline in response to both salt and drought stress (13 = Gong *et al.*, 2020; 14 = Mishra *et al.*, 2018). (D) In rice, OsHYPK serves as both a positive regulator and a substrate for the NatA complex. Comprehensive analysis of the entire N-terminal acetylome revealed that the absence of OsHYPK leads to significant alterations in the N-terminal acetylome, causing a shift in the protein turnover profile (15 = Gong *et al.*, 2022).

The accumulation of ions, particularly Na^+, causes toxic effects in the cytoplasm in most cases under salt stress. K^+ ions play an important role in many physio-chemical and metabolic process (protein synthesis, enzyme activity, stomatal movement, osmoregulation) when plants are subjected to abiotic and biotic stress (Li *et al.*, 2018). To maintain perfect balance of K^+ and Na^+ ions, several mechanisms come into play such as upregulating the expression of membrane transporters H^+-ATPases, providing H^+ power for the functioning of antiporters (NHX1, SOS1) to sequester the excess Na^+ in the vacuole and out of the cell (Deinlein *et al.*, 2014; Huang *et al.*, 2021; Chen *et al.*, 2022). Under salt stress K^+ transporters such as HKTs are upregulated (Nieves-Cordones *et al.*, 2019). Another mechanism for maintaining ion homeostasis under low concentration of K^+ ions is upregulation of *HAK5* (a high-affinity potassium ion transporter), whereby auxin-responsive factor ARF2 undergoes phosphorylation under low K^+ conditions which then reduces binding affinity for *HAK5* promoter and expression of *HAK5* is upregulated (Zhao *et al.*, 2016). The high binding competition for transporters between Na^+ and K^+ ions causes osmotic imbalance. Hence, maintaining an elevated K^+/Na^+ ratio is a strategy to reduce the toxicity induced by salt stress (Maathuis *et al.*, 2014; Chourasia *et al.*, 2021).

7.3.2 Crosstalk between reactive oxygen species and protein kinases: unravelling signalling pathways under salt stress

During salt stress, the 'SOS' pathway plays a vital role in overseeing ion homeostasis and facilitating the plant's adaptation to saline conditions. In salt-stressed plants, oxidative stress generates reactive oxygen species (ROS) mainly hydroxyl radicals (·OH), hydrogen peroxide (H_2O_2), and superoxide anion radicals (O_2^-). When ROS levels are low, they function as signals that activate downstream signalling pathways, regulating plant development and responses to both biotic and abiotic stresses. However, elevated ROS levels can significantly impede plant growth and yield (Liu *et al.*, 2022). The increased levels of ROS under salt stress in plant cells induce damage to subcellular structures and lipid peroxidation of cellular membranes, and push accumulation of toxic substances by interfering with enzymes regulating the pathways involved in nitrogen metabolism (Anjum *et al.*, 2015; Zakari *et al.*, 2020). Many protein kinases that carry out reversible phosphorylation have been elucidated and play a critical role in enhancing salt tolerance by regulating and modifying downstream target genes (Yan *et al.*, 2014; Zhang *et al.*, 2014; Yao *et al.*, 2018; Qin *et al.*, 2020). Overexpression of *MsTRX* (a thioredoxin protein-coding gene) from *Medicago sativa* improved salinity tolerance in transgenic *Nicotiana benthamiana* by scavenging ROS; upregulated genes were *Cu/Zn-SOD* (Cu/Zn superoxide dismutase), *BAM1* (barley any meristem 1), *LTP1* (barley lipid transfer protein), *Mn-SOD* (manganese superoxide dismutase), and *CBSX3* (CBS domain-containing protein 3) which regulates the thioredoxin system (Shin *et al.*, 2020). Increased MsTRX upregulated the expression of *SnRK2* (SNF1-related protein kinase 2) linking the ROS signal with ABA. The downregulated genes were *BAS1* (beta-amyrin synthase 1) and *STN7* (serine/threonine-protein kinase which phosphorylates LHCII protein 7) (Duan *et al.*, 2022). STN7 and protein phosphatase 1 (PPH1) carry out PTM of light-harvesting complex II (LHCII) antenna LHCB1 and LHCB2 and maintain redox state in thylakoid membranes in *Arabidopsis*. It seems under salinity stress that STN7 might negatively regulate ROS homeostasis (Ramakrishnan *et al.*, 2022). Calcium/calmodulin (CaM)-dependent protein kinase (CCaMK) modulates and plays a positive role in abiotic stress signalling in plants. Overexpression of the rice homolog of CCaMK (*OsDMI3*) and under H_2O_2 stress conditions led to the identification of downstream protein kinases and transcription factors. Yeast two-hybrid (Y2H) assay was performed to find the interacting protein partner of OsDMI3 supported by glutathione *S*-transferase (GST) pull down, firefly luciferase complementation imaging (LCI), and co-immunoprecipitation (Co-IP). OsDMI3 interacted with the downstream target protein OsUSX3 in cytoplasm supported by 35S-OsDMI3-YFP construct transfected in rice protoplasts. OsDMI3 is an important positive regulator in the antioxidant defence system that positively regulates OsUSX3 by phosphorylation at Ser^{245} upon oxidative stress, increasing its stability and promoting its interaction with CAT isozyme B (OsCATB), thereby increasing levels of CATB and raising CAT activity under oxidative stress (Ni *et al.*, 2022) (Fig. 7.1A). Similar to rice CCaMK, in maize ZmCCaMK phosphorylates ZmNAC84, modulating expression levels of SOD- and CAT-encoding genes and implicating its role in ABA signalling (Gong *et al.*, 2004; Chen, X. *et al.*, 2021; Xiang *et al.*, 2021) (Fig. 7.1A). CAT also modulates salt overly sensitive 2 (SOS2) (the main pathway for salt sensing and tolerance, respectively, in plants), nucleoside diphosphate kinase 1 (NDK1), calcium-dependent kinase 8 (CPPK8), no catalase activity 1 (NCA1), receptor-like kinase 1 (RLK1), and brassinosteroid-insensitive 1 (BAK1) (Verslues *et al.*, 2007; Yan *et al.*, 2014; Li *et al.*, 2015; Zou *et al.*, 2015; Zhou *et al.*, 2018; Zhang *et al.*, 2020).

7.4 Nα-Acetylation Effect on Growth Retardation and Salt Tolerance

Various sophisticated techniques like proteomics, transcriptomics, pharmacological approach, quantitative N-terminomics, and physiological assay (Aksnes *et al.*, 2015; Dinh *et al.*, 2015) have implicated the plant-specific acetyltransferases NatF and NatG in roles related to abiotic stress in the *Arabidopsis* genome.

Loss-of-function mutants of *OsHYPK* (Huntingtin yeast partner K), a gene encoding a conserved chaperone-like protein in rice, show increased tolerance to abiotic stresses. NatA binds to OsHYPK and causes the degradation of OsHYPK through NTA by the Ac/N-degron pathway induced by salt stress. This establishes an intricate mechanism of OsHYPK–NatA-mediated NTA, which plays a role in coordinating salt stress and development in rice (Gong *et al.*, 2022) (Fig. 7.1D). Recent research has shown that NTA is quite a dynamic process (previously it was thought to be static) which responds to the environment and contributes to the amelioration of biotic stress. Notably, decrease of NatA activity in plants did not result in hypersensitivity to osmotic or high-salt stress, despite the fact that NatA targets roughly twice as many substrates as NatB and *NatB* knock-out mutants showed drought-tolerant phenotypes (Linster *et al.*, 2015; Linster and Wirtz, 2018). Study revealed that NatA when downregulated showed constitutive activation of ABA. Despite previous genetic engineering studies in plants showing that depletion of NAT complexes led to diverse defects in plant development and stress responses (Pesaresi *et al.*, 2003; Linster and Wirtz, 2018; Armbruster *et al.*, 2020; Huber *et al.*, 2020), so far only a few functional substrates of NATs have been identified (Xu *et al.*, 2015). Golgi-associated NatF (NAA60) anchored to the PM has provided more information about NATs. Loss-of-function mutants (*naa60-2* and *naa60-1*) showed more sensitivity under high-salt application. The absence of NatA results in the termination of plant embryo development at the globular stage. Loss-of-function *NatB* mutants via T-DNA insertion are currently not accessible; therefore the inquiry into this matter should be approached through clustered regularly interspaced short palindromic repeats/CRISPR-associated protein 9 (CRISPR/Cas9)-mediated gene disruption in forthcoming investigations (Linster *et al.*, 2015, 2020; Linster and Wirtz, 2018; Armbruster *et al.*, 2020; Kashtwari *et al.*, 2022).

7.5 Role of Ubiquitination: Influence on Salinity Tolerance in Plants

About 1600 genes, comprising 6% of total genes in the *Arabidopsis* genome, code for the UPS (ubiquitin–26S proteasome system); this ubiquitin-facilitated protein degradation plays key roles in embryogenesis, growth, abiotic stress, hormone signalling, and senescence (Vierstra, 2009; Maathuis *et al.*, 2014). Moreover, proteomic analysis indicates that a protein is set to undergo proteasome-mediated degradation when it has modification at Lys-48. It has been elucidated that many E3 ligases mediate the degradation of salt stress-related proteins, which are involved in almost all aspects of plant signalling from development to stress response (Walsh and Sadanandom, 2014). Most of the ubiquitin E3 ligases in rice act as a negative regulator of salt-stress response. For example, rice plants overexpressing genes coding for RING finger E3 ligases, such as the salt-induced RING protein family (SIRP) that includes OsSIRP3, OsSIRP4, and OsSIRP1m, showed hypersensitivity to salt stress and reduction in root elongation and seed germination. The targets of these ligases are rice MADS-box gene 70 (OsMADS70) and an ABC domain-containing protein (OsABC1P11) (Hwang *et al.*, 2016; Li *et al.*, 2021; Al-Saharin *et al.*, 2022). Expression studies in *Arabidopsis* loss-of-function mutants of the RING finger E3 ligase called salt tolerance RING finger 1 (STRF1), a plasma-associated protein, showed enhanced salt tolerance and minimized ROS, but downstream targets are still unknown (Li *et al.*, 2021; Al-Saharin *et al.*, 2022; Wang, S. *et al.*, 2022). Currently, we do not have much information on plant U-box (PUB) E3 ligases and their roles. One PUB described in wheat (*Triticum aestivum*) is TaPUB15 and overexpression of this E3 ligase in rice and *Arabidopsis* increased salt-stress tolerance, thus acting as positive regulator; unlike TaPUB26 which acts as a negative regulator since a *Brachypodium distachyon* transgenic line overexpressing this E3 ligase had decreased salt-stress tolerance with increased ROS production and a high Na^+/K^+ ratio (Li *et al.*, 2021; Al-Saharin *et al.*, 2022; Wang, Q. *et al.*, 2022) (Fig. 7.1B). Moreover, overexpression of the zinc finger-homeodomain family transcription factor, homeobox protein 24 (HB24), causes root elongation, while loss-of-function mutants exhibit root growth inhibition under salt stress. HB24 binds to sugars and will eventually be exported by transporter 11/12 (SWEET11/12), which is regular metabolism in roots. It is hypothesized that U-box type E3 ubiquitin ligase 30 (PUB30) mediates the degradation of HB24, thereby downregulating the expression of SWEET11 and resulting in reduced sucrose supply and root growth inhibition under salt stress (Wang *et al.*, 2023) (Fig. 7.1B). It should be noted that this area needs extensive research to find out the downstream targets as well other protein–protein interactions.

7.6 SUMOylation: A Complex Post-Translational Modification Impacting Salinity Stress through its Effect on Membrane Transporters and Redox State of Plant Cells

Indications from several reports suggest that salinity stress leads to an elevation in the levels of SUMO conjugates in *Arabidopsis*. Additionally, the stress-induced SUMOylation of DELLA proteins has been observed to enhance their stability (Conti *et al.*, 2008; Bailey *et al.*, 2016). In a study demonstrating the significance of the SUMOylation type of PTM, mutants of *Arabidopsis* lacking the *SIZ1* (SUMO) gene exhibited dwarfism and hypersensitivity to salt and ABA in the early stage of seed germination, and this phenotype was successfully rescued through heterologous complementation with SlSIZ1, an E3 ligase found in the nucleus of tomato (*Solanum lycopersicum*), whereby salinity tolerance was exhibited in the early stage of seed germination (Wang, Q. *et al.*, 2022). The role of SIZ1 in salinity stress was further supported by results showing that it functions in the SUMOylation of MYB30 transcription factor, which binds to the promoter of alternative oxidase 1a gene (*AOX1a*) and upregulates its expression under salinity stress, maintaining cellular redox homeostasis of ROS through the alternative oxidase respiration route during salinity stress (Miura *et al.*, 2007, 2010; Gong *et al.*, 2020) (Fig. 7.1C). As mentioned previously, maintenance of ion homeostasis in cells and pumping out excess salt is a well-established strategy to combat salt stress. SUMOylation of the transcription factor MdbHLH104 by the E3 ligase ortholog of AtSIZ1 in *Malus domestica*, MdSIZ1, regulates the function of PM H^+-ATPases, which suggests that, because PM Na^+/H^+ antiporters like SOS1 (salt overly sensitive 1) need PM H^+-ATPases to maintain ion balance in cells and salt tolerance in plants, there might be a pathway by which SUMOylation by SIZ1 plays a key role in salt tolerance by targeting PM H^+-ATPases (Yang and Guo, 2018). Overexpression studies in *Arabidopsis* showed SIZ1 upregulated the proline biosynthesis gene *P5CS* and increased salt tolerance via the synthesis of higher amounts of osmoprotectant (Mishra *et al.*, 2018) (Fig. 7.1C). Under salt-stress treatment SUMO proteases (OST1 and OTS2) degrade SUMO–substrate complex and thus maintain SUMOylation balance in a cell. Further, *OST1* and *OST2* double mutants showed increased levels of SUMO1/2 indicating interactions between SUMO proteases and SUMO (Conti *et al.*, 2008). Understanding the role of SUMOylation and deSUMOylation will be progressive in producing salt-tolerant crops in the future and maintaining global food production. Some mentioned proteins, their subcellular localization, and how they play a key role under salinity stress are shown in Table 7.1.

Table 7.1. Diverse ubiquitin ligases for ubiquitination and E3 ligases for SUMOylation, along with their subcellular localization and functional roles in plants.

Species	E3 ligase	Pathway and E3 ligase class	Subcellular localization	Function	Reference
Triticum aestivum L.	Plant U-box 15 (TaPUB15)	UPS U-box	Nucleus and cytoplasm	Overexpressing *TaPUB15* enhances salt tolerance in crops, making it a valuable gene resource for future salt-tolerant wheat breeding	Li *et al*. (2021)
Arabidopsis thaliana	XB35 (XBAT35.2)	UPS RING	Golgi	Besides having a role in defence response, XBAT35.2 ubiquitin and ACD11 improve salt tolerance and drought stress in *Arabidopsis*	Li *et al*. (2020)
A. thaliana	Overly tolerant to salt 1 (AtOSTS1)	SUMO protease	Nucleus	Plants with elevated *OTS1* expression exhibit enhanced resistance to salt stress and a corresponding decrease in the levels of SUMOylated proteins	Kurepa *et al*. (2003); Conti *et al*. (2008)
Solanum tuberosum	Plant U-box 17 (StPUB17)	UND/PUB/ARM repeat type gene	Nucleus	StPUB17 shows broad-spectrum responses to both biotic and abiotic stresses. It plays a crucial role in enhancing late blight resistance and salt tolerance in the crop	Ni *et al*. (2010)
Oryza sativa	SAP AND MIZ 1 (OsSIZ1)	SUMO SIZ	Nucleus	Overexpression of *OsSIZ1* showed increased salt, heat, and drought tolerance through SUMOylation of stress-related substrates	Mishra *et al*. (2018)
O. sativa	Salt-induced RING protein 1 (OsSIRP1)	RING finger domain E3 ubiquitin ligase	Cytoplasm	Heterogenous overexpression of *OsSIRP1* in *Arabidopsis* negatively impacted salinity-stress tolerance during seed germination and root growth	Hwang *et al*. (2016)
O. sativa	Salt-induced RING protein 2 (OsSIRP2)	RING finger domain E3 ubiquitin ligase	Nucleus	OsSIRP2 enhances OsTKL1 degradation through ubiquitination and the 26S proteasomal pathway. Heterogeneous overexpression enhances tolerance to salinity and osmotic stress	Chapagain *et al*. (2018)
O. sativa	Salt-induced RING protein 3 (OsSIRP3)	RING finger domain E3 ubiquitin ligase	Cytoplasm	OsSIRP3 possesses E3 ligase activity attributed to its RING H2 domain. When overexpressed in *Arabidopsis*, it led to hypersensitivity phenotypes under salinity stress, indicating its role as a negative regulator of salt stress	Park *et al*. (2018)
O. sativa	Salt-induced RING protein 4 (OsSIRP4)	RING finger domain E3 ubiquitin ligase	Plasma membrane and cytosol	Overexpressing *OsSIRP4* revealed hypersensitivity to salt stress, contrasting with the responses observed in both wild-type and OsSIRP4-C269A mutated plants	Kim and Jang (2021)

O. sativa	Salt-insensitive RING-H2 type 14 (OsSIRH2-14)	RING finger domain E3 ubiquitin ligase	Golgi, cytoplasm, apparatus, and plasma membrane	OsSIRH2-14 (previously named OsRFPH2-14) positively contributes to salinity tolerance by modulating regulation of salt-related proteins, notably the HKT-type Na^+ transporter, OsHKT2	Park *et al*. (2019)
T. aestivum	PLANT U-Box 1 (PUB1)	U-box E3 ubiquitin ligase	Nucleus	Transgenic wheat with *TaPUB1* (RNAi)-mediated knockdown exhibited heightened sensitivity, while overexpression of *TaPUB1* in wheat resulted in lower cytosolic Na^+/K^+ ratios and improved antioxidant enzyme activities under salt stress. Therefore, TaPUB1 plays a positive role in regulating salt-stress tolerance in wheat	Wang *et al*. (2019)
A. thaliana	Stress associated protein 5 (SAP5)	RING finger domain E3 ubiquitin ligase	Nucleus	Plants overexpressing *AtSAP5*, which exhibits E3 ligase activity, demonstrated enhanced tolerance to various environmental challenges, such as salt stress, osmotic stress, and water deficit	Kang *et al*. (2011)
A. thaliana	Salt and drought responsiveness (SDR)	SCF (RING finger E3 ligase)	Nucleus	Plants with 35S:AntiSDR exhibit increased sensitivity to salt stress, while those with 35S:SDR demonstrate a phenotype characterized by enhanced salt tolerance	Li *et al*. (2022)
A. thaliana	PLANT U-box 30 (PUB30)	U -box E3 ubiquitin ligase	Cytoplasm, nucleus, and plasma membrane	At the germination stage, T-DNA-inserted loss-of-function mutant plants (*atpub30-1* and *atpub30-2*) exhibited enhanced tolerance to salt stress, as indicated by diverse phenotypic observations. These findings imply that AtPUB30 plays a crucial role in salt tolerance	Hwang *et al*. (2015)
A. thaliana	Salt tolerance RING finger 1 (STRF1)	RING finger domain E3 ubiquitin ligase	Plasma membrane	Seedlings with a loss-of-function mutation in *strf1-1* exhibit accelerated endocytosis in roots and changes in the expression of genes associated with membrane trafficking. Due to accumulation of antioxidant enzyme, strikingly they display enhanced tolerance to salt, ionic, and osmotic stresses	Tian *et al*. (2015)
Gossypium hirsutum	Salt-associated RING finger protein (SARP1)	C3H2C3 RING finger domain E3 ubiquitin ligase	Plasma membrane	GhSARP1 in *Arabidopsis* through heterogeneous overexpression resulted in reduced salt tolerance during both germination and post-germination stages, indicating an adverse reaction to salt stress	Liu *et al*. (2016)

7.7 Exploring Lipidation: A Novel Frontier and its Implications on Salinity Stress

Proteins belonging to the PAT (protein S-acyltransferase) family are characterized by a highly conserved cysteine-rich domain known as DHHC-CRD (Asp-His-His-Cys). These PATs play a crucial role in facilitating protein S-acylation. In plants, PATs are membrane-bound proteins with 3–6 membrane-spanning domains, and they feature diverse N- and C-terminal domains. The variability in these terminal domains is thought to be crucial for determining the substrate specificity of these proteins (Smotrys and Linder, 2004; Batistič *et al.*, 2012). Drawing insights from investigations conducted in mammalian and yeast systems, PATs operate by catalysing target proteins through a two-step ping-pong mechanism: (i) the catalytic central DHHC tetrapeptide is deprotonated and nucleophilic cysteine undergoes auto-acylation; and (ii) subsequently, the acyl group is transferred to a cysteine residue on a target protein (Jennings and Linder, 2012; Zmuda and Chamberlain, 2020). Through mutation analyses, it was discovered that the presence of the cysteine residue within the DHHC domain is essential for both the auto-acylation of PAT and its enzymatic activity (Zhou *et al.*, 2017; Li *et al.*, 2019; Chen, D. *et al.*, 2021; Jiang *et al.*, 2021; Tian *et al.*, 2022). *Arabidopsis* possesses a PAT family consisting of 24 members (Batistič, 2012). Given that the number of S-acylated proteins surpasses the number of PATs, it is probable that PATs have the capability to S-acylate multiple substrates (Hemsley, 2020).

Among the ten *Arabidopsis* CBL proteins which sense the increasing concentration of salt in a cell, the dual fatty acid modifications of N-myristoylation and S-acylation help in the cellular sorting of CBL1, SOS3/CBL4, CBL5, and CBL9. PAT10 is required for *Arabidopsis* plant development and salt tolerance via controlling the tonoplast localization of many CBLs, including CBL2, CBL3, CBL6, and CBL10, the membrane association of which is dependent on S-acylation (Qi *et al.*, 2013; Zhou *et al.*, 2013; Chai *et al.*, 2020). S-acylation of the cysteine residue next to the myristoylated glycine is considered to improve CBL protein–membrane adhesion and regulate subcellular trafficking from the ER to the PM. The CBL4–CIPK6 complex regulates AKT2's K^+ channel activity by boosting its transport from the ER to the PM (Held *et al.*, 2011). Plants with a mutation in any one of the components (CBL4 or CIPK6) exhibit delayed flowering only in short-day (SD) conditions; however, the precise molecular connection between K^+ status and flowering time has not been investigated. Further, S-acylation of SOS3 is essential for its nuclear import, influencing flowering time under both long-day (LD) and saline conditions. A mutation at the S-acylation site (SOS3-C3A) or biochemical inhibition with 2-bromotrifluoropropane (2-BrP) hinders nuclear translocation, leading to the relocalization of the SOS3–GI complex to the perinuclear ER. Yet, SOS3-C3A demonstrated partial suppression of the salt sensitivity in the *sos3-1* mutant. In contrast, the mutant with non-myristoylatable SOS3-G2A was unable to restore salt tolerance, although it supported normal flowering. This highlights the specificity of SOS3 S-acylation and its key role under salinity stress (Held *et al.*, 2011). Additionally, the critical interacting protein MdCBL1 was identified and characterized. Quick screening of PATs of confirmed S-acylated proteins in rice using an OsDHHC cDNA library and the bimolecular fluorescence complementation (BiFC) test yielded multiple PAT–substrate pairs, including OsDHHC13–OsNAC9, OsDHHC14–OsGSD1, OsDHHC18–OsNOA1, and OsDHHC30–OsCBL2. Further functional analysis of three pairs of OsDHHC30–OsCBL2 has revealed that OsDHHC30 determines the endomembranous interaction of OsCBL2 and OsCBL3 by S-acylation and contributes to salt tolerance in rice (Tian *et al.*, 2022). MdPAT16, a PAT member in apple (*M. domestica*), has been demonstrated to modulate plant resilience to salt stress and soluble sugar buildup via S-acylating MdCBL1 protein. MdCBL1 S-acylation at the N-terminus improves protein stability and PM localization (Jiang *et al.*, 2021). These findings imply that MdPAT16 palmitoylates its downstream target proteins, hence increasing their stability. This could be a missing link in the plant's salt-stress response pathway, with serious implications for fruit quality.

Many myristoylated proteins have been identified or are expected to be involved in cellular signalling pathways (Boisson *et al.*, 2003; Maurer-Stroh *et al.*, 2004; Resh, 2004) and myristoylation is frequently necessary for proper protein function. Myristoylation of the SOS3 calcium-binding protein, for example, is necessary for salt tolerance in *Arabidopsis*. Myristoylation of BON1/CPN1 is necessary for normal plant growth (Li *et al.*, 2010) and myristoylation of SnRK1 alters its catalytic activity and its role in shoot meristem development (Pierre *et al.*, 2007). SOS3 N-myristoylation at Gly-2 permits the SOS2–SOS3 complex to bind to the PM and phosphorylate SOS1 to promote Na^+ efflux. Further SOS3 is myristoylated and has the ability to bind calcium. *Arabidopsis* plants become susceptible to NaCl stress when myristoylation (G2A) or calcium binding (*sos3-1*) is disrupted. Because no myristoylation was observed when Gly-2 of SOS3 was changed to Ala, the myristyl group is most likely connected to Gly-2. The Gly-2 to Ala mutation has no effect on SOS3's calcium binding (Lindzen and Choi, 1995; Rommens *et al.*, 1995; Loh *et al.*, 1998; Ellard-Ivey *et al.*, 1999; Gao *et al.*, 2016). A potential N-myristoylation motif is found in several plant proteins; *in vitro* myristoylation of a calcium-dependent protein kinase (CDPK) from zucchini has been demonstrated but whether this is required for its function is uncertain (Ellard-Ivey *et al.*, 1999).

7.8 Glycosylation: Exploring its Role as a Key Player in the Intricate Response to Salinity Stress

N-glycosylation of proteins and lipids occurs in the ER and is considered a ubiquitous PTM in eukaryotes. Addition of preassembled core oligosaccharides (Glc3Man9GlcNac2) to the conserved sequence (in NXT/S sequence) on a nascent polypeptide chain en bloc by enzymatic activity of oligosaccharyltransferase (OT complex) leads to the formation of glycosidic linkages, enabling proper folding of proteins, cellulose synthesis, protein–protein interactions, and salt-stress tolerance in plant (Jae *et al.*, 2008). The composition of plant N-glycans is unique compared to animals and contains β1,2-xylose and core α1,3-fucose residues. Here, we throw some light on how N-glycosylation research would help in engineering of salt-tolerant plants. Defective glycosylation patterns in the *Arabidopsis* mutant *complex glycan 1* (*cgl1*), *temperature sensitive 3a* (*stt3a*), and *staurosporin* (Lane *et al.*, 2001; Jae *et al.*, 2008), which resulted in a salt-sensitive phenotype with an inhibitory effect on growth, abnormal root morphology, and callose accumulation, were pioneering to study the N-glycosylation of proteins in plants. In fact, salt stress and its impact on cell-wall synthesis have been well established through korrigan 1/radially swollen 2 (KOR1/RSW2). These findings suggest that glycosylation is a critical process in a plant's response to salt stress, offering potential strategies for improving salt tolerance in crops. However, the exact mechanisms are complex and still being explored.

7.9 Navigating the Database Landscape for Post-Translational Modifications

Due to the crucial importance of PTMs in regulation of molecular mechanisms in plants (Yang and Guo, 2018; Steinhorst *et al.*, 2022; Zhou *et al.*, 2022), to gather more information various techniques are used to quantify and profile plant PTMs, mainly mass spectrometry-based methods and proteomics (like phospho-proteomics for phosphorylation). Recently, due to advances in technologies and lower costs of sequencing entire transcriptomes, plant genomes can be acquired within no time and how can profiling and quantification of plant PTMs remain behind. Large datasets have been generated that are available at PubMed (https://pubmed.ncbi.nlm.nih.gov, accessed 20 February 2024) and annotation is done by using data taken from different online databases, like UniProt (https://www.uniprot.org, accessed 20 February 2024), TAIR (The Arabidopsis Information Resource) (https://www.arabidopsis.org, accessed 20 February 2024), RAP (The Rice Annotation Project) (https://rapdb.dna.affrc.go.jp/, accessed 20 February 2024), and IUPRED3 (Prediction of Intrinsically Unstructured Proteins) (https://iupred3.elte.hu, accessed 20 February 2024). Although many databases are available for PTMs in plants, the most recent one of its kind is qPTMplants (integrative database of quantitative Post-Translational Modifications in plants) (http://qptmplants.omicsbio.info, accessed 20 February 2024) which hosts data from 293 published articles and includes 1,242,365 experimentally PTM events (for further details, check http://qptmplants.omicsbio.info). Various databases for searching PTMs for plants in general are given in Table 7.2.

7.10 Conclusion

Developing crop cultivars resistant to salinity is prima facie of abiotic stress biology of plants, to make salinity-affected lands reusable as productive cultivation lands. Right now, to successfully develop a salt-tolerant crop cultivar, it is of utmost importance to understand in fine detail the mechanisms involved when plants are under salt stress, ranging from gene expression, to translation, to modifications at post-translation level. Elucidating the role of PTMs is indispensable, as they have a pivotal role from growth to development and in plants' response to abiotic

Table 7.2. Dedicated databases for PTMs in plants.

Database name	Release date	Available data	PTM (type)	URL locator	Reference
dbPTM 2019	2006	No. of experimental sites = 347,984 No. of putative sites = 908,917 No. of literatures = 92,648	Majority of studies include phosphorylation, acetylation, ubiquitination, succinylation, methylation, N-linked glycosylation, etc.	https://awi.cuhk.edu.cn/dbPTM/ (accessed 28 February 2024)	Huang *et al.* (2019)
SysPTM 2.0	2009	No. of experimental sites = 471,109 No. of putative sites = 52,235 No of literatures = 2420	Five modifications were taken: phosphorylation, acetylation, SUMOylation, ubiquitination, and glycosylation	http://lifecenter.sgst.cn/SysPTM/ (accessed 30 October 2023)	Li *et al.* (2014)
PhosPhAt 4.0	2013	No. of literatures = 4450	Experimental and predicted phosphorylation sites of individual proteins, kinase, or kinase family, with its substrate. Phosphorylation hotspots	https://phosphat.uni-hohenheim.de/ (accessed 28 February 2024)	Zulawski *et al.* (2013)
P3DB V4.1	2009	No. of phosphoproteins = 18179 No. of phosphosites = 71895 No. of literatures = 32	Usually focus on phosphorylation and their protein–protein interaction	https://p3db.org/home (accessed 28 February 2024)	Yao *et al.* (2014)
EPSD	2014	No. of phosphoproteins = 31,012 No of phosphosites = 82,175	Comprehensive databases contain mostly phosphosites of proteins	http://epsd.biocuckoo.cn/ (accessed 24 February 2024)	Cheng *et al.* (2014)
Plant PTM Viewer 2.0	2019	No. of phosphosites = 370,000 No. of literatures = 122	Useful for finding PTMs in a protein sequence, 19 types of PTM are included in this database	https://www.psb.ugent.be/webtools/ptm-viewer/ (accessed 24 February 2024)	Willems *et al.* (2019)

stress. Furthermore, the most investigated PTMs are phosphorylation, glycosylation, and acetylation, whereas the least studied is lipidation, which requires additional exploration to untangle the complicated processes under abiotic stress. Recent rapid progress in high-throughput proteomic techniques has enabled updates about the involvement and role of ubiquitination and SUMOylation in the PTM of proteins that are main players in pathways regulating salinity stress. Ubiquitination regulates the various transporters and minimizes ROS production under salt stresses. To better understand how SUMOylation influences a plant's ability to handle salinity and, more critically, how responses to salinity stress, including SUMOylation-dependent processes, are coordinated to maximize growth under stress, further research is required in this area. More research is needed to understand the cellular pathways concerning salt stress which are regulated by modification carried by phosphorylation, ubiquitination, SUMOylation, and other PTMs. The immediate challenge is to identify and validate these putative proteins specifically involved in salt stress. The characterization of specific substrates for modifiers, together with in-depth knowledge of subcellular compartmentalization, will lead to the design of more accurate molecular tools for salt-tolerance improvement in crops.

References

Aksnes, H., Van Damme, P., Goris, M., Starheim, K.K., Marie, M. *et al.* (2015) An organellar Nα-acetyltransferase, Naa60, acetylates cytosolic N termini of transmembrane proteins and maintains Golgi integrity. *Cell Reports* 10, 1362–1374. DOI: 10.1016/j.celrep.2015.01.053.

Al-Saharin, R., Hellmann, H. and Mooney, S. (2022) Plant E3 ligases and their role in abiotic stress response. *Cells* 11, 890.

Anjum, N.A., Sofo, A., Scopa, A., Roychoudhury, A., Gill, S.S. *et al.* (2015) Lipids and proteins – major targets of oxidative modifications in abiotic stressed plants. *Environmental Science and Pollution Research* 22, 4099–4121. DOI: 10.1007/S11356-014-3917-1.

Armbruster, L., Linster, E., Boyer, J.B., Brünje, A., Eirich, J. *et al.* (2020) NAA50 is an enzymatically active N α-acetyltransferase that is crucial for development and regulation of stress responses. *Plant Physiology* 183(4), 1502–1516.

Assaha, D.V.M., Ueda, A., Saneoka, H., Al-Yahyai, R. and Yaish, M.W. (2017) The role of Na^+ and K^+ transporters in salt stress adaptation in glycophytes. *Frontiers in Physiology* 8, 509.

Augustine, R.C. and Vierstra, R.D. (2018) SUMOylation: re-wiring the plant nucleus during stress and development. *Current Opinion in Plant Biology* 45, 143–154. DOI: 10.1016/j.pbi.2018.06.006.

Bailey, M., Srivastava, A., Conti, L., Nelis, S., Zhang, C. *et al.* (2016) Stability of small ubiquitin-like modifier (SUMO) proteases OVERLY TOLERANT TO SALT1 and -2 modulates salicylic acid signalling and SUMO1/2 conjugation in *Arabidopsis thaliana*. *Journal of Experimental Botany* 67, 353. DOI: 10.1093/jxb/erv468

Bassil, E., Tajima, H., Liang, Y.C., Ohto, M.A., Ushijima, K. *et al.* (2011) The *Arabidopsis* Na^+/H^+ antiporters NHX1 and NHX2 control vacuolar pH and K^+ homeostasis to regulate growth, flower development, and reproduction. *The Plant Cell* 23(9), 3482–3497. DOI: 10.1105/tpc.111.089581.

Batistič, O. (2012) Genomics and localization of the *Arabidopsis* DHHC-cysteine-rich domain S-acyltransferase protein family. *Plant Physiology* 160, 1597–1612. DOI: 10.1104/PP.112.203968.

Batistič, O., Rehers, M., Akerman, A., Schlücking, K., Steinhorst, L. *et al.* (2012) S-acylation-dependent association of the calcium sensor CBL2 with the vacuolar membrane is essential for proper abscisic acid responses. *Cell Research* 22, 1155–1168. DOI: 10.1038/cr.2012.71.

Boisson, B., Giglione, C. and Meinnel, T. (2003) Unexpected protein families including cell defense components feature in the N-myristoylome of a higher eukaryote. *Journal of Biological Chemistry* 278, 43418–43429. DOI: 10.1074/jbc.M307321200.

Boisson, M., Gomord, V., Audran, C., Berger, N., Dubreucq, B. *et al.* (2001) *Arabidopsis* glucosidase I mutants reveal a critical role of N-glycan trimming in seed development. *EMBO Journal* 20, 1010–1019. DOI: 10.1093/emboj/20.5.1010.

Bortesi, L. and Fischer, R. (2015) The CRISPR/Cas9 system for plant genome editing and beyond. *Biotechnology Advances* 33, 41–52.

Chai, S., Ge, F.R., Zhang, Y. and Li, S. (2020) S-acylation of CBL10/SCaBP8 by PAT10 is crucial for its tonoplast association and function in salt tolerance. *Journal of Integrative Plant Biology* 62, 718–722. DOI: 10.1111/jipb.12864.

Chapagain, S., Park, Y.C., Kim, J.H. and Jang, C.S. (2018) *Oryza sativa* salt-induced RING E3 ligase 2 (OsSIRP2) acts as a positive regulator of transketolase in plant response to salinity and osmotic stress. *Planta* 247, 925–939. DOI: 10.1007/S00425-017-2838-X.

Chen, D., Hao, F., Mu, H., Ahsan, N., Thelen, J.J. and Stacey, G. (2021) S-acylation of P2K1 mediates extracellular ATP-induced immune signaling in *Arabidopsis*. *Nature Communications* 12, 2750. DOI: 10.1038/s41467-021-22854-1.

Chen, W., Lin, F., Lin, K.-H., Chen, C., Xia, C. *et al.* (2022) Growth promotion and salt-tolerance improvement of *Gerbera jamesonii* by root colonization of *Piriformospora indica*. *Journal of Plant Growth Regulation* 41, 1219–1228.

Chen, X., Ding, Y., Yang, Y., Song, C., Wang, B. *et al.* (2021) Protein kinases in plant responses to drought, salt, and cold stress. *Journal of Integrative Plant Biology* 63, 53–78.

Cheng, H., Deng, W., Wang, Y., Ren, J., Liu, Z. and Xue, Y. (2014) dbPPT: a comprehensive database of protein phosphorylation in plants. *Database: The Journal of Biological Databases and Curation* 2014, bau121. DOI: 10.1093/database/bau121.

Chourasia, K.N., Lal, M.K., Tiwari, R.K., Dev, D., Kardile, H.B. *et al.* (2021) Salinity stress in potato: understanding physiological, biochemical and molecular responses. *Life* 11, 545. DOI: 10.3390/life11060545.

Chourasia, K.N., More, S.J., Kumar, A., Kumar, D., Singh, B. *et al.* (2022) Salinity responses and tolerance mechanisms in underground vegetable crops: an integrative review. *Planta* 255, 68. DOI: 10.1007/S00425-022-03845-y.

Conti, L., Price, G., O'Donnell, E., Schwessinger, B., Dominy, P. and Sadanandom, A. (2008) Small ubiquitin-like modifier proteases OVERLY TOLERANT TO SALT1 and -2 regulate salt stress responses in *Arabidopsis*. *The Plant Cell* 20, 2894–2908. DOI: 10.1105/tpc.108.058669.

Deinlein, U., Stephan, A.B., Horie, T., Luo, W., Xu, G. and Schroeder, J.I. (2014) Plant salt-tolerance mechanisms. *Trends in Plant Science* 19, 371–379.

Dinh, T.V., Bienvenut, W.V., Linster, E., Feldman-Salit, A., Jung, V.A. *et al.* (2015) Molecular identification and functional characterization of the first Nα-acetyltransferase in plastids by global acetylome profiling. *Proteomics* 15, 2426–2435. DOI: 10.1002/pmic.201500025.

Duan, X., Wang, Z., Zhang, Y., Li, H., Yang, M. *et al.* (2022) Overexpression of a thioredoxin-protein-encoding gene, *MsTRX*, from *Medicago sativa* enhances salt tolerance to transgenic tobacco. *Agronomy* 12, 1467.

Ellard-Ivey, M., Hopkins, R.B., White, T.J. and Lomax, T.L. (1999) Cloning, expression and N-terminal myristoylation of CpCPK1, a calcium-dependent protein kinase from zucchini (*Cucurbita pepo* L.). *Plant Molecular Biology* 39, 199–208. DOI: 10.1023/a:1006125918023.

FAO (2021) *Global Map of Salt-Affected Soils*. Food and Agriculture Organization of the United Nations, Rome. Available at: https://www.fao.org/documents/card/en/c/cb7247en (accessed 29 February 2024).

Ferrández-Ayela, A., Micol-Ponce, R., Sánchez-García, A.B., Peral, A. and Micol, M.M. (2013) Mutation of an *Arabidopsis* NatB N-alpha-terminal acetylation complex component causes pleiotropic developmental defects. *PLoS ONE* 8, e80697. DOI: 10.1371/journal.pone.0080697.

Filipčík, P., Curry, J.R. and Mace, P.D. (2017) When worlds collide – mechanisms at the interface between phosphorylation and ubiquitination. *Journal of Molecular Biology* 429, 1097–1113. DOI: 10.1016/j.jmb.2017.02.011.

Forte, G.M.A., Pool, M.R. and Stirling, C.J. (2011) N-terminal acetylation inhibits protein targeting to the endoplasmic reticulum. *PLoS Biology* 9, e1001073. DOI: 10.1371/journal.pbio.1001073.

Friso, G. and Van Wijk, K.J. (2015) Posttranslational protein modifications in plant metabolism. *Plant Physiology* 169, 1469–1487. DOI: 10.1104/pp.15.01378.

Ganie, S.A. (2020) RNA chaperones: potential candidates for engineering salt tolerance in rice. *Crop Science* 60, 530–540. DOI: 10.1002/csc2.20134.

Gao, R., Liu, P., Irwanto, N., Loh, D.R. and Wong, S.M. (2016) Upregulation of *LINC-AP2* is negatively correlated with *AP2* gene expression with *Turnip crinkle virus* infection in *Arabidopsis thaliana*. *Plant Cell Reports* 35, 2257–2267. DOI: 10.1007/S00299-016-2032-9.

Gao, W., Li, M., Yang, S., Gao, C., Su, Y. *et al.* (2022) miR2105 and the kinase OsSAPK10 co-regulate OsbZIP86 to mediate drought-induced ABA biosynthesis in rice. *Plant Physiology* 189(2), 889–905. DOI: 10.1093/plphys/kiac071.

Geiss-Friedlander, R. and Melchior, F. (2007) Concepts in sumoylation: a decade on. *Nature Reviews Molecular Cell Biology* 8, 947–956. DOI: 10.1038/nrm2293.

Gong, D., Guo, Y., Schumaker, K.S. and Zhu, J.K. (2004) The SOS3 family of calcium sensors and SOS2 family of protein kinases in *Arabidopsis*. *Plant Physiology* 134, 919–926. DOI: 10.1104/PP.103.037440.

Gong, Q., Li, S., Zheng, Y., Duan, H., Xiao, F. *et al.* (2020) SUMOylation of MYB30 enhances salt tolerance by elevating alternative respiration via transcriptionally upregulating AOX1a in *Arabidopsis*. *The Plant Journal* 102, 1157–1171. DOI: 10.1111/tpj.14689.

Gong, X., Huang, Y., Liang, Y., Yuan, Y., Liu, Y. *et al.* (2022) OsHYPK-mediated protein N-terminal acetylation coordinates plant development and abiotic stress responses in rice. *Molecular Plant* 15, 740–754. DOI: 10.1016/j.molp.2022.03.001.

Hashiguchi, A. and Komatsu, S. (2016) Impact of post-translational modifications of crop proteins under abiotic stress. *Proteomes* 4(4), 42.

Held, K., Pascaud, F., Eckert, C., Gajdanowicz, P., Hashimoto, K. *et al.* (2011) Calcium-dependent modulation and plasma membrane targeting of the AKT2 potassium channel by the CBL4/CIPK6 calcium sensor/protein kinase complex. *Cell Research* 21, 1116–1130. DOI: 10.1038/cr.2011.50.

Hemsley, P.A. (2020) S-acylation in plants: an expanding field. *Biochemical Society Transactions* 48, 529–536. DOI: 10.1042/BST20190703.

Hemsley, P.A., Weimar, T., Lilley, K.S., Dupree, P. and Grierson, C.S. (2013) A proteomic approach identifies many novel palmitoylated proteins in *Arabidopsis*. *New Phytologist* 197, 805–814. DOI: 10.1111/nph.12077.

Holmes, W.M., Mannakee, B.K., Gutenkunst, R.N. and Serio, T.R. (2014) Loss of amino-terminal acetylation suppresses a prion phenotype by modulating global protein folding. *Nature Communications* 5, 4383. DOI: 10.1038/ncomms5383.

Huang, K.Y., Lee, T.Y., Kao, H.J., Ma, C.T., Lee, C.C. *et al.* (2019) dbPTM in 2019: exploring disease association and cross-talk of post-translational modifications. *Nucleic Acids Research* 47, D298–D308. DOI: 10.1093/nar/gky1074.

Huang, L., Li, Z., Sun, C., Yin, S., Wang, B. *et al.* (2022) Genome-wide identification, molecular characterization, and gene expression analyses of honeysuckle NHX antiporters suggest their involvement in salt stress adaptation. *PeerJ* 10, e13214.

Huang, X.-Y., Wang, C.-K., Zhao, Y.-W., Sun, C.-H. and Hu, D.-G. (2021) Mechanisms and regulation of organic acid accumulation in plant vacuoles. *Horticulture Research* 8, 227.

Huber, M., Bienvenut, W.V., Linster, E., Stephan, I., Armbruster, L. *et al.* (2020) NatB-mediated N-terminal acetylation affects growth and biotic stress responses. *Plant Physiology* 182, 792–806. DOI: 10.1104/pp.19.00792.

Husnjak, K. and Dikic, I. (2012) Ubiquitin-binding proteins: decoders of ubiquitin-mediated cellular functions. *Annual Review of Biochemistry* 81, 291–322. DOI: 10.1146/annurev-biochem-051810-094654.

Hwang, J.H., Seo, D.H., Kang, B.G., Kwak, J.M. and Kim, W.T. (2015) Suppression of *Arabidopsis* AtPUB30 resulted in increased tolerance to salt stress during germination. *Plant Cell Reports* 34, 277–289. DOI: 10.1007/S00299-014-1706-4

Hwang, S.-G., Kim, J.J., Lim, S.D., Chan Park, Y., Moon, J.-C. and Jang, C.S. (2016) Molecular dissection of *Oryza sativa* salt-induced RING Finger Protein 1 (OsSIRP1): possible involvement in the sensitivity response to salinity stress. *Physiologia Plantarum* 158, 168–179. DOI: 10.1111/ppl.12459.

Jae, S.K., Frank, J., Chang, H.K., Kajiura, H., Vikram, M. *et al.* (2008) Salt tolerance of *Arabidopsis thaliana* requires maturation of N-glycosylated proteins in the Golgi apparatus. *Proceedings of the National Academy of Sciences USA* 105, 5933–5938. DOI: 10.1073/pnas.0800237105.

Jennings, B.C. and Linder, M.E. (2012) DHHC protein S-acyltransferases use similar ping-pong kinetic mechanisms but display different Acyl-CoA specificities. *Journal of Biological Chemistry* 287, 7236–7245. DOI: 10.1074/jbc.M111.337246.

Jiang, H., Ma, Q.J., Zhong, M.S., Gao, H.N., Li, Y.Y. and Hao, Y.J. (2021) The apple palmitoyltransferase MdPAT16 influences sugar content and salt tolerance via an MdCBL1–MdCIPK13–MdSUT2.2 pathway. *The Plant Journal* 106, 689–705. DOI: 10.1111/tpj.15191.

Jiang, Z., Zhou, X., Tao, M., Yuan, F., Liu, L. *et al.* (2019) Plant cell-surface GIPC sphingolipids sense salt to trigger Ca^{2+} influx. *Nature* 572, 341–346.

Johnson, K.D. and Chrispeels, M.J. (1987) Substrate specificities of N-acetylglucosaminyl-, fucosyl-, and xylosyltransferases that modify glycoproteins in the Golgi apparatus of bean cotyledons. *Plant Physiology* 84, 1301–1308.

Kang, M., Fokar, M., Abdelmageed, H. and Allen, R.D. (2011) *Arabidopsis* SAP5 functions as a positive regulator of stress responses and exhibits E3 ubiquitin ligase activity. *Plant Molecular Biology* 75, 451–466. DOI: 10.1007/S11103-011-9748-2.

Kashtwari, M., Mansoor, S., Wani, A.A., Najar, M.A., Deshmukh, R.K. *et al.* (2022) Random mutagenesis in vegetatively propagated crops: opportunities, challenges and genome editing prospects. *Molecular Biology Reports* 49, 5729–5749. DOI: 10.1007/S11033-021-06650-0.

Khandelwal, N.K., Millan, C.R., Zangari, S.I., Avila, S., Williams, D. *et al.* (2022) The structural basis for regulation of the glutathione transporter Ycf1 by regulatory domain phosphorylation. *Nature Communications* 13, 1278. DOI: 10.1038/s41467-022-28811-w.

Kim, J.H. and Jang, C.S. (2021) E3 ligase, the *Oryza sativa* salt-induced RING finger protein 4 (OsSIRP4), negatively regulates salt stress responses via degradation of the OsPEX11-1 protein. *Plant Molecular Biology* 105, 231–245. DOI: 10.1007/S11103-020-01084-X.

Kim, J.Y., Song, J.T. and Seo, H.S. (2017) Post-translational modifications of *Arabidopsis* E3 SUMO ligase AtSIZ1 are controlled by environmental conditions. *FEBS Open Bio* 7, 1622–1634.

Kim, Y.H., Khan, A.L., Waqas, M., Shim, J.K., Kim, D.H. *et al.* (2014) Silicon application to rice root zone influenced the phytohormonal and antioxidant responses under salinity stress. *Journal of Plant Growth Regulation* 33, 137–149. DOI: 10.1007/S00344-013-9356-2.

Ku, Y.S., Cheng, S.S., Cheung, M.Y., Law, C.H. and Lam, H.M. (2022) The re-localization of proteins to or away from membranes as an effective strategy for regulating stress tolerance in plants. *Membranes* 12, 1261. DOI: 10.3390/membranes12121261.

Kudla, J., Becker, D., Grill, E., Hedrich, R., Hippler, M. *et al.* (2018) Advances and current challenges in calcium signaling. *New Phytologist* 218, 414–431.

Kulathu, Y. and Komander, D. (2012) Atypical ubiquitylation – the unexplored world of polyubiquitin beyond Lys48 and Lys63 linkages. *Nature Reviews Molecular Cell Biology* 13, 508–523. DOI: 10.1038/nrm3394.

Kumar, S., Kalita, A., Srivastava, R. and Sahoo, L. (2017) Co-expression of *Arabidopsis* NHX1 and bar improves the tolerance to salinity, oxidative stress, and herbicide in transgenic mungbean. *Frontiers in Plant Science* 8, 1896. DOI: 10.3389/fpls.2017.01896.

Kurepa, J., Walker, J.M., Smalle, J., Gosink, M.M., Davis, S.J. *et al.* (2003) The small ubiquitin-like modifier (SUMO) protein modification system in *Arabidopsis*. Accumulation of SUMO1 and-2 conjugates is increased by stress. *Journal of Biological Chemistry* 278, 6862–6872. DOI: 10.1074/jbc.m209694200.

Lane, D.R., Wiedemeier, A., Peng, L., Höfte, H., Vernhettes, S. *et al.* (2001) Temperature-sensitive alleles of *RSW2* link the KORRIGAN endo-1,4-β-glucanase to cellulose synthesis and cytokinesis in *Arabidopsis*. *Plant Physiology* 126, 278–288. DOI: 10.1104/pp.126.1.278.

Laurière, M., Laurière, C., Chrispeels, M.J., Johnson, K.D. and Sturm, A. (1989) Characterization of a xylose-specific antiserum that reacts with the complex asparagine-linked glycans of extracellular and vacuolar glycoproteins. *Plant Physiology* 90, 1182–1188. DOI: 10.1104/pp.90.3.1182.

Li, B.W., Gao, S., Yang, Z.M. and Song, J.B. (2022) The F-box E3 ubiquitin ligase AtSDR is involved in salt and drought stress responses in *Arabidopsis*. *Gene* 809, 146011. DOI: 10.1016/j.gene.2021.146011.

Li, J., Jia, J., Li, H., Yu, J., Sun, H. *et al.* (2014) SysPTM 2.0: An updated systematic resource for post-translational modification. *Database: The Journal of Biological Databases and Curation* 2014, bau025. DOI: 10.1093/database/bau025.

Li, J., Liu, J., Wang, G., Cha, J.-Y., Li, G. *et al.* (2015) A chaperone function of NO CATALASE ACTIVITY1 is required to maintain catalase activity and for multiple stress responses in *Arabidopsis*. *The Plant Cell* 27, 908–925. DOI: 10.1105/tpc.114.135095.

Li, Q., Serio, R.J., Schofield, A., Liu, H., Rasmussen, S.R. *et al.* (2020) *Arabidopsis* RING-type E3 ubiquitin ligase XBAT35.2 promotes proteasome-dependent degradation of ACD11 to attenuate abiotic stress tolerance. *The Plant Journal* 104, 1712–1723. DOI: 10.1111/tpj.15032.

Li, Q., Li, B., Wang, J., Chang, X., Mao, X. and Jing, R. (2021) *TaPUB15*, a U-box E3 ubiquitin ligase gene from wheat, enhances salt tolerance in rice. *Food and Energy Security* 10, e250. DOI: 10.1002/fes3.250.

Li, W., Xu, G., Alli, A. and Yu, L. (2018) Plant HAK/KUP/KT K^+ transporters: function and regulation. *Seminars in Cell and Developmental Biology* 74, 133–141. DOI: 10.1016/j.semcdb.2017.07.009.

Li, Y., Gou, M., Sun, Q. and Hua, J. (2010) Requirement of calcium binding, myristoylation, and protein–protein interaction for the copine BON1 function in *Arabidopsis*. *Journal of Biological Chemistry* 285, 29884–29891. DOI: 10.1074/jbc.M109.066100.

Li, Y., Xu, J., Li, G., Wan, S., Batistič, O. *et al.* (2019) Protein S-acyl transferase 15 is involved in seed triacylglycerol catabolism during early seedling growth in *Arabidopsis*. *Journal of Experimental Botany* 70, 5205–5216. DOI: 10.1093/jxb/erz282.

Lindemose, S., O'Shea, C., Jensen, M.K. and Skriver, K. (2013) Structure, function and networks of transcription factors involved in abiotic stress responses. *International Journal of Molecular Sciences* 14(3), 5842–5878. DOI: 10.3390/ijms14035842.

Lindzen, E. and Choi, J.H. (1995) A carrot cDNA encoding an atypical protein kinase homologous to plant calcium-dependent protein kinases. *Plant Molecular Biology* 28, 785–797. DOI: 10.1007/BF00042065

Linster, E. and Wirtz, M. (2018) N-terminal acetylation: an essential protein modification emerges as an important regulator of stress responses. *Journal of Experimental Botany* 69, 4555–4568. DOI: 10.1093/jxb/ery241.

Linster, E., Stephan, I., Bienvenut, W.V., Maple-Grødem, J., Myklebust, L.M. *et al.* (2015) Downregulation of N-terminal acetylation triggers ABA-mediated drought responses in *Arabidopsis*. *Nature Communications* 6, 7640. DOI: 10.1038/ncomms8640.

Linster, E., Layer, D., Bienvenut, W.V., Dinh, T.V., Weyer, F.A. *et al.* (2020) The *Arabidopsis* Nα-acetyltransferase NAA60 locates to the plasma membrane and is vital for the high salt stress response. *New Phytologist* 228, 554–569. DOI: 10.1111/nph.16747.

Liu, C.C., Zhu, H.Y., Dong, X.M., Ning, D.L., Wang, H.X. *et al.* (2013) Identification and analysis of the acetylated status of poplar proteins reveals analogous N-terminal protein processing mechanisms with other eukaryotes. *PLoS ONE* 8, e58681. DOI: 10.1371/journal.pone.0058681.

Liu, L., Sun, Y., Di, P., Cui, Y., Meng, Q. *et al.* (2022) Overexpression of a *Zea mays* brassinosteroid-signaling kinase gene *ZmBSK1* confers salt stress tolerance in maize. *Frontiers in Plant Science* 13, 894710.

Liu, Y., Zhang, X., Zhu, S., Zhang, H., Li, Y. *et al.* (2016) Overexpression of *GhSARP1* encoding a E3 ligase from cotton reduce the tolerance to salt in transgenic *Arabidopsis*. *Biochemical and Biophysical Research Communications* 478(4), 1491–1496.

Loh, Y.T., Zhou, J. and Martin, G.B. (1998) The myristylation motif of Pto is not required for disease resistance. *Molecular Plant–Microbe Interactions* 11, 572–576. DOI: 10.1094/MPMI.1998.11.6.572.

Lukowitz, W., Nickle, T.C., Meinke, D.W., Last, R.L., Conklin, P.L. and Somerville, C.R. (2001) *Arabidopsis cyt1* mutants are deficient in a mannose-1-phosphate guanylyltransferase and point to a requirement of N-linked glycosylation for cellulose biosynthesis. *Proceedings of the National Academy of Sciences USA* 98, 2262–2267. DOI: 10.1073/pnas.051625798.

Lyzenga, W.J. and Stone, S.L. (2012) Abiotic stress tolerance mediated by protein ubiquitination. *Journal of Experimental Botany* 63, 599–616. DOI: 10.1093/jxb/err310.

Maathuis, F.J.M., Ahmad, I. and Patishtan, J. (2014) Regulation of Na^+ fluxes in plants. *Frontiers in Plant Science* 5, 467. DOI: 10.3389/FPLS.2014.00467.

Martinez, A., Traverso, J.A., Valot, B., Ferro, M., Espagne, C. *et al.* (2008) Extent of N-terminal modifications in cytosolic proteins from eukaryotes. *Proteomics* 8(14), 2809–2831.

Maurer-Stroh, S., Gouda, M., Novatchkova, M., Schleiffer, A., Schneider, G. *et al.* (2004) MYRbase: analysis of genome-wide glycine myristoylation enlarges the functional spectrum of eukaryotic myristoylated proteins. *Genome Biology* 5, R21. DOI: 10.1186/gb-2004-5-3-r21.

Millar, A.H., Heazlewood, J.L., Giglione, C., Holdsworth, M.J., Bachmair, A. and Schulze, W.X. (2019) The scope, functions, and dynamics of posttranslational protein modifications. *Annual Review of Plant Biology* 70, 119–151. DOI: 10.1146/annurev-arplant-050718-100211.

Mishra, N., Srivastava, A.P., Esmaeili, N., Hu, W. and Shen, G. (2018) Overexpression of the rice gene *OsSIZ1* in *Arabidopsis* improves drought-, heat-, and salt-tolerance simultaneously. *PLoS ONE* 13, e0201716. DOI: 10.1371/journal.pone.0201716.

Miura, K., Jin, J.B. and Hasegawa, P.M. (2007) Sumoylation, a post-translational regulatory process in plants. *Current Opinion in Plant Biology* 10, 495–502. DOI: 10.1016/J.PBI.2007.07.002.

Miura, K., Lee, J., Miura, T. and Hasegawa, P.M. (2010) SIZ1 controls cell growth and plant development in *Arabidopsis* through salicylic acid. *Plant & Cell Physiology* 51, 103–113. DOI: 10.1093/PCP/PCP171.

Molotoks, A., Smith, P. and Dawson, T.P. (2021) Impacts of land use, population, and climate change on global food security. *Food and Energy Security* 10, e261.

Nazir, R., Mandal, S., Mitra, S., Ghorai, M., Das, N. *et al.* (2022) Clustered regularly interspaced short palindromic repeats (CRISPR)/CRISPR-associated genome-editing toolkit to enhance salt stress tolerance in rice and wheat. *Physiologia Plantarum* 174, e13642. DOI: 10.1111/ppl.13642.

Ni, L., Wang, Q., Chen, C., Wang, S., Shen, T. *et al.* (2022) OsDMI3-mediated OsUXS3 phosphorylation improves oxidative stress tolerance by modulating OsCATB protein abundance in rice. *Journal of Integrative Plant Biology* 64, 1087–1101. DOI: 10.1111/jipb.13255.

Ni, X., Tian, Z., Liu, J., Song, B., Li, J. *et al.* (2010) StPUB17, a novel potato UND/PUB/ARM repeat type gene, is associated with late blight resistance and NaCl stress. *Plant Science* 178, 158–169. DOI: 10.1016/j.plantsci.2009.12.002.

Nickle, T.C. and Meinke, D.W. (1998) A cytokinesis-defective mutant of *Arabidopsis* (cyt1) characterized by embryonic lethality, incomplete cell walls, and excessive callose accumulation. *The Plant Journal* 15, 321–332. DOI: 10.1046/j.1365-313X.1998.00212.x.

Nieves-Cordones, M., Lara, A., Ródenas, R., Amo, J., Rivero, R.M. *et al.* (2019) Modulation of K^+ translocation by AKT1 and AtHAK5 in *Arabidopsis* plants. *Plant, Cell & Environment* 42, 2357–2371.

Ogunniyi, A.I., Mavrotas, G., Olagunju, K.O., Fadare, O. and Adedoyin, R. (2020) Governance quality, remittances and their implications for food and nutrition security in Sub-Saharan Africa. *World Development* 127, 104752. DOI: 10.1016/j.worlddev.2019.104752.

Park, Y.C., Moon, J.C., Chapagain, S., Oh, D.G., Kim, J.J. and Jang, C.S. (2018) Role of salt-induced RING finger protein 3 (OsSIRP3), a negative regulator of salinity stress response by modulating the level of its target proteins. *Environmental and Experimental Botany* 155, 21–30.

Park, Y.C., Lim, S.D., Moon, J.C. and Jang, C.S. (2019) A rice really interesting new gene H 2-type E 3 ligase, OsSIRH2-14, enhances salinity tolerance via ubiquitin/26 S proteasome-mediated degradation of salt-related. *Plant, Cell & Environment* 42, 3061–3076. DOI: 10.1111/pce.13619.

Pesaresi, P., Gardner, N.A., Masiero, S., Dietzmann, A., Eichacker, L. *et al.* (2003) Cytoplasmic N-terminal protein acetylation is required for efficient photosynthesis in *Arabidopsis*. *The Plant Cell* 15, 1817–1832. DOI: 10.1105/tpc.012377.

Pierre, M., Traverso, J.A., Boisson, B., Domenichini, S., Bouchez, D. *et al.* (2007) N-myristoylation regulates the SnRK1 pathway in *Arabidopsis*. *The Plant Cell* 19(9), 2804–2821.

Qi, B., Doughty, J. and Hooley, R. (2013) A Golgi and tonoplast localized S-acyl transferase is involved in cell expansion, cell division, vascular patterning and fertility in *Arabidopsis*. *New Phytologist* 200, 444–456. DOI: 10.1111/nph.12385.

Qin, X., Duan, Z., Zheng, Y., Liu, W.-C., Guo, S. *et al.* (2020) ABC1K10a, an atypical kinase, functions in plant salt stress tolerance. *BMC Plant Biology* 20, 720.

Ramakrishnan, M., Papolu, P.K., Satish, L., Vinod, K.K., Wei, Q. *et al.* (2022) Redox status of the plant cell determines epigenetic modifications under abiotic stress conditions and during developmental processes. *Journal of Advanced Research* 42, 99–116.

Resh, M.D. (2004) Membrane targeting of lipid modified signal transduction proteins. *Subcellular Biochemistry* 37, 217–232. DOI: 10.1007/978-1-4757-5806-1_6.

Rommens, C.M.T., Salmeron, J.M., Baulcombe, D.C. and Staskawicz, B.J. (1995) Use of a gene expression system based on potato virus X to rapidly identify and characterize a tomato Pto homolog that controls fenthion sensitivity. *The Plant Cell* 7, 249–257. DOI: 10.1105/tpc.7.3.249.

Rudashevskaya, E.L., Ye, J., Jensen, O.N., Fuglsang, A.T. and Palmgren, M.G. (2012) Phosphosite mapping of P-type plasma membrane H^+-ATPase in homologous and heterologous environments. *Journal of Biological Chemistry* 287, 4904–4913. DOI: 10.1074/jbc.M111.307264.

Scott, D.C., Monda, J.K., Bennett, E.J., Harper, J.W. and Schulman, B.A. (2011) N-terminal acetylation acts as an avidity enhancer within an interconnected multiprotein complex. *Science (New York, N.Y.)* 334, 674–678. DOI: 10.1126/SCIENCE.1209307.

Scott, D.C., Hammill, J.T., Min, J., Rhee, D.Y., Connelly, M. *et al.* (2017) Blocking an N-terminal acetylation-dependent protein interaction inhibits an E3 ligase. *Nature Chemical Biology* 13, 850–857. DOI: 10.1038/nchembio.2386.

Shi, H., Ishitani, M., Kim, C. and Zhu, J.K. (2000) The *Arabidopsis thaliana* salt tolerance gene *SOS1* encodes a putative Na^+/H^+ antiporter. *Proceedings of the National Academy of Sciences USA* 97, 6896–6901. DOI: 10.1073/pnas.120170197.

Shi, H., Quintero, F.J., Pardo, J.M. and Zhu, J.-K. (2002) The putative plasma membrane Na^+/H^+ antiporter SOS1 controls long-distance Na^+ transport in plants. *The Plant Cell* 14, 465–477.

Shin, J.S., So, W.M., Kim, S.Y., Noh, M. *et al.* (2020) CBSX3-Trxo-2 regulates ROS generation of mitochondrial complex II (succinate dehydrogenase) in *Arabidopsis*. *Plant Science* 294, 110458. DOI: 10.1016/j.plantsci.2020.110458.

Smotrys, J.E. and Linder, M.E. (2004) Palmitoylation of intracellular signaling proteins: regulation and function. *Annual Review of Biochemistry* 73, 559–587. DOI: 10.1146/annurev.biochem.73.011303.073954.

Song, L. and Luo, Z.Q. (2019) Post-translational regulation of ubiquitin signaling. *Journal of Cell Biology* 218, 1776–1786. DOI: 10.1083/jcb.201902074.

Srivastava, A.K., Zhang, C., Yates, G., Bailey, M., Brown, A. and Sadanandom, A. (2016) SUMO is a critical regulator of salt stress responses in rice. *Plant Physiology* 170, 2378–2391. DOI: 10.1104/pp.15.01530.

Srivastava, M. and Sadanandom, A. (2020) An insight into the factors influencing specificity of the SUMO system in plants. *Plants (Basel)* 9, 1788. DOI: 10.3390/plants9121788.

Starheim, K.K., Gevaert, K. and Arnesen, T. (2012) Protein N-terminal acetyltransferases: when the start matters. *Trends in Biochemical Sciences* 37(4), 152–161. DOI: 10.1016/j.tibs.2012.02.003.

Steinhorst, L. and Kudla, J. (2019) How plants perceive salt. *Nature* 572, 318–320.

Steinhorst, L., He, G., Moore, L.K., Schültke, S., Schmitz-Thom, I. *et al.* (2022) A Ca^{2+}-sensor switch for tolerance to elevated salt stress in *Arabidopsis*. *Developmental Cell* 57(17), 2081–2094.

Stewart Gillmor, C., Poindexter, P., Lorieau, J., Palcic, M.M. and Somerville, C. (2002) α-Glucosidase I is required for cellulose biosynthesis and morphogenesis in *Arabidopsis*. *Journal of Cell Biology* 156, 1003–1013. DOI: 10.1083/jcb.200111093.

Su, Y., Luo, W., Lin, W., Ma, L. and Kabir, M.H. (2015) Model of cation transportation mediated by high-affinity potassium transporters (HKTs) in higher plants. *Biological Procedures Online* 17, 1. DOI: 10.1186/s12575-014-0013-3.

Tian, M., Lou, L., Liu, L., Yu, F., Zhao, Q. *et al.* (2015) The RING finger E3 ligase STRF1 is involved in membrane trafficking and modulates salt-stress response in *Arabidopsis thaliana*. *The Plant Journal* 82, 81–92. DOI: 10.1111/tpj.12797.

Tian, Y., Zeng, H., Wu, J., Huang, J., Gao, Q. *et al.* (2022) Screening DHHCs of S-acylated proteins using an OsDHHC cDNA library and bimolecular fluorescence complementation in rice. *The Plant Journal* 110, 1763–1780. DOI: 10.1111/tpj.15769.

Van Wijk, K.J., Friso, G., Walther, D. and Schulze, W.X. (2014) Meta-analysis of *Arabidopsis thaliana* phospho-proteomics data reveals compartmentalization of phosphorylation motifs. *The Plant Cell* 26(6), 2367–2389. DOI: 10.1105/tpc.114.125815.

Verslues, P.E., Batelli, G., Grillo, S., Agius, F., Kim, Y.-S. *et al.* (2007) Interaction of SOS2 with nucleoside diphosphate kinase 2 and catalases reveals a point of connection between salt stress and H_2O_2 signaling in *Arabidopsis thaliana*. *Molecular and Cellular Biology* 27, 7771–7780. DOI: 10.1128/mcb.00429-07.

Vierstra, R.D. (2009) The ubiquitin–26S proteasome system at the nexus of plant biology. *Nature Reviews Molecular Cell Biology* 10, 385–397. DOI: 10.1038/nrm2688.

Walsh, C.K. and Sadanandom, A. (2014) Ubiquitin chain topology in plant cell signaling: a new facet to an evergreen story. *Frontiers in Plant Science* 5, 122. DOI: 10.3389/FPLS.2014.00122.

Wang, Q., Lei, S., Qian, J., Zheng, M. and Hsu, Y.F. (2022) Overexpression of tomato SIZ2 in *Arabidopsis* improves plant salinity tolerance. *Journal of Plant Growth Regulation* 42, 1893–1904. DOI: 10.1007/S00344-022-10667-5.

Wang, S., Lv, X., Zhang, J. Chen, D., Chen, S. *et al.* (2022) Roles of E3 ubiquitin ligases in plant responses to abiotic stresses. *International Journal of Molecular Sciences* 23, 2308. DOI: 10.3390/ijms23042308.

Wang, W., Wang, W., Wu, Y., Li, Q., Zhang, G. *et al.* (2019) The involvement of wheat U-box E3 ubiquitin ligase TaPUB1 in salt stress tolerance. *Journal of Integrative Plant Biology* 62, 631–651. DOI: 10.1111/jipb.12842.

Wang, Y., Liu, Z., Cheng, H., Gao, T., Pan, Z. *et al.* (2014) EKPD: a hierarchical database of eukaryotic protein kinases and protein phosphatases. *Nucleic Acids Research* 42(D1), D496–D502. DOI: 10.1093/nar/gkt1121.

Wang, Y., Zhao, H., Xu, L., Zhang, H., Xing, H., Fu, Y. and Zhu, L. (2023) PUB30-mediated downregulation of the HB24-SWEET11 module is involved in root growth inhibition under salt stress by attenuating sucrose supply in *Arabidopsis*. *New Phytologist* 237, 1667–1683. DOI: 10.1111/nph.18635.

Willems, P., Horne, A., van Parys, T., Goormachtig, S., de Smet, I. *et al.* (2019) The Plant PTM Viewer, a central resource for exploring plant protein modifications. *The Plant Journal* 99, 752–762. DOI: 10.1111/tpj.14345.

Xiang, Y., Bian, X., Wei, T., Yan, J., Sun, X. *et al.* (2021) ZmMPK5 phosphorylates ZmNAC49 to enhance oxidative stress tolerance in maize. *New Phytologist* 232, 2400–2417. DOI: 10.1111/nph.17761.

Xu, F., Huang, Y., Li, L., Gannon, P., Linster, E. *et al.* (2015) Two N-terminal acetyltransferases antagonistically regulate the stability of a nod-like receptor in *Arabidopsis*. *The Plant Cell* 27, 1547–1562. DOI: 10.1105/tpc.15.00173.

Yamauchi, S., Fusada, N., Hayashi H., Utsumi, T., Uozumi, N. *et al.* (2010) The consensus motif for N-myristoylation of plant proteins in a wheat germ cell-free translation system. *The FEBS Journal* 277(17), 3596–3607. DOI: 10.1111/j.1742-4658.2010.07768.x.

Yan, H., Jia, H., Chen, X., Hao, L., An, H. and Guo, X. (2014) The cotton WRKY transcription factor GhWRKY17 functions in drought and salt stress in transgenic *Nicotiana benthamiana* through ABA signaling and the modulation of reactive oxygen species production. *Plant & Cell Physiology* 55, 2060–2076. DOI: 10.1093/pcp/pcu133.

Yan, J., Kim, Y.J. and Somers, D.E. (2021) Post-translational mechanisms of plant circadian regulation. *Genes (Basel)* 12, 325. DOI: 10.3390/genes12030325.

Yang, D., Fang, Q., Wang, M., Ren, R., Wang, H. *et al.* (2013) Nα-acetylated Sir3 stabilizes the conformation of a nucleosome-binding loop in the BAH domain. *Nature Structural & Molecular Biology* 20, 1116–1118. DOI: 10.1038/nsmb.2637.

Yang, Y. and Guo, Y. (2018) Elucidating the molecular mechanisms mediating plant salt-stress responses. *New Phytologist* 217, 523–539. DOI: 10.1111/nph.14920.

Yang, Y., Wu, Y., Ma, L., Yang, Z., Dong, Q. *et al.* (2019) The Ca^{2+} sensor SCaBP3/CBL7 modulates plasma membrane H^+-ATPase activity and promotes alkali tolerance in *Arabidopsis*. *The Plant Cell* 31, 1367–1384.

Yao, Q., Ge, H., Wu, S., Zhang, N., Chen, W. *et al.* (2014) P^3DB 3.0: from plant phosphorylation sites to protein networks. *Nucleic Acids Research* 42, D1206–D1213. DOI: 10.1093/nar/gkt1135.

Yao, W., Zhao, K., Cheng, Z., Li, X., Zhou, B. and Jiang, T. (2018) Transcriptome analysis of poplar under salt stress and over-expression of transcription factor NAC57 gene confers salt tolerance in transgenic *Arabidopsis*. *Frontiers in Plant Science* 9, 1121.

Zakari, S.A., Asad, M.-A.-U., Han, Z., Zhao, Q. and Cheng, F. (2020) Relationship of nitrogen deficiency-induced leaf senescence with ROS generation and ABA concentration in rice flag leaves. *Journal of Plant Growth Regulation* 39, 1503–1517.

Zhang, A., Liu, Y., Wang, F., Li, T., Chen, Z. *et al.* (2019) Enhanced rice salinity tolerance via CRISPR/Cas9-targeted mutagenesis of the *OsRR22* gene. *Molecular Breeding* 39, 47.

Zhang, D., Jiang, S., Pan, J., Kong, X., Zhou, Y., Liu, Y. and Li, D. (2014) The overexpression of a maize mitogen-activated protein kinase gene (ZmMPK5) confers salt stress tolerance and induces defence responses in tobacco. *Plant Biology* 16, 558–570. DOI: 10.1111/plb.12084.

Zhang, S., Li, C., Ren, H., Zhao, T., Li, Q. *et al.* (2020) BAK1 mediates light intensity to phosphorylate and activate catalases to regulate plant growth and development. *International Journal of Molecular Sciences* 21, 1437. DOI: 10.3390/ijms21041437.

Zhao, S., Zhang, M.-L., Ma, T.-L. and Wang, Y. (2016) Phosphorylation of ARF2 relieves its repression of transcription of the K^+ transporter gene *HAK5* in response to low potassium stress. *The Plant Cell* 28, 3005–3019.

Zhou, B., Lin, J.Z., Peng, D., Yang, Y.Z., Guo, M. *et al.* (2017) Plant architecture and grain yield are regulated by the novel DHHC-type zinc finger protein genes in rice (*Oryza sativa* L.). *Plant Science* 254, 12–21. DOI: 10.1016/j.plantsci.2016.08.015.

Zhou, H., Lin, H., Chen, S., Becker, K., Yang, Y. *et al.* (2014) Inhibition of the *Arabidopsis* salt overly sensitive pathway by 14-3-3 proteins. *The Plant Cell* 26, 1166–1182.

Zhou, L.Z., Li, S., Feng, Q.N., Zhang, Y.L., Zhao, X. *et al.* (2013) PROTEIN S-ACYL TRANSFERASE10 is critical for development and salt tolerance in *Arabidopsis*. *The Plant Cell* 25, 1093–1107. DOI: 10.1105/tpc.112.108829.

Zhou, X., Li, J., Wang, Y., Liang, X., Zhang, M. *et al.* (2022) The classical SOS pathway confers natural variation of salt tolerance in maize. *New Phytologist* 236, 479–494.

Zhou, Y.B., Liu, C., Tang, D.Y., Yan, L., Wang, D. *et al.* (2018) The receptor-like cytoplasmic kinase STRK1 phosphorylates and activates CatC, thereby regulating H_2O_2 homeostasis and improving salt tolerance in rice. *The Plant Cell* 30, 1100–1118. DOI: 10.1105/tpc.17.01000.

Zhu, J.K. (2016) Abiotic stress signaling and responses in plants. *Cell* 167(2), 313–324. DOI: 10.1016/j.cell.2016.08.029.

Zmuda, F. and Chamberlain, L.H. (2020) Regulatory effects of post-translational modifications on zDHHC S-acyltransferases. *Journal of Biological Chemistry* 295, 14640–14652. DOI: 10.1074/jbc.REV120.014717.

Zou, J.-J., Li, X.-D., Ratnasekera, D., Wang, C., Liu, W.-X. *et al.* (2015) *Arabidopsis* CALCIUM-DEPENDENT PROTEIN KINASE8 and CATALASE3 function in abscisic acid-mediated signaling and H_2O_2 homeostasis in stomatal guard cells under drought stress. *The Plant Cell* 27, 1445–1460. DOI: 10.1105/tpc.15.00144.

Zulawski, M., Braginets, R. and Schulze, W.X. (2013) PhosPhAt goes kinases – searchable protein kinase target information in the plant phosphorylation site database PhosPhAt. *Nucleic Acids Research* 41, D1176–D1184. DOI: 10.1093/nar/gks1081.

Zybailov, B., Rutschow, H., Friso, G., Rudella, A., Emanuelsson, O. *et al.* (2008) Sorting signals, N-terminal modifications and abundance of the chloroplast proteome. *PLoS ONE* 3, e1994. DOI: 10.1371/journal.pone.0001994.

8 Integration of Genomics-Assisted and Speed Breeding for Enhancement of Plant Salt Tolerance

Farwa Basit[1], Xianzhong Feng[2*] and Javaid Akhter Bhat[2*]
[1]*Institute of Crop Sciences, College of Agriculture and Biotechnology, Zhejiang University, Hangzhou, China;* [2]*Zhejiang Lab, Hangzhou, China*

Abstract

Salt stress is known as the foremost abiotic stress and is presenting a serious problem in the agricultural world because it causes a noticeable inhibition in the growth and developmental attributes of various crops. However, molecular breeding is opening a new way to improve the performance of cultivars by developing novel breeding techniques via exposing molecular genetic markers for genetic evolution, even under salinity stress. These molecular techniques have changed the whole scenario concerning crop development in both model as well as non-model crop species. To this end, speed breeding can gear up the number of generations produced per year, which, in turn, substantially minimizes the time interval for the development of improved crop cultivars. It has been demonstrated that an integrated approach of genomics-assisted breeding (GAB) and speed breeding tremendously increases the speed for the development of salt-tolerant crop cultivars. By combining these techniques, the genetic gains in the improvement of complex traits can be obtained within a short time period, and with less cost, under a salt-stressed environment. In this chapter, we discuss the role of GAB and speed breeding in harnessing the true potential for the accelerated development of salt-tolerant plant varieties.

8.1 Introduction

The sessile nature of plants has led them to be highly exposed to environmental stresses, both biotic (such as diseases and insect pests) and abiotic (such as salinity, drought, heat, cold, and flooding) (Lamalakshmi Devi *et al.*, 2017). The occurrence of these environmental stresses is further enhanced by global climate change (Xiao *et al.*, 2022). These stresses result in considerable losses in the yield, overall production, and quality of crop plants. In this regard, salinity stress is one of the major abiotic constraints leading to substantial losses in agricultural production annually. It has been estimated that ~20% and ~50% of the world's cultivated and irrigated lands, respectively, are affected by salt-stress conditions (Gengmao *et al.*, 2015). Cultivation of crop plants in these saline areas is a tough job for farmers, as the yield is very low (Gengmao *et al.*, 2015). Hence, there is an immediate prerequisite to develop salt-tolerant crop cultivars to avoid yield losses in the salt-affected areas. In this context, recent advances in genomics-assisted breeding (GAB) and speed breeding have provided the best alternative to

*Email: fengxianzhong@iga.ac.cn; javid.akhter69@gmail.com

DOI: 10.1079/9781800623033.0008

hasten the development of improved salt-tolerant cultivars with high accuracy and precision (Rana *et al.*, 2019).

GAB has emerged as an important approach that utilizes genotype/marker data to select the desirable crop genotype/cultivar. This approach upsurges the proficiency of a breeding programme by reducing the selection time, cost, as well as labour used (Lorenz *et al.*, 2011). Marker-assisted selection (MAS) and genomic selection (GS) are the two main approaches of GAB that use molecular markers to select salt-tolerant crop genotypes instead of phenotypic traits (Bhat and Yu, 2021). MAS, which also includes marker-assisted backcrossing (MABC) and marker-assisted recurrent selection (MARS), is effective for the major-effect genes/quantitative trait loci (QTLs) having a considerable effect on trait expression of salt tolerance (Bhat *et al.*, 2016). These approaches can be used to select the major gene variation. However, for the complex quantitative traits governed by the minor genes/QTLs, GS has emerged as the most powerful genomic approach to breed for complex crop traits (Irving *et al.*, 2012). Most of the variation in salt resistance in plants is regulated by minor gene variation. Hence, the GS approach has been suggested to have great potential for breeding salt-tolerant crop cultivars (Liu *et al.*, 2020). Different studies have used the GS approach in various plant species such as apple (Liu *et al.*, 2020), rice (Bartholomé *et al.*, 2022), as well as wheat (Javid *et al.*, 2022), and these studies showed GS as a potential approach for breeding salt-tolerant plant cultivars.

To cope with the demands posed by rapid growth of the world's population, there is a need to develop some efficient techniques to accomplish the food requirements of the growing population in the era of severe global climate change. Although gene editing technologies and functional genomics have provided new tools for breeding plant cultivars having improved salt tolerance in an accurate and efficient manner (Song *et al.*, 2022), it is time consuming to grow these plants traditionally in the field for assessing their performance. A botanist documented a hundred years ago that plants can grow better and quicker in an artificially controlled environment (Pfeiffer, 1926). Since then it has been observed that flowering is persuaded by constant light intensities in various plants such as pulses, herbs, weed species, cereals, vegetables, and ornamentals, etc. (Davis and Burns, 2016). Afterwards, the term 'speed breeding' was introduced by The University of Queensland (Australia), the John Innes Centre (UK), and collaborators for quick plant development (Watson *et al.*, 2018). This technique lowered the plant generation time fivefold compared to plants grown with traditional methods in fields as well as being 2.5 times quicker than plants grown in greenhouses (Watson *et al.*, 2018). Speed breeding has been used for the breeding of salt-tolerant plant cultivars such as in rice (Rana *et al.*, 2019) and the great ability of speed breeding to accelerate the development of plant cultivars has been revealed (Table 8.1).

In this chapter, we discuss the adverse effects induced by salt stress on crop growth, as well as how GAB and speed breeding will provide the best platform to develop the salt-tolerant crop cultivars at higher speed as well as higher accuracy that are required to meet the needs of the growing global population and maintain global food security.

8.2 Effect of Salt Stress on Plant Growth

Globally, almost 20% of the agricultural (cultivated) lands, and nearly 50% of the irrigated lands, are affected by the poisonous effects of salt toxicity (Gengmao *et al.*, 2015). In China, ~ 34.6 million ha of land are polluted by salinity. A wide range of crop varieties, also including medicinal plant types that have a vital role in human disease treatments, are endangered by salt stress.

Environmental conditions are changing with time, which is causing dramatic effects on plant growth and yield development. These changed environmental behaviours are known as abiotic stresses (Begum *et al.*, 2019). Salt stress is considered one of the major factors which restrict plants' development in their natural environment. Plants show various dynamic as well as complex responses towards salt stress, which include both reversible and non-reversible alterations. For instance, the stress-induced changes cause numerous transcriptional modifications comprising cell and tissue specificities (Prittesh *et al.*, 2020).

Table 8.1. Speed breeding technique for the rapid generation of various plant species.

Species type	Generations per year	Technique	Conditions (temperature, light)	References
Wheat (spring)	6	Photoperiod	22/17°C, bright/dark	Watson *et al.* (2018)
Chickpea	6	Photoperiod and temperature	22/17°C, bright/dark	Watson *et al.* (2018)
Durum wheat	6	Photoperiod	22/17°C, bright/dark	Watson *et al.* (2018)
Barley	6	Photoperiod	22/17°C, bright/dark	Watson *et al.* (2018)
Pea	6	Photoperiod	22/17°C, bright/dark	Watson *et al.* (2018)
Canola	4	Photoperiod	22/17°C, bright/dark	Watson *et al.* (2018)
Amaranth	6	Photoperiod and temperature	–	Stetter *et al.* (2016)
Barley	9	Photoperiod temperature, soil fertility, immature seed germination, and embryo rescue	20°C, bright/dark	Zheng *et al.* (2013)
Faba bean	7	Plant hormones, photoperiod, light intensity, and immature seeds	22/18°C, bright/dark	Mobini *et al.* (2015)
Sugarcane	4	Photoperiod and temperature	25°C, day/dark	Hale *et al.* (2017)
Pea	5	Plant hormones, photoperiod, and immature seed germination	21/16°C, bright/dark	Mobini and Warkentin (2016)
Sorghum	6	Photoperiod, temperature, and immature seed germination	30/25°C, bright/dark	Forster *et al.* (2014)
Soybean	5	Photoperiod, temperature, and immature seed germination	28°C, day/dark	Jähne *et al.* (2020)
Rice	4	Photoperiod, temperature, and high-density planting	30/25°C, bright/dark	Collard *et al.* (2017)
Pea	5	Photoperiod and temperature	30/25°C, bright/dark	Ochatt *et al.* (2002)
Lentil	8	Plant hormones, photoperiod, light intensity, and immature seeds	22/18°C, bright/dark	Mobini *et al.* (2015)

Salt accumulation in plants is dependent on environmental factors like pH, temperature, nutrient availability, etc., and these environmental factors significantly determine the salt accumulation in plants (Aghaei and Komatsu, 2013). Munns and Tester (2008) demonstrated that ionic as well as osmotic stress is induced by salinity stress via the uptake of excessive amounts of particular ions such as Na^+ and Cl^-. The stress induced by salinity imposes adverse effects on the cellular membrane which further cause reductions in the efficacy of photosynthetic attributes, crop productivity, and plant growth traits as well (Ganie, 2020; Gamalero and Glick, 2022). According to earlier investigations, plants represent the stress responses in the form of reactive oxygen species (ROS) induction and use them as signalling molecules in the cellular redox mechanism. However, ROS are generated in an excessive amount under severely stressed conditions, which causes cellular oxidative damage (Ondrasek *et al.*, 2022). To overcome this salt-induced stress condition, plants have developed stress-tolerance mechanisms. For example, the regulation of anthocyanin biosynthesis genes occurs under salt stress by ensuing ROS signalling and anthocyanins, which are triggered by stress conditions. This mechanism further supports the ROS scavenging, and reduces the ionic and osmotic injuries (Ben Abdallah *et al.*, 2016). Xu and Rothstein (2018) have documented that stimulation of the anthocyanin biosynthesis genes, which activates the production of anthocyanin, thus directs ROS scavenging. Moreover, other authors have suggested that anthocyanin biosynthesis genes (*PAL* and

CHS) can be considered candidate genes for molecular breeding to increase the salt-stress resistance in plants (Dehghan *et al.*, 2014). As per the above discussion, many plants take up higher concentrations of Na^+ and Cl^- ions under salt excess conditions, and because of carbon and also water uptake restriction, over-accumulation of these ions inside plant cells may cause ion homeostasis disruption and/or nutritional imbalance. Subsequently, it causes ROS overgeneration, which inhibits the chlorophyll pigment contents, photosynthetic ratio, and plant growth. Against salt stress, the plant also develops a specific defence mechanism known as an antioxidant enzymatic and non-enzymatic defence system by stimulating the anthocyanin synthesis. Likewise, the stress-tolerance mechanism by stimulating antioxidant activities is documented in different plants, i.e. rice (Kibria *et al.*, 2017), wheat (Ahanger and Agarwal, 2017), maize (Ren *et al.*, 2020), and soybean (Soliman *et al.*, 2020). Additionally, the overexpression of *AtDFR*, *AmDEL*, and *UFGT2* genes confirmed the stimulation of antioxidant activities and proline contents, which increased the rate of water uptake, photosynthetic attributes, and ultimately plant growth rate under salt-induced damages (Khan *et al.*, 2017; Zhang *et al.*, 2019). On the other hand, the antioxidant defence mechanism (superoxide dismutase (SOD), peroxidase (POD), catalase (CAT), ascorbate peroxidase (APX), glutathione reductase (GR), glutathione disulfide (GSSG), glutathione (GSH), etc.) as well as proline contents improve the ROS-scavenging capability if anthocyanins located in the plant's organs are not enough to restrain them (Zhang *et al.*, 2019; Ganie, 2021).

Previously it was stated that anthocyanin-augmented plants have greater stress tolerance while the tolerant plants do not have a need for greater anthocyanin contents to acclimatize to the stress environment. The foremost responsibility of anthocyanin is to prevent ion toxicity and elevate the chelating ions, consequently enhancing the ROS scavenging under salt stress, which assists in the modulation of ion homeostasis, water uptake, osmotic balance, and photosynthetic level (Zahra *et al.*, 2012; Naing and Kim, 2021). Taking the damage by salinity stress to crop plants into consideration, it is required to develop a potential strategy by using modern technologies (genome selection, as well as speed breeding) to increase the salt-stress tolerance in plants. Development of salt-tolerant crop varieties through modern breeding techniques is an environmentally safer and effective approach to mitigate the losses resulting from salt stress (Bhat and Yu, 2021).

8.3 Genomics-Assisted Breeding for Development of Salt-Tolerant Crop Cultivars

Using integrated genomic technologies with traditional breeding can significantly improve the effectiveness of responding to present and upcoming environmental concerns. In these circumstances, there must be access to germplasm, genetic, and genomic resources for all plant species in order to use decision support systems to quickly improve their genetic productivity (Zheng *et al.*, 2022). The development of mapping populations with the purpose of identifying markers associated to QTLs for crop traits is the first step in the development of first-generation molecular breeding procedures (MAB and MARS). Most economically significant features, including biomass accumulation, fertility improvement, yield production, and salinity-stress tolerance, are polygenic in nature and regulated by numerous genes with minor effects (Meuwissen *et al.*, 2001). The latest breeding strategy, GS, can be used to enhance complex traits like salinity tolerance and to improve quantitative traits especially by employing genome-wide marker data without requiring the identification of markers linked to QTLs for desired attributes (Rutkoski *et al.*, 2011). For the purpose of training a prediction model to determine genomic estimated breeding values (GEBVs), GS uses a 'training population' of individuals who have both been genotyped and phenotyped. GEBVs for untested individuals from a 'candidate population' can then be estimated using this approach, and selection candidates (SCs) for making crosses or for advanced yield trials can be identified. GEBVs are a great selection criterion, despite the fact that they do not reveal the function of the underlying QTLs or genes for the trait of interest (Jannink *et al.*, 2010; Rutkoski *et al.*, 2011).

GAB approaches include MAS, MABC, MARC, and GS (Bhat and Yu, 2021; Naing and

Kim, 2021). The main benefit of GAB is that the genotypic data can be utilized at an early plant growth stage (i.e. seedling) to select the phenotypic performance of the traits that appear at the maturity stage. Therefore, it can noticeably save money, time, as well as the labour required for the massive phenotypic selection in numerous environments (Bhat and Yu, 2021). Besides, it will increase the precision as well as accuracy in the selection of the crop cultivars because GAB approaches are independent of the environmental influence. MAS exploits the molecular markers that have been previously reported to be linked with particular genomic loci or genes of interest. The specific markers will be used to trace the desirable allele of the gene during the selection of the improved cultivars possessing the superior trait phenotype (Das *et al.*, 2015). However, only the genes with major effects on the trait can be selected through MAS. This approach has been extensively used in plant breeding for the development of high-yielding and stress tolerance/resistance cultivars (as reviewed by Hasan *et al.*, 2021). Rana *et al.* (2019) used the integrated approach of MAS for the development of salt-tolerant rice cultivars. Their study suggested that MAS is the best alternative to accelerate the development of salt-tolerant crop cultivars (Rana *et al.*, 2019).

MABC is a type of GAB through which one or a few major genomic loci governing the yield, quality, or stress tolerance is/are incorporated from the donor parent (usually a non-commercial cultivar) into the recipient elite parent (commercial cultivar) (Hasan *et al.*, 2015). In the last two decades, this approach has been successfully used to develop biotic and abiotic stress-tolerant crop cultivars (as reviewed by Janaki Ramayya *et al.*, 2021). Until now crop cultivars with improved quality, yield, abiotic/salt tolerance, and disease resistance have been produced by using MABC technology (Kim *et al.*, 2022). For example, Moniruzzaman *et al.* (2012) have utilized MABC to identify and develop salt-tolerant rice lines in a reduced time period. Moreover, this approach has been used to transfer the Saltol QTL (a major genomic locus regulating salt tolerance) from the 'FL478' salt-tolerant donor parent to the 'Bacthom 7' recipient rice cultivar (Vu *et al.*, 2012); and the study showed that introgression lines can be directly used as salinity-tolerant cultivars. Similarly, this approach has been used by other researchers to develop salt-tolerant rice lines such as salt-tolerant 'IR64' (Ho *et al.*, 2016), salt-tolerant 'Pusa Basmati 1121' (Babu *et al.*, 2017), and 'Pusa Basmati 1509' (Yadav *et al.*, 2020). These investigations demonstrated the potential of MABC in the development of salinity-tolerant crop varieties.

MARC is another approach of MAB, in which several genome regions are targeted by using numerous molecular markers associated to the trait of interest for the selection of desirable crop genotype. MARC allows selection of the best-performing salt-tolerant cultivar/genotypes based on the marker genotype, followed by intercrossing among the selected genotypes for the next selection cycle (Ye and Smith, 2008). MARC has the capability to improve the breeding efficiency and assists to integrate the numerous promising genes or QTLs from diverse sources by recurrent selection grounded on a multi-parent population. It also allows the forward breeding of identified genes to be performed as well as the pyramiding of multiple QTLs. This approach has been successfully used to improve crown rot resistance in wheat (Rahman *et al.*, 2020). Bankole *et al.* (2017) successfully used MARS for the improvement of yield and its related traits under drought stress in maize. These studies demonstrated the efficacious application of MARS in improving stress tolerance in crop plants, thus suggesting the use of MARS for breeding salt-tolerant crop varieties.

All the above three approaches of GAB, i.e. MAS, MABC, and MARS, are applicable for the major gene variation. However, in the case of minor genes that govern the most part of the variation for complex traits such as salt tolerance, the more recent approach of GS has great promise and applications (Bhat and Yu, 2021; Bartholomé *et al.*, 2022). The role of GS in breeding salt-tolerant cultivars is discussed in the following section.

8.4 Genomic Selection for Breeding Salt Tolerance

Crop traits are of two types based on the inheritance pattern, i.e. quantitative and qualitative traits. The qualitative traits are regulated by one or a few major gene/s, although the quantitative traits are regulated by many minor genes present across the whole crop genome. Salt tolerance in crop plants is a complex quantitative trait that possesses low heritability and is greatly affected by environmental factors. The complex nature of salt tolerance has resulted in slower improvement via conventional breeding approaches (Bhat *et al.*, 2016). Furthermore, the earlier MAB techniques, MABC and MARS, were unable to deal with salt resistance in crop plants because these approaches only apply to major genes and are insufficient to handle the relatively minor gene variation that accounts for the majority of the salt tolerance variation (Bhat *et al.*, 2016; Zhang *et al.*, 2023). Therefore, these approaches of MAB are observed to perform inferior to phenotypic selection when most of the salt tolerance variation of crop plants is controlled by minor genes (Bhat *et al.*, 2016). Hence, researchers have developed a better alternative for breeding the complex traits such as salt tolerance, and this approach is known as genomic selection. GS uses high-density markers across the whole genome to estimate the net genetic worth of an individual, that is in contrast with MAS utilizing only a few selected markers. This approach first generates the prediction model dependent on the phenotypic as well as the genotypic data of the training population (TP), which is then used to estimate the GEBVs of the individuals of the breeding population (BP) from their genotypic data only (Meuwissen *et al.*, 2001) (Fig. 8.1). In this way the desirable individual is selected either for its use as a parent in hybridization or directly used as a variety for cultivation.

Hence, in the GS analysis, genome-wide markers disseminated across the entire genome are utilized to develop a model that predicts the performance of the breeding individual based on the genetic effects of all loci governing the trait, regardless how many are minor or major (Bhat *et al.*, 2016). In this way, this approach prevents the loss of variation controlled by minor genes and incorporates this variation in the selection process. Multiple studies have revealed that GS possesses great potential for the improvement of complex traits such as yield, biotic and abiotic stresses (see details in Bhat and Yu, 2021) (Tables 8.2 and 8.3). Besides, GS has the advantage of using the genotypic data collected from the early plant growth stage, such as seed or seedling stage, and predicts the phenotypic performance of the traits that express late at the maturity stage; hence greatly speeding up the selection process of superior individuals. Therefore, GS has been used in multiple crops for various traits, and has shown great successes for breeding the complex traits such as yield and agronomic traits in wheat (Lozada *et al.*, 2019), seed composition as well as yield traits in soybean (Stewart-Brown *et al.*, 2019), yield in maize (Beyene *et al.*, 2015), drought resistance in pea (Annicchiarico *et al.*, 2017) and chickpea (Li, Y. *et al.*, 2018), as well as disease resistance in barley (Tiede and Smith, 2018). Moreover, Liu *et al.* (2020) recently also demonstrated the great potential of genomic-assisted prediction for the breeding of salt-tolerant apple rootstocks. Hence, all these studies suggest a great future for GS in the breeding of salt-tolerant crop varieties.

8.5 Genomics-Assisted Breeding for Salt Tolerance in the Era of Genotyping based on High-Throughput Next-Generation Sequencing

In plant selection via GAB, it is essential to distinguish the precise location and foremost effect of the gene/QTL on the trait of interest. Hence, the genotyping technique based on next-generation sequencing (NGS) has a great role to play in improving the gene-mapping resolution as well as the labelling of the gene/QTL near the linked maker (Bhat *et al.*, 2016). This is because NGS provides huge marker density for genotyping a large population size of crop plants in the case of genome-wide association studies (GWAS), thereby improving the mapping resolution as well as refining the tight association between the marker and the QTL/gene. Using a varied and large population in the GWAS analysis permits more recombination breakpoints to be identified, which in turn assists in the identification of specific candidate genes with greater accuracy (Zargar *et al.*, 2015). Numerous investigations have used NGS-based genotyping in various crop species for different traits in conducting GWAS. It has been reported that NGS-based high-throughput genotyping has increased the resolution as well as the mapping of

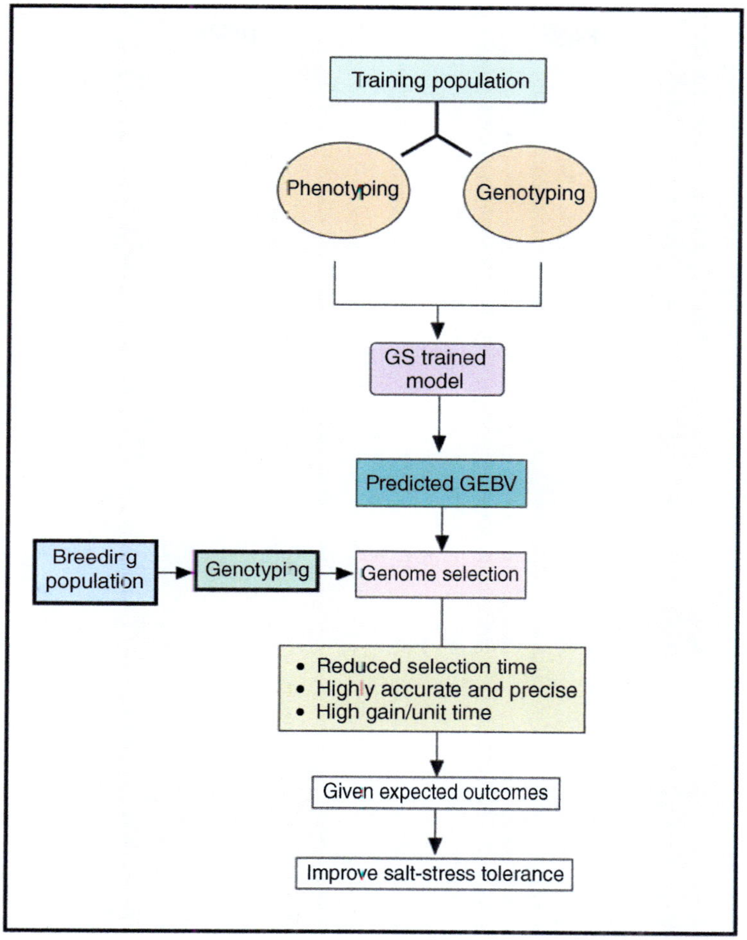

Fig. 8.1. Diagrammatic presentation of GS steps and its usage for crop improvement.

the underlying candidate genes for various traits including salt stress (Oyiga *et al.*, 2018). This approach has been effectively used to elucidate the genetic basis of salt tolerance at high resolution and mapping accuracy in different crop species such as camelina (Luo *et al.*, 2020), sesame (Li, D. *et al.*, 2018), wheat (Alotaibi *et al.*, 2022), and rice (Yuan *et al.*, 2020). Moreover, the NGS-based genotyping technique allowed researchers to identify candidate genes for nitrogen fixation (Torkamaneh *et al.*, 2021) and to improve plant height (Borah *et al.*, 2018), disease resistance (Mwando *et al.*, 2020), and protein content regulation (Sui and Wang, 2020). To this end, the use of bulk segregant analysis (BSA) and its amended techniques in trait mapping has been made successful by the use of the NGS approaches (Gu *et al.*, 2018; Majeed *et al.*, 2022). For example, the BSA-seq/QTL-seq has been used in rice to identify the major loci and candidate gene regulating salt tolerance at the bud burst stage (Lei *et al.*, 2020). Similarly, Zhang *et al.* (2022) used BSA-seq for the identification of candidate genes regulating salt tolerance in sorghum.

Therefore, as compared to the prior marker system, NGS technology is considered as highly efficient and precise for gene mapping as well as candidate gene identification (Bhat and Yu, 2021). Besides, it has the capability to simultaneously conduct discovery, genotyping, and validation of

Table 8.2. Genome selection-mediated marker–trait prediction as well as associations in different plant species.

Species type	High-throughput genotype platform[a]	Trait	Population size	Model[b]	Total SNPs	References
Wheat	GBS	Rust resistance in stem	365	G-BLUP	4,040	Rutkoski *et al.* (2014)
Soybean	SNP-Chip	Yield improvement	483	RR-BLUP	2,647	Stewart-Brown *et al.* (2019)
Rice	GBS	Flowering; grain yield improvement	363	RR-BLUP	73,147	Spindel *et al.* (2015)
Soybean	GBS	Grain yield improvement	139	G-BLUP	2,395	Jarquín *et al.* (2014)
Wheat	GBS	Grain yield improvement	254	BLUP	41,371	Poland *et al.* (2012)
Wheat	DArTseq	Grain yield improvement	803	G-BLUP	–	Saint Pierre *et al.* (2016)
Wheat	GBS	Grain yield; Fusarium head	470	BLUP	4,858	Hoffstetter *et al.* (2016)
Wheat	GBS	Heat; drought stress	10,819	G-BLUP	40,000	Crossa *et al.* (2016)
Soybean	GBS	Improve amino acid contents	249	RR-BLUP	23,279	Qin *et al.* (2019)
Soybean	SNP-Chip	Yield-related traits	235	RR-BLUP	5,361	Ma *et al.* (2016)
Soybean	SNP-Chip	Grain yield improvement	1,248	G-BLUP	4,141	Duhnen *et al.* (2017)
Soybean	SNP-Chip	Yield-related traits improvement	324	eBLUP	4,947	Matei *et al.* (2018)
Soybean	SNP-Chip	Yield and protein contents	227	RR-BLUP and Bayesian models	3,000	Đorđević *et al.* (2019)
Soybean	SNP-Chip	Nematode resistance	234	G-BLUP	3,782	Ravelombola *et al.* (2019)
Soybean	SNP-Chip	Seed weight enhancement	309	RR-BLUP	31,045	Zhang *et al.* (2016)
Soybean	GBS	Improved plant height; maturity	301	G-BLUP	8,140	Jarquín *et al.* (2019)
Maize	GBS	Drought-stress tolerance	3,273	PGBLUP, PRKHS	58,731	Zhang *et al.* (2015, 2017)
Maize	DArTseq	Ear rot disease resistance	238	PGBLUP, PRKHS	235,265	Dos Santos *et al.* (2016)
Lucerne (alfalfa)	GBS	Biomass yield enhancement	190	BLUP	10,000	Li *et al.* (2015)
Maize	GBS	Grain yield; anthesis date	504	PGBLUP, PRKHS	158,281	Crossa *et al.* (2016)
Lucerne (alfalfa)	GBS	Biomass augmentation	278	SVR	10,000	Annicchiarico *et al.* (2015)
Maize	GBS	Grain yield; anthesis date	296	PGBLUP, PRKHS	235,265	Crossa *et al.* (2013)
Perennial ryegrass	GBS	Plant herbage dry weight	211	RR-BLUP	10,885	Faville *et al.* (2016)
Miscanthus	RADseq	Phenology; cell-wall composition; biomass accumulation	138	BLUP	20,000	Slavov *et al.* (2014)
Intermediate wheat grass	GBS	Yield and other agronomic traits	1,126	RR-BLUP	3,883	Zhang *et al.* (2017)

[a]SNP-Chip, single nucleotide polymorphism-chip; DArTseq, diversity array technology sequencing; RADseq, restriction site-associated DNA sequencing.
[b]RR-BLUP, ridge-regression best linear unbiased prediction; G-BLUP, genomic best linear unbiased prediction; BLUP, best linear unbiased prediction; eBLUP, empirical best linear unbiased prediction; PGBLUP, genomic best linear unbiased predictor; PRKHS, pedigree reproducing kernel Hilbert spaces; SVR, support vector regression.

Table 8.3. Transcriptional modifications and their mode of action in various crop plants.

Crop	Transcriptional modifications (transcription factors/genes/proteins)	Functional modifications	References
Solanum chilense	WRKY1, WRKY3, and WRKY72	Salinity-stress tolerance	Kashyap *et al.* (2020)
Arabidopsis thaliana	Transgenic overexpression of OsBTBZ1 in Arabidopsis	Salt-stress tolerance	Saputro *et al.* (2023)
S. chilense and *Solanum lycopersicum*	ScbZIP and SlbZIP	Salt-stress tolerance	Zhu *et al.* (2018); Kashyap *et al.* (2020)
Brassica rapa	Genome-wide identification and characterization of WRKYs against salt stress	Salinity-stress tolerance	Kayum *et al.* (2015)
S. lycopersicum and *A. thaliana*	Tomato DREB2	Salt tolerance	Hichri *et al.* (2016)
S. lycopersicum	Tomato ERF84, ERF5	Salt-stress tolerance	Pan *et al.* (2012)
Raphanus sativus	Radish WRKY	Salinity-stress tolerance	Karanja *et al.* (2017)

single-nucleotide polymorphisms (SNPs), by such means as genotyping by sequencing (GBS), in a mapping or mutant population; that in turn reduces the ascertainment bias (Bhat *et al.*, 2016). Furthermore, NGS-based approaches possess the prodigious aptitude to determine genome duplication concerns in an intricate as well as paleopolyploid crop (Oyiga *et al.*, 2018). Moreover, the more advanced approach of GAB known as GS was almost impossible before the emergence of NGS-based genotyping. Because the GS approach requires genome-wide markers at higher density to achieve accuracy in the genomic prediction of the desirable cultivar, in this regard the emergence of NGS has greatly facilitated use of the GS approach in crop development for both model and non-model crop species (Liu *et al.*, 2020). In conclusion, NGS-based genotyping has played an enormous role in increasing the efficiency and precision of the GAB approach for crop improvement, that was impossible to achieve before the emergence of NGS.

8.6 Speed Breeding Applications in Salt-Stress Mitigation

Salt stress restricts plant growth and thus yield by affecting ~20% of agricultural land every year worldwide. The exposure to salinity causes severe morpho-physiological as well as molecular damage by inducing ionic and osmotic stress (Ondrasek *et al.*, 2022). Plants develop some tolerance mechanisms to resist low or moderate salt-induced toxicity via reprogramming the various physiological, biochemical, molecular, and developmental processes (Yi *et al.*, 2022). However, under constant/severe salinity stress, plants' defence mechanism becomes dysfunctional; thus, it is required to develop a specific mechanism for sustainable crop production under salt stress. By getting an understanding of the genetic as well as molecular mechanisms against salt tolerance, scientists are trying to develop some crop improvement techniques according to the molecular signals towards cellular responses (Riyazuddin *et al.*, 2020; Ganie *et al.*, 2021). Speed breeding can open new vistas by applying the precise plant growth environment of the greenhouse to develop quicker generation advancements and evaluations of phenotypes under salt stress. The foremost benefit of using speed breeding under controlled environments is to lower the risk of experimental error and to minimize the effects of interactions among the environment and genotype on phenotypic traits (G × E interactions). Controlled conditions have the capability to monitor the stress initiation level such as the increase of NaCl concentration in water in the case of salt stress (Ahmad *et al.*, 2021). In contrast, a controlled environment does not mimic the natural environmental conditions, nor the cost-effective veracity compared to the field conditions, because the crops are grown in pots. However, salt-stress tolerance experiments can be conducted in a greenhouse and speed breeding can be integrated as a regular component of this application. Advanced research is focusing on lowering salt stress by incorporating various molecular techniques, i.e. transgenics, GS, genome editing, MAB, etc., at numerous plant growth stages to improve salt tolerance (Wanga *et al.*, 2021). Speed breeding provides a definite stress target against different environmental conditions, enabling calibration of screens for stress resistance, identification of superior parents, exposure of various mechanisms of salt-stress tolerance, identification as well as classification of genes conferring tolerance, pyramiding diverse tolerance in elite lines by MAB and breeding line assessment under the target environmental stress, etc. (Rana *et al.*, 2019). Consequently, speed breeding can offer a better scope to increase tolerance to salt-induced stress and to cultivate improved genotypes in different crops like rice, oilseeds, cereals, pulses, and vegetables at a faster rate.

8.7 Integrated Approach of Speed Breeding and Genomics-Assisted Breeding for Development of Salt-Tolerant Cultivars

A lot of development and progress has been made in the field of GAB. The availability of high-quality and high-density DNA markers distributed across the whole genome can be easily offered by cost-efficient and advanced sequencing procedures (Wang *et al.*, 2018). The high-throughput technologies have made plant breeding more advanced; for example, the application of GS in crop breeding has changed the entire paradigm for the development of amended crop cultivars. In this regard, the integration of speed breeding with GAB will greatly accelerate the process for the advancement of superior crop cultivars because speed breeding will allow more generations of crop plants to be produced in much less time as compared to the conventional crop cultivation approaches. For instance, Watson *et al.* (2019) verified that speed breeding allows the production of six generations per year for pea, barley, durum, chickpea, and spring wheat and four generations per year for the crop canola, which otherwise produce only two or three generations per year under normal glasshouse conditions. Hence, these authors suggested that combining speed breeding with the modern breeding approaches of GAB and genome editing will tremendously gear up the production of modified crop cultivars. Rana *et al.* (2019) coupled speed breeding with MAS for the development of salt-tolerant rice cultivars. They demonstrated 'YNU31-2-4' as a potential candidate for a new salt-tolerant cultivar in rice especially at the seedling and reproductive stages. Under salt stress at the reproductive stage, the 'YNU31-2-4' plants displayed an enhanced phenotype with noticeably higher net CO_2 assimilation and less yield drop than the wild type. 'YNU31-2-4' is a prospective, new rice cultivar that has high salt-stress tolerance during the seedling and reproductive stages and that may be able to maintain yields in the face of uncertain climatic conditions (Rana *et al.*, 2019). Chimmili *et al.* (2022) elaborated that an integrated breeding approach which contains novel methods such as GAB, speed breeding, and GS enables quick generation and evaluation of inbred lines. Additionally, this suggests lessening the generation cycle in a number of crops, including rice, to speed up generation advancement; and furthermore, it accelerates the breeding cycle, enabling even faster generational growth. An integrated breeding strategy could include fast line derivation, phenotyping of yield-related factors, and indirect phenotypic as well as multivariate genomic selection for pertinent traits before field evaluation in a controlled environment.

In wheat, the combined approach of genome selection and speed breeding has been revealed to improve the yield considerably (Watson *et al.*, 2019). In addition, the combined approaches of genome selection and speed breeding have also been utilized to evaluate the recombinant breeding lines and unintended phenotypical assortment to increase traits such as height, biomass, flowering time, grain quality, and quantity as well (Watson *et al.*, 2019). Moreover, Dr Lee Hickey and colleagues from The University of Queensland, Australia have established a novel methodology that integrates two different technologies, i.e. speed

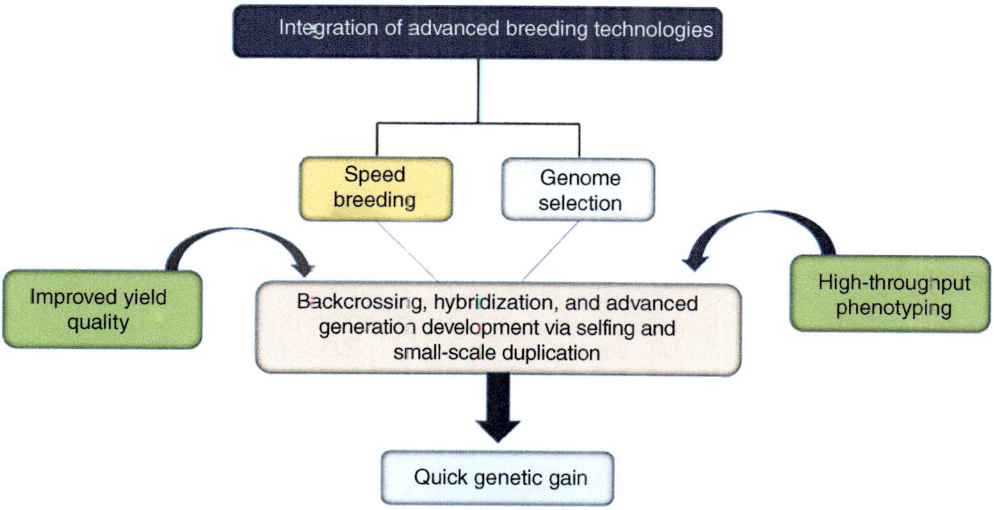

Fig. 8.2. Integration of genome selection and speed breeding in advanced breeding technologies for quick genetic gains.

breeding and genome selection, to gain higher quality and yield (Ahmad *et al*., 2021). Overall, the integrated approach of GAB and speed breeding has tremendous future potential to overcome the challenges of global food security (Fig. 8.2).

8.8 Conclusion

Until now, an important contribution of GAB has been noticed in the field of molecular breeding for the production of improved cultivars. Thus, GAB especially GS has emerged as a potential technique to breed crop plants for complex, economically important traits. The success of GAB and GS goes back to the NGS technology, which has provided low-cost high-density SNP genotyping platforms. To this end, speed breeding has provided the best alternative to attain a greater annual number of generations for crop plants; that in turn has accelerated a higher rate of improvement in agricultural varieties. The speed breeding has opened ways for the full exploitation of GAB in crop improvement. Moreover, the integrated approach of speed breeding and GAB has been demonstrated to have tremendous potential for the production of cultivars that are tolerant of stress, such as salt-tolerant crop cultivars. Hence, GAB and speed breeding will be the future for crop breeding and maintaining global food security.

References

Aghaei, K. and Komatsu, S. (2013) Crop and medicinal plants proteomics in response to salt stress. *Frontiers in Plant Science* 4, 8.

Ahanger, M.A. and Agarwal, R. (2017) Salinity stress induced alterations in antioxidant metabolism and nitrogen assimilation in wheat (*Triticum aestivum* L.) as influenced by potassium supplementation. *Plant Physiology and Biochemistry* 115, 449–460.

Ahmad, S., Makhmale, S., Bosamia, T.C., Sangh, C. and Nawade, B. (2021) Speed breeding: a potential tool for mitigating abiotic stresses. In: Rai, A.C., Rai, A., Rai, K.K., Rai, V.P. and Kumar, A. (eds) *Stress Tolerance in Horticultural Crops: Challenges and Mitigation Strategies*. Elsevier, Cambridge, pp. 51–61.

Alotaibi, F.S., Al-Qthanin, R.N., Aljabri, M., Shehzad, T., Albaqami, M. and Abou-Elwafa, S.F. (2022) Identification of genomic regions associated with agronomical traits of bread wheat under two levels of salinity using GWAS. *Plant Molecular Biology Reporter* 40, 595–609.

Annicchiarico, P., Nazzicari, N., Li, X., Wei, Y., Pecetti, L. and Brummer, E.C. (2015) Accuracy of genomic selection for alfalfa biomass yield in different reference populations. *BMC Genomics* 16, 1020.

Annicchiarico, P., Nazzicari, N., Pecetti, L., Romani, M., Ferrari, B. *et al.* (2017) GBS-based genomic selection for pea grain yield under severe terminal drought. *The Plant Genome* 10, plantgenome2016.07.0072.

Babu, N.N., Krishnan, S.G., Vinod, K.K., Krishnamurthy, S.L., Singh, V.K. *et al.* (2017) Marker aided incorporation of *Saltol*, a major QTL associated with seedling stage salt tolerance, into *Oryza sativa* 'Pusa Basmati 1121'. *Frontiers in Plant Science* 8, 41.

Bankole, F., Menkir, A., Olaoye, G., Crossa, J., Hearne, S. *et al.* (2017) Genetic gains in yield and yield related traits under drought stress and favorable environments in a maize population improved using marker assisted recurrent selection. *Frontiers in Plant Science* 8, 808.

Bartholomé, J., Frouin, J., Brottier, L., Cao, T.V., Boisnard, A. *et al.* (2023) Genomic selection for salinity tolerance in japonica rice. *PLoS ONE* 18, e0291833.

Begum, N., Qin, C., Ahanger, M.A., Raza, S., Khan, M.I. *et al.* (2019) Role of arbuscular mycorrhizal fungi in plant growth regulation: Implications in abiotic stress tolerance. *Frontiers in Plant Science* 10, 1068.

Ben Abdallah, S., Aung, B., Amyot, L., Lalin, I., Lachâal, M. *et al.* (2016) Salt stress (NaCl) affects plant growth and branch pathways of carotenoid and flavonoid biosyntheses in *Solanum nigrum*. *Acta Physiologiae Plantarum* 38, 1–13.

Beyene, Y., Semagn, K., Mugo, S., Tarekegne, A., Babu, R. *et al.* (2015) Genetic gains in grain yield through genomic selection in eight bi-parental maize populations under drought stress. *Crop Science* 55, 154–163.

Bhat, J.A. and Yu, D. (2021) High-throughput NGS-based genotyping and phenotyping: role in genomics-assisted breeding for soybean improvement. *Legume Science* 3, e81.

Bhat, J.A., Ali, S., Salgotra, R.K., Mir, Z.A., Dutta, S. *et al.* (2016) Genomic selection in the era of next generation sequencing for complex traits in plant breeding. *Frontiers in Genetics* 7, 221.

Borah, P., Das, A., Milner, M.J., Ali, A., Bentley, A.R. and Pandey, R. (2018) Long non-coding RNAs as endogenous target mimics and exploration of their role in low nutrient stress tolerance in plants. *Genes (Basel)* 9, 459.

Chimmili, S.R., Kanneboina, S., Hanjagi, P.S., Basavaraj, P.S., Sakhare, A.S. *et al.* (2022) Integrating advanced molecular, genomic, and speed breeding methods for genetic improvement of stress tolerance in rice. In: Gowdra Mallikarjuna, M., Nayaka, S.C. and Kaul, T. (eds) *Next-Generation Plant Breeding Approaches for Stress Resilience in Cereal Crops*. Springer, Singapore, pp. 263–283.

Collard, B.C., Beredo, J.C., Lenaerts, B., Mendoza, R., Santelices, R. *et al.* (2017) Revisiting rice breeding methods – evaluating the use of rapid generation advance (RGA) for routine rice breeding. *Plant Production Science* 20, 337–352.

Crossa, J., Beyene, Y., Kassa, S., Pérez, P., Hickey, J.M. *et al.* (2013) Genomic prediction in maize breeding populations with genotyping-by-sequencing. *G3: Genes, Genomes, Genetics* 3, 1903–1926.

Crossa, J., Jarquín, D., Franco, J., Pérez-Rodríguez, P., Burgueño, J. *et al.* (2016) Genomic prediction of gene bank wheat landraces. *G3: Genes, Genomes, Genetics* 6, 1819–1834.

Das, S., Upadhyaya, H.D., Srivastava, R., Bajaj, D., Gowda, C.L.L. *et al.* (2015) Genome-wide insertion–deletion (InDel) marker discovery and genotyping for genomics-assisted breeding applications in chickpea. *DNA Research* 22, 377–386.

Davis, P.A. and Burns, C. (2016) Photobiology in protected horticulture. *Food and Energy Security* 5, 223–238.

Dehghan, S., Sadeghi, M., Pöppel, A., Fischer, R., Lakes-Harlan, R. *et al.* (2014) Differential inductions of phenylalanine ammonia-lyase and chalcone synthase during wounding, salicylic acid treatment, and salinity stress in safflower, *Carthamus tinctorius*. *Bioscience Reports* 34, 273–282.

Đorđević, V., Ćeran, M., Miladinović, J., Balešević-Tubić, S., Petrović, K. *et al.* (2019) Exploring the performance of genomic prediction models for soybean yield using different validation approaches. *Molecular Breeding* 39, 74.

Dos Santos, J.P.R., Pires, L.P.M., de Castro Vasconcellos, R.C., Pereira, G.S., Von Pinho, R.G. and Balestre, M. (2016) Genomic selection to resistance to *Stenocarpella maydis* in maize lines using DArTseq markers. *BMC Genetics* 17, 86.

Duhnen, A., Gras, A., Teyssèdre, S., Romestant, M., Claustres, B. *et al.* (2017) Genomic selection for yield and seed protein content in soybean: a study of breeding program data and assessment of prediction accuracy. *Crop Science* 57, 1325–1337.

Faville, M.J., Ganesh, S., Moraga, R., Easton, H.S., Jahufer, M.Z.Z. *et al.* (2016) Development of genomic selection for perennial ryegrass. In: Roldán-Ruiz, I., Baert, J. and Reheul, D. (eds) *Breeding in a World of Scarcity: Proceedings of the 2015 Meeting of the Section 'Forage Crops and Amenity Grasses' of Eucarpia*. Springer, Cham, Switzerland, pp. 139–143.

Forster, B.P., Till, B.J., Ghanim, A.M.A., Huynh, H.O.A., Burstmayr, H. and Caligari, P.D.S. (2015) Accelerated plant breeding. *CABI Reviews* 9, 043.

Gamalero, E. and Glick, B.R. (2022) Recent advances in bacterial amelioration of plant drought and salt stress. *Biology* 11, 437.

Ganie, S.A. (2020) RNA chaperones: potential candidates for engineering salt tolerance in rice. *Crop Science* 60(2), 530–540.

Ganie, S.A. (2021) Amino acids other than proline and their participation in abiotic stress tolerance. In: Wani, S.H., Gangola, M.P. and Ramadoss, B.R. (eds) *Compatible Solutes Engineering for Crop Plants Facing Climate Change*. Springer, Cham, Switzerland, pp. 47–96.

Ganie, S.A., Wani, S.H., Henry, R. and Hensel, G. (2021) Improving rice salt tolerance by precision breeding in a new era. *Current Opinion in Plant Biology* 60, 101996.

Gengmao, Z., Yu, H., Xing, S., Shihui, L., Quanmei, S. and Changhai, W. (2015) Salinity stress increases secondary metabolites and enzyme activity in safflower. *Industrial Crops and Products* 64, 175–181.

Gu, X.H., Jiang, D.L., Huang, Y., Li, B.J., Chen, C.H. *et al.* (2018) Identifying a major QTL associated with salinity tolerance in Nile tilapia using QTL-Seq. *Marine Biotechnology* 20, 98–107.

Hale, A.L., White, P.M., Webber, C.L. 3rd and Todd, J.R. (2017) Effect of growing media and fertilization on sugarcane flowering under artificial photoperiod. *PLoS ONE* 12, e0181639.

Hasan, M.M., Rafii, M.Y., Ismail, M.R., Mahmood, M., Rahim, H.A. *et al.* (2015) Marker-assisted backcrossing: a useful method for rice improvement. *Biotechnology and Biotechnological Equipment* 29, 237–254.

Hasan, N., Choudhary, S., Naaz, N., Sharma, N. and Laskar, R.A. (2021) Recent advancements in molecular marker-assisted selection and applications in plant breeding programmes. *Journal of Genetic Engineering and Biotechnology* 19, 128.

Hichri, I., Muhovski, Y., Clippe, A., Žižková, E., Dobrev, P.I. *et al.* (2016) SlDREB2, a tomato dehydration-responsive element-binding 2 transcription factor, mediates salt stress tolerance in tomato and *Arabidopsis*. *Plant, Cell & Environment* 39, 62–79.

Ho, V.T., Thomson, M.J. and Ismail, A.M. (2016) Development of salt tolerant IR64 near isogenic lines through marker-assisted breeding. *Journal of Crop Science and Biotechnology* 19, 373–381.

Hoffstetter, A., Cabrera, A. and Sneller, C. (2016) Identifying quantitative trait loci for economic traits in an elite soft red winter wheat population. *Crop Science* 56, 547–558.

Irving, H., Riveron, J., Ibrahim, S., Lobo, N. and Wondji, C. (2012) Positional cloning of *rp2* QTL associates the *P450* genes *CYP6Z1, CYP6Z3* and *CYP6M7* with pyrethroid resistance in the malaria vector *Anopheles funestus*. *Heredity* 109, 383–392.

Jähne, F., Hahn, V., Würschum, T. and Leiser, W.L. (2020) Speed breeding short-day crops by LED-controlled light schemes. *Theoretical and Applied Genetics* 133, 2335–2342.

Janaki Ramayya, P., Vinukonda, V.P., Singh, U.M., Alam, S., Venkateshwarlu, C. *et al.* (2021) Marker-assisted forward and backcross breeding for improvement of elite Indian rice variety Naveen for multiple biotic and abiotic stress tolerance. *PLoS ONE* 16, e0256721.

Jannink, J.L., Lorenz, A.J. and Iwata, H. (2010) Genomic selection in plant breeding: from theory to practice. *Briefings in Functional Genomics* 9, 166–177.

Jarquín, D., Kocak, K., Posadas, L., Hyma, K., Jedlicka, J. *et al.* (2014) Genotyping by sequencing for genomic prediction in a soybean breeding population. *BMC Genomics* 15, 740.

Jarquín, D., Howard, R., Graef, G. and Lorenz, A. (2019) Response surface analysis of genomic prediction accuracy values using quality control covariates in soybean. *Evolutionary Bioinformatics Online* 15, 1176934319831307.

Javid, S., Bihamta, M.R., Omidi, M., Abbasi, A.R., Alipour, H. and Ingvarsson, P.K. (2022) Genome-wide association study (GWAS) and genome prediction of seedling salt tolerance in bread wheat (*Triticum aestivum* L.). *BMC Plant Biology* 22, 581.

Karanja, B.K., Fan, L., Xu, L., Wang, Y., Zhu, X. *et al.* (2017) Genome-wide characterization of the *WRKY* gene family in radish (*Raphanus sativus* L.) reveals its critical functions under different abiotic stresses. *Plant Cell Reports* 36, 1757–1773.

Kashyap, S., Kumari, N., Mishra, P., Moharana, D.P., Aamir, M. *et al.* (2020) Transcriptional regulation-mediating ROS homeostasis and physio-biochemical changes in wild tomato (*Solanum chilense*) and cultivated tomato (*Solanum lycopersicum*) under high salinity. *Saudi Journal of Biological Sciences* 27, 1999–2009.

Kayum, M.A., Jung, H.J., Park, J.I., Ahmed, N.U., Saha, G. *et al.* (2015) Identification and expression analysis of WRKY family genes under biotic and abiotic stresses in *Brassica rapa*. *Molecular Genetics and Genomics* 290, 79–95.

Khan, A.L., Waqas, M., Asaf, S., Kamran, M., Shahzad, R. *et al.* (2017) Plant growth-promoting endophyte *Sphingomonas* sp. LK11 alleviates salinity stress in *Solanum pimpinellifolium*. *Environmental and Experimental Botany* 133, 58–69.

Kibria, M.G., Hossain, M., Murata, Y. and Hoque, M.A. (2017) Antioxidant defense mechanisms of salinity tolerance in rice genotypes. *Rice Science* 24, 155–162.

Kim, M.-S., Yu, J.-K., Ko, S.-R., Kim, K.-J., Ji, H. *et al.* (2022) Marker-assisted backcrossing (MABc) to improve eating quality with thin seed coat and aleurone layer of non-glutinous japonica variety in rice. *Genes (Basel)* 13, 210.

Lamalakshmi Devi, E., Kumar, S., Basanta Singh, T., Sharma, S.K., Beemrote, A. *et al.* (2017) Adaptation strategies and defence mechanisms of plants during environmental stress. In: Ghorbanpour, M. and Varma, A. (eds) *Medicinal Plants and Environmental Challenges*. Springer, Cham, Switzerland, pp. 359–413.

Lei, L., Zheng, H., Bi, Y., Yang, L., Liu, H. *et al.* (2020) Identification of a major QTL and candidate gene analysis of salt tolerance at the bud burst stage in rice (*Oryza sativa* L.) using QTL-Seq and RNA-Seq. *Rice* 13, 55.

Li, D., Dossa, K., Zhang, Y., Wei, X., Wang, L. *et al.* (2018) GWAS uncovers differential genetic bases for drought and salt tolerances in sesame at the germination stage. *Genes (Basel)* 9, 87.

Li, X., Wei, Y., Acharya, A., Hansen, J.L., Crawford, J.L. *et al.* (2015) Genomic prediction of biomass yield in two selection cycles of a tetraploid alfalfa breeding population. *The Plant Genome* 8, plantgenome2014.12.0090.

Li, Y., Ruperao, P., Batley, J., Edwards, D., Khan, T. *et al.* (2018) Investigating drought tolerance in chickpea using genome-wide association mapping and genomic selection based on whole-genome resequencing data. *Frontiers in Plant Science* 9, 190.

Liu, J., Shen, F., Xiao, Y., Fang, H., Qiu, C. *et al.* (2020) Genomics-assisted prediction of salt and alkali tolerances and functional marker development in apple rootstocks. *BMC Genomics* 21, 550.

Lorenz, A.J., Chao, S., Asoro, F.G., Heffner, E.L., Hayashi, T. *et al.* (2011) Genomic selection in plant breeding: knowledge and prospects. *Advances in Agronomy* 110, 77–123.

Lozada, D.N., Mason, R.E., Sarinelli, J.M. and Brown-Guedira, G. (2019) Accuracy of genomic selection for grain yield and agronomic traits in soft red winter wheat. *BMC Genetics* 20, 82.

Luo, Z., Szczepanek, A. and Abdel-Haleem, H. (2020) Genome-wide association study (GWAS) analysis of camelina seedling germination under salt stress condition. *Agronomy* 10, 1444.

Ma, Y., Reif, J.C., Jiang, Y., Wen, Z., Wang, D. *et al.* (2016) Potential of marker selection to increase prediction accuracy of genomic selection in soybean (*Glycine max* L.). *Molecular Breeding* 36, 113.

Majeed, A., Johar, P., Raina, A., Salgotra, R.K., Feng, X. and Bhat, J.A. (2022) Harnessing the potential of bulk segregant analysis sequencing and its related approaches in crop breeding. *Frontiers in Genetics* 13, 944501.

Matei, G., Woyann, L.G., Milioli, A.S., de Bem Oliveira, I., Zdziarski, A.D. *et al.* (2018) Genomic selection in soybean: accuracy and time gain in relation to phenotypic selection. *Molecular Breeding* 38, 117.

Meuwissen, T.H., Hayes, B.J. and Goddard, M.E. (2001) Prediction of total genetic value using genome-wide dense marker maps. *Genetics* 157, 1819–1829.

Mobini, S.H. and Warkentin, T.D. (2016) A simple and efficient method of in vivo rapid generation technology in pea (*Pisum sativum* L.). *In Vitro Cellular & Developmental Biology – Plant* 52, 530–536.

Mobini, S.H., Lulsdorf, M., Warkentin, T.D. and Vandenberg, A. (2015) Plant growth regulators improve in vitro flowering and rapid generation advancement in lentil and faba bean. *In Vitro Cellular & Developmental Biology – Plant* 51, 71–79.

Moniruzzaman, M., Alam, M.S., Rashid, J.A., Begum, S.N. and Islam, M.M. (2012) Marker-assisted backcrossing for identification of salt tolerant rice lines. *International Journal of Agricultural Research, Innovation and Technology* 2(2), 1–8.

Munns, R. and Tester, M. (2008) Mechanisms of salinity tolerance. *Annual Review of Plant Biology* 59, 651–681.

Mwando, E., Han, Y., Angessa, T.T., Zhou, G., Hill, C.B. *et al.* (2020) Genome-wide association study of salinity tolerance during germination in barley (*Hordeum vulgare* L.). *Frontiers in Plant Science* 11, 118.

Naing, A.H. and Kim, C.K. (2021) Abiotic stress-induced anthocyanins in plants: their role in tolerance to abiotic stresses. *Physiologia Plantarum* 172, 1711–1723.

Ochatt, S.J., Sangwan, R.S., Marget, P., Ndong, Y.A., Rancillac, M. *et al.* (2002) New approaches towards the shortening of generation cycles for faster breeding of protein legumes. *Plant Breeding* 121, 436–440.
Ondrasek, G., Rathod, S., Manohara, K.K. Gireesh, C., Anantha, M.S. *et al.* (2022) Salt stress in plants and mitigation approaches. *Plants* 11, 717.
Oyiga, B.C., Sharma, R.C., Baum, M., Ogbonnaya, F.C., Léon, J. and Ballvora, A. (2018) Allelic variations and differential expressions detected at quantitative trait loci for salt stress tolerance in wheat. *Plant, Cell & Environment* 41, 919–935.
Pan, Y., Seymour, G.B., Lu, C., Hu, Z., Chen, X. and Chen, G. (2012) An ethylene response factor (ERF5) promoting adaptation to drought and salt tolerance in tomato. *Plant Cell Reports* 31, 349–360.
Pfeiffer, N.E. (1926) Microchemical and morphological studies of effect of light on plants. *Botanical Gazette* 81, 173–195.
Poland, J., Endelman, J., Dawson, J., Rutkoski, J., Wu, S. *et al.* (2012) Genomic selection in wheat breeding using genotyping-by-sequencing. *The Plant Genome* 5, 103–113.
Prittesh, P., Avnika, P., Kinjal, P., Jinal, H.N., Sakthivel, K. and Amaresan, N. (2020) Amelioration effect of salt-tolerant plant growth-promoting bacteria on growth and physiological properties of rice (*Oryza sativa*) under salt-stressed conditions. *Archives of Microbiology* 202, 2419–2428.
Qin, J., Shi, A., Song, Q., Li, S., Wang, F. *et al.* (2019) Genome wide association study and genomic selection of amino acid concentrations in soybean seeds. *Frontiers in Plant Science* 10, 1445.
Rahman, M., Davies, P., Bansal, U., Pasam, R., Hayden, M. and Trethowan, R. (2020) Marker-assisted recurrent selection improves the crown rot resistance of bread wheat. *Molecular Breeding* 40, 28.
Rana, M.M., Takamatsu, T., Baslam, M., Kaneko, K., Itoh, K. *et al.* (2019) Salt tolerance improvement in rice through efficient SNP marker-assisted selection coupled with speed-breeding. *International Journal of Molecular Sciences* 20, 2585.
Ravelombola, W.S., Qin, J., Shi, A., Nice, L. Bao, Y. *et al.* (2019) Genome-wide association study and genomic selection for soybean chlorophyll content associated with soybean cyst nematode tolerance. *BMC Genomics* 20, 904.
Ren, J., Ye, J., Yin, L., Li, G., Deng, X. and Wang, S. (2020) Exogenous melatonin improves salt tolerance by mitigating osmotic, ion, and oxidative stresses in maize seedlings. *Agronomy* 10, 663.
Riyazuddin, R., Verma, R., Singh, K., Nisha, N., Keisham, M. *et al.* (2020) Ethylene: a master regulator of salinity stress tolerance in plants. *Biomolecules* 10, 959.
Rutkoski, J.E., Heffner, E.L. and Sorrells, M.E. (2011) Genomic selection for durable stem rust resistance in wheat. *Euphytica* 179, 161–173.
Rutkoski, J.E., Poland, J.A., Singh, R.P., Huerta-Espino, J., Bhavani, S. *et al.* (2014) Genomic selection for quantitative adult plant stem rust resistance in wheat. *The Plant Genome* 7, plantgenome2014.02.0006.
Saint Pierre, C., Burgueño, J., Crossa, J., Fuentes Dávila, G., Figueroa López, P. *et al.* (2016) Genomic prediction models for grain yield of spring bread wheat in diverse agro-ecological zones. *Scientific Reports* 6, 27312.
Saputro, T.B., Jakada, B.H., Chutimanukul, P., Comai, L., Buaboocha, T. and Chadchawan, S. (2023) *OsBTBZ1* confers salt stress tolerance in *Arabidopsis thaliana*. *International Journal of Molecular Sciences* 24, 14483.
Slavov, G.T., Nipper, R., Robson, P., Farrar, K., Allison, G.G. *et al.* (2014) Genome-wide association studies and prediction of 17 traits related to phenology, biomass and cell wall composition in the energy grass *Miscanthus sinensis*. *New Phytologist* 201, 1227–1239.
Soliman, M.H., Abdulmajeed, A.M., Alhaithloul, H., Alharbi, B.M., El-Esawi, M.A. *et al.* (2020) Saponin biopriming positively stimulates antioxidants defense, osmolytes metabolism and ionic status to confer salt stress tolerance in soybean. *Acta Physiologiae Plantarum* 42, 114.
Song, Y., Duan, X., Wang, P., Li, X., Yuan, X. *et al.* (2022) Comprehensive speed breeding: a high-throughput and rapid generation system for long-day crops. *Plant Biotechnology Journal* 20, 13–15.
Spindel, J., Begum, H., Akdemir, D., Virk, P., Collard, B. *et al.* (2015) Genomic selection and association mapping in rice (*Oryza sativa*): effect of trait genetic architecture, training population composition, marker number and statistical model on accuracy of rice genomic selection in elite, tropical rice breeding lines. *PLoS Genetics* 11, e1004982.
Stetter, M.G., Zeitler, L., Steinhaus, A., Kroener, K., Biljeck, M. and Schmid, K.J. (2016) Crossing methods and cultivation conditions for rapid production of segregating populations in three grain amaranth species. *Frontiers in Plant Science* 7, 816.
Stewart-Brown, B.B., Song, Q., Vaughn, J.N. and Li, Z. (2019) Genomic selection for yield and seed composition traits within an applied soybean breeding program. *G3: Genes, Genomes, Genetics* 9, 2253–2265.

Sui, D. and Wang, B. (2020) Transcriptome analysis reveals complex defensive mechanisms in salt-tolerant and salt-sensitive shrub willow genotypes under salinity stress. *International Journal of Genomics* 2020, 6870157.

Tiede, T. and Smith, K.P. (2018) Evaluation and retrospective optimization of genomic selection for yield and disease resistance in spring barley. *Molecular Breeding* 38, 55.

Torkamaneh, D., Laroche, J., Boyle, B., Hyten, D.L. and Belzile, F. (2021) A bumper crop of SNPs in soybean through high-density genotyping-by-sequencing (HD-GBS). *Plant Biotechnology Journal* 19, 860.

Vu, H.T.T., Le, D.D., Ismail, A.M. and Le, H.H. (2012) Marker-assisted backcrossing (MABC) for improved salinity tolerance in rice (*Oryza sativa* L.) to cope with climate change in Vietnam. *Australian Journal of Crop Science* 6, 1649–1654.

Wang, X., Xu, Y., Hu, Z. and Xu, C. (2018) Genomic selection methods for crop improvement: current status and prospects. *Crop Journal* 6, 330–340.

Wanga, M.A., Shimelis, H., Mashilo, J. and Laing, M.D. (2021) Opportunities and challenges of speed breeding: a review. *Plant Breeding* 140, 185–194.

Watson, A., Ghosh, S., Williams, M.J., Cuddy, W.S., Simmonds, J. *et al.* (2018) Speed breeding is a powerful tool to accelerate crop research and breeding. *Nature Plants* 4, 23–29.

Watson, A., Hickey, L.T., Christopher, J., Rutkoski, J., Poland, J. and Hayes, B.J. (2019) Multivariate genomic selection and potential of rapid indirect selection with speed breeding in spring wheat. *Crop Science* 59, 1945–1959.

Xiao, X., Xu, Y., Zhang, X., Wang, F., Lu, X. *et al.* (2022) Amplified upward trend of the joint occurrences of heat and ozone extremes in China over 2013–20. *Bulletin of the American Meteorological Society* 103, E1330–E1342.

Xu, Z. and Rothstein, S.J. (2018) ROS-induced anthocyanin production provides feedback protection by scavenging ROS and maintaining photosynthetic capacity in *Arabidopsis*. *Plant Signaling & Behavior* 13(3), 1364–1377.

Yadav, A.K., Kumar, A., Grover, N., Ellur, R.K., Krishnan, S.G. *et al.* (2020) Marker aided introgression of 'Saltol', a major QTL for seedling stage salinity tolerance into an elite Basmati rice variety 'Pusa Basmati 1509'. *Scientific Reports* 10, 13877.

Ye, G. and Smith, K.F. (2008) Marker-assisted gene pyramiding for inbred line development: practical applications. *International Journal of Plant Breeding* 2, 11–22.

Yi, Y., Peng, Y., Song, T., Lu, S., Teng, Z. *et al.* (2022) NLP2-NR module associated NO is involved in regulating seed germination in rice under salt stress. *Plants* 11, 795.

Yuan, J., Wang, X., Zhao, Y., Khan, N.U., Zhao, Z. *et al.* (2020) Genetic basis and identification of candidate genes for salt tolerance in rice by GWAS. *Scientific Reports* 10, 9958.

Zahra, H., Naser, K., Masoud, M. and Saeed, M. (2012) Enhancement of compatible solute and secondary metabolites production in *Plantago ovata* Forsk. by salinity stress. *Journal of Medicinal Plant Research: Planta Medica* 6, 3495–3500.

Zargar, S.M., Raatz, B., Sonah, H., Bhat, J.A., Dar, Z.A. *et al.* (2015) Recent advances in molecular marker techniques: insight into QTL mapping, GWAS and genomic selection in plants. *Journal of Crop Science and Biotechnology* 18, 293–308.

Zhang, A., Liu, Y., Wang, F., Li, T., Chen, Z. *et al.* (2019) Enhanced rice salinity tolerance via CRISPR/Cas9-targeted mutagenesis of the *OsRR22* gene. *Molecular Breeding* 39, 47.

Zhang, G., Zhou, J., Peng, Y., Tan, Z., Zhang, Y. *et al.* (2023) High-throughput phenotyping-based quantitative trait loci mapping reveals the genetic architecture of the salt stress tolerance of *Brassica napus*. *Plant, Cell & Environment* 46, 549–566.

Zhang, J., Song, Q., Cregan, P.B. and Jiang, G.L. (2016) Genome-wide association study, genomic prediction and marker-assisted selection for seed weight in soybean (*Glycine max*). *Theoretical and Applied Genetics* 129, 117–130.

Zhang, K., Zhang, Z., Lu, F., Wang, Y., Ke, F. *et al.* (2022) Bulked segregant analysis-sequencing identification of candidate genes for salt tolerance at the seedling stage of sorghum (*Sorghum bicolor*). *Plant Breeding* 141, 366–378.

Zhang, X., Pérez-Rodríguez, P., Semagn, K., Beyene, Y., Babu, R. *et al.* (2015) Genomic prediction in biparental tropical maize populations in water-stressed and well-watered environments using low-density and GBS SNPs. *Heredity* 114, 291–299.

Zhang, X., Pérez-Rodríguez, P., Burgueño, J., Olsen, M., Buckler, E. *et al.* (2017) Rapid cycling genomic selection in a multiparental tropical maize population. *G3: Genes, Genomes, Genetics* 7, 2315–2326.

Zheng, J., Fan, J., Zhou, M., Zhang, F., Liao, Z. *et al*. (2022) Ridge–furrow plastic film mulching enhances grain yield and yield stability of rainfed maize by improving resources capture and use efficiency in a semi-humid drought-prone region. *Agricultural Water Management* 269, 107654.

Zheng, Z., Wang, H.B., Chen, G.D., Yan, G.J. and Liu, C.J. (2013) A procedure allowing up to eight generations of wheat and nine generations of barley per annum. *Euphytica* 191, 311–316.

Zhu, M., Meng, X., Cai, J., Li, G., Dong, T. and Li, Z. (2018) Basic leucine zipper transcription factor SlbZIP1 mediates salt and drought stress tolerance in tomato. *BMC Plant Biology* 18, 83.

9 Advances in Plant Phenotyping for Enhanced Salt Tolerance

Kehinde A. Adeboye[1,*], Oyeboade A. Oyetunde[2], Victor J. Bamisaye[1] and Emmanuel O. Idehen[3]

[1]Department of Agricultural Technology, Ekiti State Polytechnic, Isan, Nigeria; [2]Department of Horticulture, Lagos State University of Science and Technology, Ikorodu, Nigeria; [3]Department of Plant Breeding and Seed Technology, Federal University of Agriculture, Abeokuta, Nigeria

Abstract

Salinity stress is one of the most severe abiotic restrictions on crop productivity around the world. It is fundamentally caused by the accumulation of sodium, which may be natural as a result of geological and meteorological processes or caused as a result of human activities. Salt stress has a substantial impact on plant development and growth from the seed germination stage to maturity and, in extreme cases, may lead to plant death. Crop plants have evolved with various morphological and physiological characteristics that allow them to cope with and ameliorate the effects of salinity stress. The variability observed in these traits in response to salinity stress paved the way for breeding-based crop improvement strategies. However, phenotyping of plants' observable characteristics is a major constraint among others to efficient breeding programmes. This chapter explicates the technological advances in precision phenotyping of germplasm, and the strategies for integrating high-throughput phenotyping with genetic studies for enhanced salt tolerance in crops are discussed.

9.1 Introduction

Salinity stress is a major challenge to crop productivity worldwide, as it affects approximately 10% of the world's total cropland area (FAO, 2021). It occurs when high levels of salts, especially sodium, accumulate in the soil, either naturally or due to human activities such as irrigation with salty water or the use of fertilizers containing high levels of salts (Rouached *et al.*, 2015). The accumulation of salts in the soil affects the plant's ability to absorb water, leading to dehydration, osmotic stress, ion toxicity, and ultimately plant death (Safdar *et al.*, 2019; Ganie, 2020). Salinity stress can occur at any stage of crop development, from seed germination to maturity, and can significantly reduce crop yield (Panta *et al.*, 2014; Majeed and Muhammad, 2019).

Crop plants have developed different morphological and physiological traits to cope with salinity stress (Acosta-Motos *et al.*, 2017). These include reduced leaf area, thickened cuticles, enhanced root growth, and changes in stomatal behaviour, which allow the plant to conserve water, limit salt uptake, and maintain ion homeostasis in its cells (Munns and Tester, 2008; Gupta and Huang, 2014; Ganie, 2021). However, the variability in these traits among different crop species and varieties presents a challenge to breeding for salt tolerance.

*Email: adeboye@daad-alumni.de

© CAB International 2024. *The Genetics of Salt Tolerance in Plants: A Central Dogma Perspective and Strategies for Enhancement* (S. Ganie and S.H. Wani eds)
DOI: 10.1079/9781800623033.0009

Breeding programmes that aim to develop salt-tolerant crop varieties require identifying and selecting desirable traits associated with salt tolerance. However, this process is hampered by the lack of efficient and accurate methods for phenotyping plant traits (Tattaris *et al.*, 2016). Phenotyping is the process of observing and measuring a plant's physical and physiological traits, and it is essential in identifying salt-tolerance traits for breeding programmes (Passioura 2012; Paez-Garcia *et al.*, 2015). While valuable, traditional phenotyping methods such as visual scoring, biomass measurements, and yield assessments are often subjective, time-consuming, and require large amounts of resources (Furbank and Tester, 2011; Araus and Cairns, 2014; Chen *et al.*, 2014). Therefore, the development of high-throughput phenotyping technologies has become essential in improving the accuracy and efficiency of screening for salt tolerance.

Recent advances in precision phenotyping have revolutionized the field of plant breeding by improving the accuracy and efficiency of plant trait measurement (Fahlgren *et al.*, 2015). These technologies allow for the non-destructive and rapid assessment of plant traits, providing breeders with valuable data that can be used to select plants with desirable traits (Deery and Jones, 2021), such as salt tolerance. Furthermore, the combination of high-throughput phenotyping with genetic studies can enable the identification of genes and molecular mechanisms underlying salt tolerance (Chen *et al.*, 2014) and accelerate the development of salt-tolerant crop varieties.

This chapter discusses the importance of phenotyping in genetic studies of salt tolerance by highlighting the available methods of screening crop germplasm to enhance salt tolerance. We examine technological advances in phenotyping germplasm and discuss the strategies for integrating high-throughput phenotyping with genetic studies to breed and enhance crops for salt tolerance.

9.2 Importance of Phenotyping in Genetic Studies of Salt Tolerance

Phenotyping is a crucial aspect of genetic studies for enhancing salt tolerance in crops as it provides valuable information about plant performance under salt-stress conditions (Morton *et al.*, 2019; Afzal *et al.*, 2023). It involves the measurement and evaluation of various plant traits, such as growth, morphology, and physiological and biochemical profiles, to identify the best-performing genotypes that can thrive under salinity-stress conditions (Gosa *et al.*, 2019; Thoday-Kennedy *et al.*, 2021; Masarmi *et al.*, 2023). Phenotyping helps plant breeders to identify the most effective breeding strategies to enhance salt tolerance and develop salt-tolerant varieties that can sustain productivity under stress conditions (Afzal *et al.*, 2023).

There are several mediums for screening crop germplasm for salt tolerance. Crop germplasm can be screened for salt tolerance using the hydroponic method, which involves growing plants in a nutrient solution without any soil substrate (Shavrukov *et al.*, 2012; Tavakkoli *et al.*, 2012; Atzori *et al.*, 2019). This method allows for precise control of salinity levels and the ability to rapidly screen large numbers of plant lines. Additionally, hydroponic screening can be carried out in a controlled environment, which reduces the environmental variation in salt concentrations and allows for more accurate measurement of plant responses, especially for root traits (Balliu *et al.*, 2021). Also, screening of crop germplasm can be done directly in the field, involving evaluating plant growth and development in soil with varying salinity levels in the farmer's field (Krishnamurthy *et al.*, 2016). As described by Kwon *et al.* (2015), this method is more realistic and relevant to field conditions than hydroponic screening. However, it can be challenging to control environmental factors, such as temperature, light, and precipitation. On the other hand, screening of germplasm using soil salinity involves the addition of salt to the soil to create the desired salinity levels. This method is more controlled than field screening but can be time-consuming and labour-intensive (Mustafa and Akhtar, 2019).

Furthermore, germplasm screening for salt tolerance can also be achieved by exposing plants to varying concentrations of sodium chloride in a controlled environment. This method is highly reproducible and allows for precise control of salt concentrations, but it may not accurately reflect field conditions (Iqbal *et al.*, 2020). Other salt sources, such as calcium chloride or magnesium sulfate, can be used to evaluate plant responses to different types of salinity (Kalhori *et al.*, 2018). This method can provide insights into the specific mechanisms of salt tolerance in plants and may identify genotypes with specific types of salt tolerance.

Additionally, the timing of the phenotyping is crucial for identifying the best-performing genotypes. The most vulnerable period of plant growth, when the crop is most susceptible to salt stress, is the early vegetative stage (Tarchoun *et al.*, 2022). Phenotyping during this stage can help identify genotypes that can tolerate salinity stress and continue to grow and produce even under stressful conditions. However, it is also important to note that phenotyping should be performed at different stages of plant development to ensure that the selected genotypes can maintain their performance and productivity throughout the crop growth cycle, as noted by Maier *et al.* (2022).

9.3 Advances in Precision Phenotyping of Salt Tolerance

Plant phenotyping has long been a critical component of plant breeding for developing salt-tolerant crops. Traditional methods of phenotyping germplasm for salt tolerance include visual scoring, chlorophyll fluorescence imaging, leaf water potential measurement, ion content measurement, and growth measurement (Zhou *et al.*, 2018; Arief *et al.*, 2023). Visual scoring is a simple method to assess plant growth and development, but it is subjective and may not be accurate enough for precise evaluation of salt tolerance. Chlorophyll fluorescence imaging measures the amount of chlorophyll and the photosynthetic capacity of plants, while leaf water potential measurement assesses the plant's ability to maintain turgor pressure under salinity stress (Arief *et al.*, 2023; Hu and Schmidhalter, 2023). Ion content measurement involves analysing the concentration of ions, such as sodium, potassium, and calcium, which play a critical role in salt-stress response. Growth measurement, including shoot and root growth, is also a crucial phenotyping method, providing information on the plant's overall performance under salt stress. For instance, Urbanavičiūtė *et al.* (2023), Pour-Aboughadareh *et al.* (2021), and Gandullo *et al.* (2021) employed such parameters to investigate the performance of wheat, barley, and tomato genotypes, respectively, under salt stress.

In recent years, precision phenotyping has emerged as a more efficient and accurate approach to assess plant growth and development under salinity stress. It involves the use of high-throughput phenotyping technologies to accurately and rapidly measure various plant traits in response to salinity stress (Xiao *et al.*, 2022). High-throughput phenotyping methods include imaging technologies such as hyperspectral imaging, thermal imaging, and fluorescence imaging, which can visualize and quantify plant traits related to salinity stress like leaf area, chlorophyll content, and photosynthetic efficiency (Li *et al.*, 2010). Sensors and probes can measure plant physiological parameters such as water potential, stomatal conductance, and ion concentrations in real time, providing insights into plant responses to salinity stress and aiding in the identification of salt-tolerant genotypes (Koman *et al.*, 2017).

Robotics can automate the process of phenotyping large numbers of plants, performing tasks such as watering, fertilizing, and imaging plants, enabling researchers to collect vast amounts of data quickly (Czedik-Eysenberg *et al.*, 2018). Microfluidic devices can screen large numbers of plant samples for specific traits related to salt tolerance, rapidly sorting and analysing samples based on their physical and chemical properties (Rafiq *et al.*, 2020). Furthermore, remote sensing, involving the use of satellites or unmanned aerial vehicles (UAVs), can collect data on plant growth and development, monitor large numbers of plants, and identify genotypes that are tolerant to salt stress (Ivushkin *et al.*, 2019). This enables the collection of continuous and repetitive data over time, allowing for the observation of changes in plant traits and responses to environmental factors.

Moreover, computer vision technologies such as deep learning algorithms and artificial intelligence (AI) can be used to analyse large datasets generated by high-throughput phenotyping and remote-sensing platforms. These technologies can automate data analysis, identifying complex patterns and relationships that would be difficult to detect using traditional statistical methods (Richardson *et al.*, 2018).

These techniques can help to overcome some of the bottlenecks associated with breeding crops for salt tolerance, such as the difficulty in phenotyping salt-tolerance traits. High-throughput phenotyping technologies allow for the measurement of multiple traits simultaneously and with high accuracy, enabling breeders to identify and select plants with desirable salt-tolerance traits (Morton *et al.*, 2019). Moreover, precision phenotyping can help to identify novel salt-tolerance traits that may not have been previously observed using conventional phenotyping methods. For example, precision phenotyping can identify traits related to root morphology, architecture, and physiology, which are critical for salt tolerance but are difficult to observe using traditional phenotyping methods (Awlia *et al.*, 2016).

9.4 Integrating High-Throughput Phenotyping with Genetic Studies for Enhanced Salt Tolerance in Crops

Salt tolerance is a complex trait influenced by multiple genes and environmental variables, making it difficult to improve through traditional breeding methods alone (Afzal *et al.*, 2023). However, with the advancement of genetic studies, such as marker technology, it has become possible to identify the specific mechanism that controls salt tolerance and develop new breeding strategies to enhance salt tolerance in crops. Marker-assisted breeding involves the use of genome-wide molecular markers to track specific genes and genomic regions associated with salt tolerance. Based on genome-wide data, the marker–trait association analysis helps to locate and map genomic regions or loci associated with salt-tolerance traits in a defined mapping population. Depending on the mapping population, association analysis may be conducted in a defined family using their linkage map (Adeboye *et al.*, 2021) or among diverse individuals based on linkage disequilibrium (Bhat *et al.*, 2022a,b). Both involve integrating genetic information with phenotypic data to track associations using various statistical procedures, implying that precision phenotyping is crucial to enhancing the efficiency of these genetic approaches.

By accurately measuring and quantifying these traits, researchers can better understand the genetic basis of salt tolerance and identify key genetic markers that are associated with salt tolerance. For example, precision phenotyping can be used to visualize and quantify plant traits associated with salinity stress, including leaf area, chlorophyll content, and photosynthetic efficiency (Li *et al.*, 2010). This information can be integrated with genetic data obtained through genome-wide marker scanning in a linkage disequilibrium and linkage mapping analysis to identify quantitative trait loci (QTLs) or genomic regions that are associated with salt tolerance (Islam *et al.*, 2019; Luo *et al.*, 2019; Han *et al.*, 2020; Jahan *et al.*, 2020; Guo *et al.*, 2021). Another application of precision phenotyping is genomic selection. In genomic selection, precision phenotyping can be used to accurately measure the traits of interest for selection in breeding programmes. The accuracy of phenotypic data is critical for the success of genomic selection, as accurate phenotypic data can lead to the identification of relevant genes associated with salt tolerance. In combination with genotyping, precision phenotyping can be used to develop genomic prediction models that can predict the performance of untested genotypes based on their genotypic and phenotypic information (Liu *et al.*, 2020). This approach can significantly reduce the time and resources needed for breeding programmes, as it allows for the selection of desirable genotypes at an earlier stage of the breeding process.

In genome editing, precision phenotyping can be used to validate the effectiveness of the edited gene or trait. For instance, after editing a gene associated with salt tolerance, precision phenotyping can be used to confirm the impact of the edit on the phenotype of the plant (Nascimento *et al.*, 2023). This can be done by measuring the plant's response to salinity stress under controlled conditions. The accuracy of the phenotypic data is essential in determining the efficiency of genome editing technology in enhancing salt tolerance.

Also, precision phenotyping is crucial in transcriptomics, proteomics, and metabolomics analyses to select genotypes with contrasting salt-tolerance traits, which can be used to identify differentially expressed genes, proteins, and metabolites associated with salt tolerance (Niron *et al.*, 2021; Naik *et al.*, 2023). This approach can provide insights into the molecular mechanisms underlying salt tolerance, leading to the identification of candidate genes or proteins for further investigation. In addition, precision phenotyping can be used to validate the function of the identified genes, proteins, and metabolites and their impact on the plant's salt tolerance.

The integration of tools like transcriptomics, proteomics, and metabolomics with high-throughput phenotyping and genetic studies can further enhance salt-tolerance breeding. For example, combining transcriptomic analysis with QTL mapping (Lei *et al.*, 2020) can help to identify the genetic basis of salt tolerance and reveal the key genes involved in salt-stress response. Proteomic analysis can be used to identify proteins associated with salt-stress response and can provide a better understanding of the molecular mechanisms involved in salt tolerance. Metabolomic analysis can identify key metabolites involved in salt-stress response and can provide insights into the metabolic pathways involved in salt tolerance. These strategies can be used to identify genetic loci associated with salt tolerance, develop genetic markers for marker-assisted selection, and develop predictive models of plant performance under different stress conditions (Ebersbach *et al.*, 2022).

9.5 Conclusion

Salinity stress presents a significant threat to global crop productivity and growth. The development of crop varieties with improved salt tolerance through breeding programmes has been impeded by the difficulty in accurately phenotyping plant traits. However, the use of high-throughput techniques for precise phenotyping of germplasm provides a promising solution to this challenge. It enables rapid and efficient screening of a large number of genotypes for salt tolerance, which can subsequently be analysed genetically to develop salt-tolerant crops. By leveraging these technological advancements, we can enhance the efficiency and efficacy of breeding-based strategies aimed at improving crop salt tolerance, ultimately contributing to food security and promoting sustainable agriculture worldwide.

References

Acosta-Motos, J., Ortuño, M., Bernal-Vicente, A., Diaz-Vivancos, P., Sanchez-Blanco, M. and Hernandez, J. (2017) Plant responses to salt stress: adaptive mechanisms. *Agronomy* 7(1), 18. DOI: 10.3390/agronomy7010018.

Adeboye, K.A., Semon, M., Oyetunde, O.A., Oduwaye, O.A., Adebambo, A.O. *et al.* (2021) Diversity array technology (DArT)-based mapping of phenotypic variations among recombinant inbred lines of WAB638-1/PRIMAVERA under drought stress. *Euphytica* 217(6), 130.

Afzal, M., Hindawi, S.E.S., Alghamdi, S.S., Migdadi, H.H., Khan, M.A. *et al.* (2023) Potential breeding strategies for improving salt tolerance in crop plants. *Journal of Plant Growth Regulation* 42, 3365–3387. DOI: 10.1007/s00344-022-10797-w.

Araus, J.L. and Cairns, J.E. (2014) Field high-throughput phenotyping: the new crop breeding frontier. *Trends in Plant Science* 19(1), 52–61.

Arief, M.A.A., Kim, H., Kurniawan, H., Nugroho, A.P., Kim, T. and Cho, B.-K. (2023) Chlorophyll fluorescence imaging for early detection of drought and heat stress in strawberry plants. *Plants (Basel)* 12, 1387. DOI: 10.3390/plants12061387.

Atzori, G., Mancuso, S. and Masi, E. (2019) Seawater potential use in soilless culture: a review. *Scientia Horticulturae* 249, 199–207.

Awlia, M., Nigro, A., Fajkus, J., Schmoeckel, S.M., Negrão, S. *et al.* (2016) High-throughput non-destructive phenotyping of traits that contribute to salinity tolerance in *Arabidopsis thaliana*. *Frontiers in Plant Science* 7, 1414. DOI: 10.3389/fpls.2016.01414.

Balliu, A., Zheng, Y., Sallaku, G., Fernández, J.A., Gruda, N.S. and Tuzel, Y. (2021) Environmental and cultivation factors affect the morphology, architecture and performance of root systems in soilless grown plants. *Horticulturae* 7, 243. DOI: 10.3390/horticulturae7080243.

Bhat, J.A., Adeboye, K.A., Ganie, S.A., Barmukh, R., Hu, D. *et al.* (2022a) Genome-wide association study, haplotype analysis, and genomic prediction reveal the genetic basis of yield-related traits in soybean (*Glycine max* L.). *Frontiers in Genetics* 13, 953833. DOI: 10.3389/fgene.2022.953833.

Bhat, J.A., Karikari, B., Adeboye, K.A., Ganie, S.A., Barmukh, R. *et al.* (2022b) Identification of superior haplotypes in a diverse natural population for breeding desirable plant height in soybean. *Theoretical and Applied Genetics* 135(7), 2407–2422.

Chen, D., Neumann, K., Friedel, S., Kilian, B., Chen, M. *et al.* (2014) Dissecting the phenotypic components of crop plant growth and drought responses based on high-throughput image analysis. *The Plant Cell* 26, 4636–4655.

Czedik-Eysenberg, A., Seitner, S., Güldener, U., Koemeda, S., Jež, J. *et al.* (2018) The 'PhenoBox', a flexible, automated, open-source plant phenotyping solution. *New Phytologist* 219, 808–823.

Deery, D.M. and Jones H.G. (2021) Field phenomics: will it enable crop improvement? *Plant Phenomics* 2021, 9871989. DOI: 10.34133/2021/9871989.

Ebersbach, J., Khan, N.A., McQuillan, I., Higgins, E.E., Horner, K. *et al.* (2022) Exploiting high-throughput indoor phenotyping to characterize the founders of a structured *B. napus* breeding population. *Frontiers in Plant Science* 12, 780250. DOI: 10.3389/fpls.2021.780250.

Fahlgren, N., Gehan, M.A. and Baxter, I. (2015) Lights, camera, action: high-throughput plant phenotyping is ready for a close-up. *Current Opinion in Plant Biology* 24, 93–99. DOI: 10.1016/j.pbi.2015.02.006.

FAO (2021) *Global Map of Salt-Affected Soils*. Food and Agriculture Organization of the United Nations, Rome, p. 20.

Furbank, R.T. and Tester, M. (2011) Phenomics – technologies to relieve the phenotyping bottleneck. *Trends in Plant Science* 16, 635–644.

Gandullo, J., Ahmad, S., Darwish, E., Karlova, R. and Testerink, C. (2021) Phenotyping tomato root developmental plasticity in response to salinity in soil rhizotrons. *Plant Phenomics* 2021, 2760532. DOI: 10.34133/2021/2760532.

Ganie, S.A. (2020) RNA chaperones: potential candidates for engineering salt tolerance in rice. *Crop Science* 60(2), 530–540.

Ganie, S.A. (2021) Amino acids other than proline and their participation in abiotic stress tolerance. In: Wani, S.H., Gangola, M.P. and Ramadoss, B.R. (eds) *Compatible Solutes Engineering for Crop Plants Facing Climate Change*. Springer, Cham, Switzerland, pp. 47–96.

Gosa, S.C., Lupo, Y. and Moshelion, M. (2019) Quantitative and comparative analysis of whole-plant performance for functional physiological traits phenotyping: new tools to support pre-breeding and plant stress physiology studies. *Plant Science* 282, 49–59.

Guo, A.H., Su, Y., Huang, Y., Wang, Y.M., Nie, H.S. *et al.* (2021) QTL controlling fiber quality traits under salt stress in upland cotton (*Gossypium hirsutum* L.). *Theoretical and Applied Genetics* 134, 661–685.

Gupta, B. and Huang, B. (2014) Mechanism of salinity tolerance in plants: physiological, biochemical, and molecular characterization. *International Journal of Genomics* 2014, 701596. DOI: 10.1155/2014/701596.

Han, J.H., Shin, N.H., Moon, J.H., Yi, C., Yoo, S.C. and Chin, J.H. (2020) Genetic and phenotypic characterization of rice backcrossed inbred sister lines of *Saltol* in temperate saline reclaimed area. *Plant Breeding and Biotechnology* 8(1), 58–68.

Hu, Y. and Schmidhalter, U. (2023) Opportunity and challenges of phenotyping plant salt tolerance. *Trends in Plant Science* 28(5), 552–566. DOI: 10.1016/j.tplants.2022.12.010.

Iqbal, S., Hussain, S., Abdul Qayyaum, M., Ashraf, M. and Saifullah (2020) The response of maize physiology under salinity stress and its coping strategies. In: Hossain, A. (ed.) *Plant Stress Physiology*. IntechOpen, London. 25 pp. DOI: 10.5772/intechopen.92213.

Islam, M.S., Ontoy, J. and Subudhi, P.K. (2019) Meta-analysis of quantitative trait loci associated with seedling-stage salt tolerance in rice (*Oryza sativa* L.). *Plants (Basel)* 8(2), 33.

Ivushkin, K., Bartholomeus, H.M., Bregt, A.K., Pulatov, A.S., Franceschini, M.H. *et al.* (2019) UAV based soil salinity assessment of cropland. *Geoderma* 338, 502–512.

Jahan, N., Zhang, Y.U., Lv, Y., Song, M., Zhao, C. *et al.* (2020) QTL analysis for rice salinity tolerance and fine mapping of a candidate locus qSL7 for shoot length under salt stress. *Plant Growth Regulation* 90, 307–319.

Kahori, N., Ying, T., Nulit, R., Sahebi, M., Abiri, R. and Atabaki, N. (2018) Effect of four different salts on seed germination and morphological characteristics of *Oryza sativa* L. cv. MR219. *International Journal of Advanced Research in Botany* 4(1), 29–45. DOI: 10.20431/2455-4316.0401005.

Koman, V.B., Lew, T.T., Wong, M.H., Kwak, S.Y., Giraldo, J.P. and Strano, M.S. (2017) Persistent drought monitoring using a microfluidic-printed electro-mechanical sensor of stomata in planta. *Lab on a Chip* 17(23), 4015–4024.

Krishnamurthy, S.L., Gautam, R.K., Sharma, P.C. and Sharma, D.K. (2016) Effect of different salt stresses on agro-morphological traits and utilisation of salt stress indices for reproductive stage salt tolerance in rice. *Field Crops Research* 190, 26–33. DOI: 10.1016/j.fcr.2016.02.018.

Kwon, T., Kim, K., Yoon, H., Lee, S., Kim, B. and Siddiqui, Z.S. (2015) Phenotyping of plants for drought and salt tolerance using infra-red thermography. *Plant Breeding and Biotechnology* 3, 299–307.

Lei, L., Zheng, H., Bi, Y., Yang, L., Liu, H. *et al.* (2020) Identification of a major QTL and candidate gene analysis of salt tolerance at the bud burst stage in rice (*Oryza sativa* L.) using QTL-Seq and RNA-Seq. *Rice* 13, 55.

Li, G., Wan, S., Zhou, J., Yang, Z. and Qin, P. (2010) Leaf chlorophyll fluorescence, hyperspectral reflectance, pigments content, malondialdehyde and proline accumulation responses of castor bean (*Ricinus communis* L.) seedlings to salt stress levels. *Industrial Crops and Products* 31, 13–19.

Liu, J., Shen, F., Xiao, Y., Fang, H., Qiu, C. *et al.* (2020) Genomics-assisted prediction of salt and alkali tolerances and functional marker development in apple rootstocks. *BMC Genomics* 21(1), 550.

Luo, M., Zhang, Y., Chen, K., Kong, M., Song, W. *et al.* (2019) Mapping of quantitative trait loci for seedling salt tolerance in maize. *Molecular Breeding* 39, 64.

Maier, C.R., Chen, Z.-H., Cazzonelli, C.I., Tissue, D.T. and Ghannoum, O. (2022) Precise phenotyping for improved crop quality and management in protected cropping: a review. *Crops* 2, 336–350. DOI: 10.3390/crops2040024.

Majeed, A. and Muhammad, Z. (2019) Salinity: a major agricultural problem – causes, impacts on crop productivity and management strategies. In: Hasanuzzaman, M., Hakeem, K., Nahar, K. and Alharby, H. (eds) *Plant Abiotic Stress Tolerance*. Springer, Cham, Switzerland, pp. 83–99. DOI: 10.1007/978-3-030-06118-0_3.

Masarmi, A.G., Solouki, M., Fakheri, B., Kalaji, H.M., Mahgdingad, N. *et al.* (2023) Comparing the salinity tolerance of twenty different wheat genotypes on the basis of their physiological and biochemical parameters under NaCl stress. *PLoS ONE* 18(3), e0282606. DOI: 10.1371/journal.pone.0282606.

Morton, M.J., Awlia, M., Al-Tamimi, N., Saade, S., Pailles, Y. *et al.* (2019) Salt stress under the scalpel – dissecting the genetics of salt tolerance. *The Plant Journal* 97(1), 148–163.

Munns, R. and Tester, M. (2008) Mechanisms of salinity tolerance. *Annual Review of Plant Biology* 59, 651–681.

Mustafa, G. and Akhtar, M.S. (2019) Crops and methods to control soil salinity. In: Akhtar, M. (ed.) *Salt Stress, Microbes, and Plant Interactions: Mechanisms and Molecular Approaches*. Springer, Singapore, pp. 237–251. DOI: 10.1007/978-981-13-8805-7_11.

Naik, B., Kumar, V., Rizwanuddin, S., Chauhan, M., Choudhary, M. *et al.* (2023) Genomics, proteomics, and metabolomics approaches to improve abiotic stress tolerance in tomato plant. *International Journal of Molecular Sciences* 24, 3025. DOI: 10.3390/ijms24033025.

Nascimento, F.D.S., Rocha, A.J., Soares, J.M.D.S., Mascarenhas, M.S., Ferreira, M.D.S. *et al.* (2023) Gene editing for plant resistance to abiotic factors: a systematic review. *Plants (Basel)* 12(2), 305. DOI: 10.3390/plants12020305.

Niron, H., Barlas, N., Salih, B. and Türet, M. (2021) Comparative transcriptome, metabolome, and ionome analysis of two contrasting common bean genotypes in saline conditions. *Frontiers in Plant Science* 11, 599501. DOI: 10.3389/fpls.2020.599501.

Paez-Garcia, A., Motes, C.M., Scheible, W.R., Chen, R., Blancaflor, E.B. and Monteros, M.J. (2015) Root traits and phenotyping strategies for plant improvement. *Plants (Basel)* 4(2), 334–355.

Panta, S., Flowers, T.J., Lane, P., Doyle, R., Haros, G. and Shabala, S. (2014) Halophyte agriculture: success stories. *Environmental and Experimental Botany* 107, 71–83.

Passioura, J.B. (2012) Phenotyping for drought tolerance in grain crops: when is it useful to breeders? *Functional Plant Biology* 39, 851–859. DOI: 10.1071/FP12079.

Pour-Aboughadareh, A., Sanjani, S., Nikkhah-Chamanabad, H., Reza, M., Asadi, M.A. and Amini, A. (2021) Identification of salt-tolerant barley genotypes using multiple-traits index and yield performance at the early growth and maturity stages. *Bulletin of National Research Center* 45, 117. DOI: 10.1186/s42269-021-00576-0.

Rafiq, K., Akram, M.S., Shahid, M., Qaisar, U. and Rashid, N. (2020) Enhancement of salt tolerance in maize (*Zea mays* L.) using locally isolated *Bacillus* spp. SR-2-1/1. *Biologia* 75, 1425–1436.

Richardson, A., Mulder, T.V. and Vehbi, T. (2018) *Nowcasting New Zealand GDP Using Machine Learning Algorithms*. CAMA Working Paper 47/2018. Centre for Applied Macroeconomic Analysis, Crawford School of Public Policy, The Australian National University, Canberra.

Rouached, H., Pal, S., Rachmilevitch, S., Libault, M. and Tran, L.P. (2015) Plants coping abiotic and biotic stresses: a tale of diligent management. *BioMed Research International* 2015, 754754.

Safdar, H., Amin, A., Shafiq, Y., Ali, A., Yasin, R. *et al.* (2019) A review: impact of salinity on plant growth. *Nature and Science* 17(1), 34–40.

Shavrukov, Y., Genc, Y. and Hayes, J. (2012) The use of hydroponics in abiotic stress tolerance research. In: Asao, T. (ed.) *Hydroponics – A Standard Methodology for Plant Biological Researches*. IntechOpen, London, pp. 39–66. DOI: 10.5772/35206.

Tarchoun, N., Saadaoui, W., Mezghani, N., Pavli, O.I., Falleh, H. and Petropoulos, S.A. (2022) The effects of salt stress on germination, seedling growth and biochemical responses of Tunisian squash (*Cucurbita maxima* Duchesne) germplasm. *Plants (Basel)* 11(6), 800. DOI: 10.3390/plants11060800.

Tattaris, M., Reynolds, M.P. and Chapman, S.C. (2016) A direct comparison of remote sensing approaches for high-throughput phenotyping in plant breeding. *Frontiers in Plant Science* 7, 1131.

Tavakkoli, E., Fatehi, F., Rengasamy, P. and McDonald, G.K. (2012) A comparison of hydroponic and soil-based screening methods to identify salt tolerance in the field in barley. *Journal of Experimental Botany* 63, 3853–3867.

Thoday-Kennedy, E., Joshi, S., Daetwyler, H.D., Hayden, M., Hudson, D. *et al.* (2021) Digital phenotyping to delineate salinity response in safflower genotypes. *Frontiers in Plant Science* 12, 662498. DOI: 10.3389/fpls.2021.662498.

Urbanavičiūtė, I., Bonfiglioli, L. and Pagnotta, M.A. (2023) Phenotypic and genotypic diversity of roots response to salt in durum wheat seedlings. *Plants (Basel)* 12, 412. DOI: 10.3390/plants12020412.

Xiao, Q., Bai, X., Zhang, C. and He, Y. (2022) Advanced high-throughput plant phenotyping techniques for genome-wide association studies: a review. *Journal of Advanced Research* 35, 215–230.

Zhou, J., Chen, H., Zhou, J., Fu, X., Ye, H. and Nguyen, H.T. (2018) Development of an automated phenotyping platform for quantifying soybean dynamic responses to salinity stress in greenhouse environment, *Computers and Electronics in Agriculture* 151, 319–330. DOI: 10.1016/j.compag.2018.06.016

10 *De Novo* Domestication of Wild and Halophilic Plants for Designing Salt-Tolerant Crops: Acceleration through CRISPR

Balasundari Dharbaranyam, Kausalya Sakthivel and Gayatri Venkataraman*
Plant Molecular Biology Laboratory, M.S. Swaminathan Research Foundation, Chennai, India

Abstract

Salinity is a major abiotic stress factor affecting crop yield worldwide. Significant advances have been made in understanding the molecular basis of plant salinity tolerance, many involving transporter genes governing sodium and/or potassium homeostasis. Salt exclusion has been a premier trait targeted in most crop breeding programmes, utilizing intraspecific variation or trait introgression from close crop wild relatives (CWRs). There is also evidence to suggest that, as a 'by-product' of domestication, many crop species have lost naturally existing salt-tolerance traits. In this scenario, salt-tolerant CWRs (primary, secondary, or tertiary gene pools) can be targeted for '*de novo*' domestication using knowledge-based breeding strategies. Domestication traits in crop species constitute a shared repertoire of phenotypic features. A prerequisite for *de novo* domestication of CWRs is a pipeline of tools and strategies: availability of a high-quality reference genome, efficient plant transformation processes, information on genes involved in domestication, and the ability to 'edit' these genes in the CWR using CRISPR (clustered regularly interspaced short palindromic repeats) or prime editing. Incorporation of focused identification of germplasm (involving CWRs) and rapid generation advancement techniques (speed breeding, CRISPR-Combo) into the above pipeline is also recommended to accelerate development of *de novo* domesticated CWRs with pre-existing traits for salt tolerance.

10.1 Introduction

Soil salinization has been identified as a major cause of land degradation. Primary soil salinization arises due to geochemical cycling processes while anthropogenic activities, including modern agricultural practices, contribute to secondary salinization. Soil surveys conducted by the Food and Agriculture Organization of the United Nations (FAO) as far back as the 1970s–1980s estimated that approximately 6.5% of the world's arable and marginal soils were either saline or sodic (FAO, 2016). Four decades of data on spatiotemporal variation in topsoil salinity suggests that the regions of the world with the highest salt-affected areas occur on three continents: Asia, Africa, and Australia (Hassani *et al.*, 2020). Given that the projected world population will be approximately 9.9 billion by 2050, it is imperative that global food production be

*Email: gayatri@mssrf.res.in

 The Genetics of Salt Tolerance in Plants: A Central Dogma Perspective and Strategies for Enhancement (S. Ganie and S.H. Wani eds)
DOI: 10.1079/9781800623033.0010

increased by 57% to match consumption (UN Report, 2013). Therefore, addressing the problem of salinity is crucial to food security.

10.2 Plants and Salinity

Soil contains chlorides/sulfates of sodium, calcium, and magnesium and their relative quantities are the basis of classifying soils as saline, sodic, or saline sodic (United States Salinity Laboratory Staff, 1954; Choudhary and Kharche, 2018) (Table 10.1). Electrical conductance (EC) units, sodium adsorption ratio (SAR), and the exchangeable sodium percentage (ESP) are three parameters used together to classify the soil types mentioned above.

Salinity and sodicity affect soil properties differently. At lower limits, salinity affects soil physical properties positively (via soil pH) by binding soil particles through flocculation, permitting soil aeration, and promoting root growth and penetration. When soil exchangeable sodium content exceeds calcium and magnesium, soils turn sodic. Sodic soils are also alkaline (pH 8.5–10). A combination of high pH and high concentration of sodium ions under wet conditions leads to soil dispersion. At lower moisture levels, however, clay aggregations plug soil pores, reducing soil water permeability and hydraulic conductivity.

Salinity has multiple effects on plant growth. The immediate 'osmotic response' to salinity stress (following perception), involving stomatal closure and inhibition of shoot growth, precedes Na^+ accumulation in the plant (Munns and Tester, 2008; Ganie, 2021). Toxicity of Na^+ ions in the 'ionic phase' is dealt with using a combination of avoidance and tolerance strategies, involving a range of transporters expressed at the plasma membrane and/or tonoplast (mentioned below). When present in high quantities in the root zone, Na^+ is taken up by non-selective cation channels (NSCCs) in roots (Demidchik and Tester, 2002) that cannot discriminate between K^+ and Na^+ due to similar hydrated ionic radii (Benito *et al.*, 2014). While Na^+ is not an essential mineral for plant growth (except in halophytes, mentioned below) (Kronzucker and Britto, 2011; Shabala, 2013; Maathuis, 2014; Flowers and Colmer, 2015; Wu *et al.*, 2018), K^+ is an essential macronutrient for plant growth (10% of plant dry weight) (White and Karley, 2010), crucial for a range of basic metabolic processes (Jegadeeson *et al.*, 2019). Within wide variations in soil K^+ concentration (0.01–20 mM), plant cells maintain a relatively constant cytoplasmic K^+ concentration (80–100 mM) (Rodríguez-Navarro, 2000), accumulating large amounts in vacuoles that controls cell turgor, driving cell expansion and growth. Increased Na^+ uptake under salinity reduces K^+ content in the cytosol of plant cells (Wu *et al.*, 2018; Rubio *et al.*, 2020), with major consequences for efficient metabolic functioning

Halophytes are plant species belonging to diverse genera that have the ability to survive and complete a life cycle at a salinity of 200 mM NaCl (Colmer and Flowers, 2008). Most halophytes accumulate sodium in shoot tissues as osmoticum and are said to show 'tissue tolerance' to sodium. Glycophytes, in contrast, are plants that have evolved under natural selective pressures in low soil sodium ecosystems 'excluding sodium' from aboveground tissues (Cheeseman, 2015). Thus, tissue tolerance in halophytes versus shoot exclusion of sodium in glycophytes are defining traits that have evolved in relation to soil sodium content. In addition, halophytes differ from glycophytes in other important traits like K^+ retention under salinity, finer steady-state control of reactive oxygen species (ROS) and cell-specific cytosolic Ca^{2+} signatures (second messengers generated in response to salinity stress within plant cells), increased water-use efficiency (WUE), alterations in stomatal traits, increased leaf succulence, and reverse sap flow under hypersaline conditions. The next section briefly lists ion transporters related to sodium homeostasis in plants.

Table 10.1. Salinity parameters and associated soil types. (Source: United States Salinity Laboratory Staff, 1954; public domain.)

Soil type	Soil pH	EC	SAR	ESP
Un-salinized soil	<8.5	<4.0	<13	<15
Saline	<8.5	>4.0	<13	<15
Sodic	<8.5	<4.0	>13	>15
Saline sodic	<8.5	>4.0	>13	>15

10.3 Ion Transporters in Sodium Homeostasis

Radial transport of sodium into roots can be controlled at different points: (i) epidermis; (ii) cortical tissues; (iii) root endodermis; and

(iv) stele. Ion movement in plants can be symplastic (cell to cell via plasmodesmata), apoplastic (extracellular), or transcellular (coupled transporter function across cells) (Barberon and Geldner, 2014). Since cell/tissue-specific transporter expression can vary widely between glycophytic and halophytic species and transporter function is also under extensive post-translational control, sodium homeostasis in plants is complex and shows species/genotype-specific variation (Assaha *et al.*, 2017).

NSCCs mediate sodium uptake in roots but do not strongly differentiate between K^+ and Na^+ and can also mediate divalent cation (Ca^{2+}) uptake. Based on pharmacological inhibitors/activators, NSCCs are divided into four classes: (i) constitutive; (ii) ROS-activated; (iii) ligand-activated; and (iv) mechanosensitive (Demidchik, 2006; Demidchik and Maathuis, 2007). While candidate NSCC-encoding genes have been identified in plant genomes, establishing a direct cause-and-effect relationship between candidate genes and *in vivo* function is a grey area in plant research. More recent research implicates aquaporins, cyclic nucleotide-gated channels, ionotropic glutamate receptors, and plant PQ loop proteins (PQLs) as putative NSCCs (Demidchik and Maathuis, 2007; Alqahtani *et al.*, 2021).

Within the cell, sodium can accumulate in vacuoles. Plant Na^+/H^+ antiporters (NHX) occur as multigene families in plants. Stringent organelle-specific compartmentation (vacuoles) of Na^+ under salinity was thought to be the primary function of intracellular NHX transporters (Apse *et al.*, 1999; Blumwald *et al.*, 2000; Pardo *et al.*, 2006; Apse and Blumwald, 2007; Zhang *et al.*, 2008). Knock-out data in *Arabidopsis*, however, suggests that intracellular NHX-type transporters have a crucial role in regulation of vacuolar and/or vesicular pH controlling K^+ homeostasis, driving cell expansion processes. These transporters also have a role under salinity as silencing of endosomal or vesicular NHX isoforms in *Arabidopsis* results in salt-sensitive plants (Rodríguez-Rosales *et al.*, 2008; Bassil *et al.*, 2011). In rice, *OsNHX2* is associated with salinity tolerance and *OsNHX1* with biomass under drought conditions in genome-wide association studies (GWAS) (Qu *et al.*, 2020; Chen, Q. *et al.*, 2021; Nayyeripasand *et al.*, 2021). Tissue tolerance in halophytes is attributed to the ability of vacuolar compartment(s) in leaf tissues to accumulate sodium, partly as a cheap osmoticum to drive growth under salinity instead of potassium (Shabala, 2013).

AtNHX7/SOS1, a plasma-membrane Na^+/H^+ antiporter, extrudes Na^+ from the cell after phosphorylation by SOS2, a serine-threonine kinase. SOS2 is recruited to the plasma membrane after interaction with SOS3 (calcium-dependent manner) to activate SOS1 (Shi *et al.*, 2000, 2002). Depending on tissue-specific expression, SOS1 may extrude sodium from root tip epidermal cells or contribute to loading of Na^+ into xylem, thus controlling long-distance transport to the shoot (Shi *et al.*, 2000). The SOS1–SOS3 module is also conserved in rice (Martínez-Atienza *et al.*, 2007). Reducing/knocking out *SOS1* expression in tomato or rice confers sensitivity to salinity (Olias *et al.*, 2009; El Mahi *et al.*, 2019) and *SOS1* knock-outs also show perturbations in K^+ homeostasis that may be related to higher Na^+/K^+ ratios in root cortical cells, inhibiting K^+ uptake mechanisms (El Mahi *et al.*, 2019). In halophytic *Thellungiella halophila*, SOS1 is implicated in sodium extrusion (root elongation zone) and also in xylem loading of sodium (Oh *et al.*, 2009). However, in halophytic *Oryza coarctata*, inhibitor-based electrophysiological studies have ruled out a role for SOS1 in sodium extrusion at the root surface (Ishikawa *et al.*, 2022). Species-specific alterations in expression, coupled with post-translational checkpoints, may be important determinants of SOS1 function (Ganie, 2020). Since halophytes show more precise temporal control of xylem loading of sodium compared to glycophytic species, comparing SOS1 function in halophytes and glycophytes would estimate relative contributions towards epidermal tissue-specific sodium extrusion under salinity and/or xylem loading.

Cation chloride co-transporter (CCC) family members mediate coupled movement of Cl^- with K^+ and/or Na^+ cations across the plasma membrane in animals (Hübner and Rust, 2007). *AtCCC* is a single-copy gene in *Arabidopsis thaliana* (Colmenero-Flores *et al.*, 2007) while two or more genes occur in other species (Henderson *et al.*, 2018). While both *AtCCC* and grapevine CCC both localize to the trans-Golgi network (Henderson *et al.*, 2015), *OsCCC1* localizes to the plasma membrane. The complex as well as pleiotropic nature of growth defects in *AtCCC* and *OsCCC1* mutants under normal growth conditions suggests a pivotal cellular role for CCC transporters in plant growth. However, the role of this transporter in salt tolerance needs further investigation, given divergent subcellular localizations reported so far in dicots and monocots. While an intracellular localization would confer tissue tolerance, a plasma-membrane localization would be more in sync with a role in xylem loading/unloading of sodium under salinity.

Type I high-affinity K^+ transporters (HKT) are associated with Na^+ homeostasis in plants. *AtHKT1*, the first Type I HKT isolated from plants, is sodium specific and expressed at the xylem–xylem parenchyma interface in the root under salinity, sequestering sodium from root xylem parenchyma and excluding it from leaves/shoot (Uozumi *et al.*, 2000; Mäser *et al.*, 2002; Berthomieu *et al.*, 2003). Independent genetic screens subsequently reinforced the role of *AtHKT1* in conferring leaf Na^+ exclusion (Berthomieu *et al.*, 2003; Gong *et al.*, 2004). Type I HKTs in monocot crop species (*HKT1;5* in most cases and *HKT1;4* in wheat and rice) are also important determinants for cereal crop salinity tolerance in the field (Lindsay *et al.*, 2004; Ren *et al.*, 2005; Huang *et al.*, 2006; Byrt *et al.*, 2007; Munns *et al.*, 2012; Houston *et al.*, 2020). *AtHKT1* orthologs in rice and wheat (including allelic variants: *OsHKT1*;5-*SKC1*/*saltol*; *TaHKT1*;5D; *TmHKT1*;5-A) also show similar localization profiles (root stele) (Venkataraman *et al.*, 2021). Recent data from Wang, Z. *et al.* (2021) suggests that, at least in cultivated rice, root-specific expression of *OsHKT1*;5 under salinity is under extensive transcriptional control. Finally, sodium retrieval from the xylem by *HKT1;5* and xylem loading of sodium by *SOS1* seem to be under reciprocal control that also ties in with potassium nutrition in ways so far not understood in plant species. Unravelling controls underpinning this *HKT1;5–SOS1* interplay and intersection with potassium nutrition under salinity in glycophytes and halophytes is crucial to understanding salinity tolerance.

10.4 Crop Domestication and Loss of Halophytism

The transition from a hunter-gatherer to a more settled mode of existence during the Neolithic Age was possible, in part, because of the domestication of crop species from crop wild relatives (CWRs). Domestication of plants is a significant milestone in human history, with cultivation of domesticated crops allowing for a more settled lifestyle and the expansion of human civilizations (Stetter *et al.*, 2020). Archaeobotanical evidence suggests that domestication was gradual, dispersed, and occurred multiple times in many crop lineages (Gross and Olsen, 2010) and is also supported by current genetic and genomic studies (Bohra *et al.*, 2022). Domestication of crops was a co-evolutionary process, driven in part by conscious human choice (e.g. colour, taste). Many traits associated with domestication (e.g. shattering, seed size, dormancy, germination synchrony), however, were the result of a natural transition from a free-living mode of existence (CWRs) to a 'managed' mode of cultivation and were driven by unconscious human selection (Purugganan, 2019). While domestication resulted in substantial phenotypic changes relative to CWRs, many of the phenotypic traits resulting in domestication are shared across crop species, resulting in either parallel or convergent evolution (Pickersgill, 2018). This shared repertoire of phenotypic traits across plant species with similar uses constitute the 'domestication syndrome' (DS) (Hammer, 1984). Abbo *et al.* (2014) have distinguished between 'core' DS traits that lead to defined CWR–domesticated crop dimorphism and other DS traits that show a phenotypic continuum between wild and domesticated crops, suggesting the latter represent post-domestication diversification. Thus, most 'core' DS traits are mono- or digenic (usually recessive) while DS traits showing a continuum are polygenic in nature (Table 10.2).

With certain traits, it is possible to define the order of appearance through archaeobotanical records. Thus, increase in grain size in cereals pre-dates loss of shattering (Fuller *et al.*, 2007) and increase in seed size is evident earlier in *Poaceae* species relative to *Fabaceae* (Bohra *et al.*, 2022). In contrast to sexually reproducing crop species (cereals, legumes), domestication in vegetatively propagated crop species is not so well documented. Denham *et al.* (2020) have proposed some convergent traits were co-opted in early domestication of vegetatively propagated field crops (reproduction mode, edible component yield, harvesting ease, defensive adaptations, timing of production, and plant architecture).

Table 10.2. Crop domestication traits. (Content created by the authors and based on references cited.)

Crop	Domestication trait	QTL(s)/gene(s) associated with the trait	Significance	References
Rice	Absence of abscission layer at spikelet base (non-shattering spikelet)	*qSH1*, *sh4*, *SHAT1*	Loss of natural seed dispersal	Oka and Morishima (1971); Harlan *et al.* (1973); Fuller *et al.* (2007); Wedger and Olsen (2018); Xu and Sun (2021)
	Increased grain size or weight	*GW2*	Increased seed vigour	
	Erect growth habit	*PROG1* and *PROG7*	Increased grain yield and recovery percentage	
	Compact plant architecture	*TAC1* and *TIG1*		
	Shorter or no awns	*An-1*, *LABA1* and *GAD1*	Loss of natural dispersal aids	
	White pericarp and straw yellow hull	*Bh4* and *Rc*	Reduced seed dormancy	
	Increased spikelet number per panicle	*NOG1*, *FZP*, *NAL1*, and *APO1*	Increased grain yield	
	Reduced grain dormancy	*OsG1, Sdr4*, and *OsGA20ox2*	Synchronized germination	
Wheat	Less hairiness	*Hg1* and *Hg2*	Reduced seed dispersal	Fuller *et al.* (2007); Sakuma *et al.* (2018); Würschum *et al.* (2018); DeWitt *et al.* (2019); Rahman *et al.* (2020); DeHaan *et al.* (2020); Wu *et al.* (2021)
	Shorter or no awns	*B1*, *B2*, and *Hd*		
	Plump-shaped spikelets (wider and short)	*GS3*, *GW2*, and *SW5*		
	Non-brittle rachis (non-shattering)	*Br1*, *Br2*, and *Br3*	Loss of natural seed dispersal	
	Soft glumes	*Tg1* and *sog*	Easy to harvest	
	Free threshing	*Q*		
	Increased grain number per spikelet	*GNI1*	Improved yield	
	Increased grain size and weight	*GW2* and *GASR7*	Germination under deep burial	
	Reduced plant height	*Rht-B1b* and *Rht-D1b*	Lodging resistance	
	Reduced flowering time	*ppd-1* and *vrn-1*	Survives under varied conditions and increased yield	
Barley	Non-shattering semi-tough rachis	*Btr1* and *Btr2*	Loss of natural seed dispersal	Hillman and Davies (1990); Fuller *et al.* (2007); Pourkheirandish and Komatsuda (2007)
	Six-rowed spike	*Vrs1* and *Int-c*	Increased seed yield	
	Naked caryopsis	*Nud*	Easy to process and dietary use	
	Reduced seed dormancy	*SD1* and *SD2*	Enhanced seed germination and seed dispersal	
	No or reduced requirement for vernalization	*Vrn-H1*, *Vrn-H2*, and *Vrn-H3*	Season-independent cultivation	
	Photoperiod insensitivity	*Ppd-H1* and *ppd-H2*	Adaptation to wider geographic conditions	

Continued

Table 10.2. Continued.

Crop	Domestication trait	QTL(s)/gene(s) associated with the trait	Significance	References
Maize	Fewer and shorter branches	*tb1*	Increased harvest index	Harlan *et al.* (1973); Dorweiler *et al.* (1993); Gepts (2010); Stitzer and Ibarra (2018); Chen, T. *et al.* (2021)
	Fewer ears	*gt1*		
	Non-disarticulating rachis (non-shattering)	*ZmSh1-1*, *ZmSh1-5.1* + *ZmSh1-5.2*	Loss of seed dispersal	
	Polystichous ear phyllotaxy	*zfl2*, *ids1*, *UB3*	Increased yield potential	
	Less branched inflorescence	*ra1*		
	Pistillate terminal lateral inflorescence	*tb1*, *tru1*		
	Naked grained cobs	*tga1*	Easy to harvest	
Pearl millet	Loss of abscission layer at the base of involucre	*al6*	Loss of natural seed shedding	Poncet *et al.* (2002); Fuller *et al.* (2007); Remigereau *et al.* (2011)
	Reduced plant height	*d2*	Lodging resistance	
	Compact plant architecture	*Pgtb1*	Increased seed recovery percentage	
	Increased spike weight	*wes2*	Increased seed vigour and yield	
	Increased seed size	–		
	Club-shaped and apically thicker seeds	–		
Sorghum	Determinate growth	–	Increased seed yield and recovery percentage	Harlan *et al.* (1973); Lenser and Theißen (2013); Hao *et al.* (2021)
	Heavy and compact head	–		
	Shorter internodes	–		
	Reduced seed dormancy	*qDor7*	Increased seed vigour	
	Waxy endosperm	*SbGBSSI*	Improved dietary value	
	Non-shattering	*Sh1*	Increased seed retention	
	Reduced plant height	*dw3*	Lodging resistance	
	Increased grain size and weight	*qGW1*	Increased yield potential and seed vigour	
	Reduced flowering time	*sbPPR37*	Wider adaptability	
Tomato	Increased fruit weight	*fw2.2*, *LC* and *FA*	Increase fruit yield	Bai and Lindhout (2007); Paran and Van Der Knaap (2007)
	Pear-shaped elongated fruit	*Ovate*	Amenable for mechanical harvesting	
	Separation of fruit at proximal end	*j* and *j2*		
	Determinate plant growth	*SP*		
	Fruit colour	*r*, *tangerine*, *beta*, *old-gold*, *high pigment-1*, and *high pigment-2*	Improved fruit quality	

Continued

Table 10.2. Continued.

Crop	Domestication trait	QTL(s)/gene(s) associated with the trait	Significance	References
Soybean	Non-dehiscent pod	*GmSHAT1-5, qPDH1*	Loss of natural seed dispersal	Lu, C. *et al.* (2022)
	Reduced seed dormancy	*G*	Synchronized germination	
	Thin seed coat	*GmHs1-1, GmGH9B8*	Breaking seed dormancy	
	Shiny seed coat	*B1*	Reduced allergen levels	
	Increased seed oil content	*GmSWEET39*	Increased seed size and yield	
	Photoperiod insensitivity	*Tof12*	Wider adaptability to different geographic conditions	

The process of crop domestication has resulted in a loss of genetic diversity and has also had undesirable impacts on beneficial traits (plant immunity, nutritional quality/flavour, adaptation) (Meyer *et al.*, 2012; Byrne *et al.*, 2020; Singh and Knaap, 2022). In addition, since domestication was focused on desirable aboveground parts, beneficial root traits found in CWRs have been excluded in many modern crop varieties. 'Omics' technologies offer a deeper understanding of the genetic and molecular components of the domestication process to link gene(s) to phenotypes. The availability of high-resolution genome sequences allows for annotation of quantitative trait locus (QTL) hotspots with multiple genes for structural and regulatory variation (Barmukh *et al.*, 2022). Similar approaches coupled with genome editing strategies can be used to understand the genetic basis of indirect effects of domestication (antagonistic pleiotropic effects, negative epistasis, and inheritance of undesirable alleles due to linkage drag) (Singh and Knapp, 2022).

While the Green Revolution conferred food sufficiency, the need to use resource-intensive agriculture to increase yield has had unintended manifold consequences (Hazell, 2002). This, coupled with the reliance on a few 'mega-varieties' for food security in the context of climate change and increasing land salinization, requires a reassessment of strategies used to breed saline-tolerant crops. Thus, the premier trait targeted with regard to salinity tolerance has so far been 'sodium exclusion', with breeding for salinity tolerance involving phenotyping plant crosses (between tolerant and sensitive donors) for 'exclusion of sodium from the shoot'. This has resulted in parallel selection of *HKT1;5* in conferring the 'sodium exclusion from shoot' trait, at least in graminaceous crop species (Venkataraman *et al.*, 2021).

There is also evidence to suggest that naturally existing salt-tolerance traits (associated with many transporters mentioned in Section 10.3) in CWRs were lost during domestication. Thus, a specific haplotype of *SlHAK20*, encoding a clade IV HAK/KUP/KT transporter, is associated with lowered Na^+/K^+ in roots and salinity tolerance of wild tomato (*Solanum pimpinellifolium*) but not domesticated or improved accessions of cultivated tomato (*Solanum lycopersicum* var. *cerasiforme*/*S. lycopersicum*) (Wang *et al.*, 2020). *SlHAK20* occurs in chromosomal 4 regions of tomato associated with a domestication sweep, indicating that the sensitive haplotype was inadvertently selected during the domestication-based selection for larger fruits. Single-nucleotide polymorphisms (SNPs) in key *cis*-acting stress-responsive elements in the promoter of the plasma-membrane antiporter (*SlSOS1*) and a 53 bp insertion of an additional repressor (ABI4) binding B-element in the promoter of the *SlSOS2* gene (leading to reduced *SlSOS2* expression) in cultivated tomato varieties leads to loss of salinity tolerance relative to wild tomato (Wang, Z. *et al.*, 2021; Hong *et al.*, 2023). A naturally occurring non-synonymous SNP (SNP947-G) in *ZmHKT1;2* (*ZmNC2*) enhances Na^+ transport, promoting shoot Na^+ exclusion and salt tolerance in maize (Zhang *et al.*, 2023). SNP947-G occurs at a higher frequency in wild maize (teosinte) relative to cultivated maize, suggesting that genomic regions flanking this SNP have probably undergone negative selection during domestication or post-domestication dispersal of maize. A natural variation in the promoter region of the sulfate transporter gene *ZmSULTR3;4* of cultivated maize inbred lines during domestication leads to shorter lateral root lengths making them more sensitive to salt stress as compared to the wild-type teosinte (Zhang *et al.*, 2022). An ancestral haplotype of *GmSalt3* (encoding a cation/H^+ (CHX)-type exchanger) in cultivated soybean is associated with salinity tolerance and is under positive selection in soybean cultivars growing in saline regions (Guan *et al.*, 2014). However, there is also evidence for positive selection for salinity-tolerance traits from CWRs during domestication. As an example, non-synonymous variation (SNP389) in *ZmHKT2* (*qKC3*) derived from wild maize reduces K^+ transport activity in roots, thus increasing K^+ in xylem sap, shoot K^+ content, and salt tolerance, and was positively selected during maize domestication (Cao *et al.*, 2019).

10.5 Crop Wild Relatives as Sources of Traits for Salinity Tolerance

CWRs are wild plant species genetically related to present-day domesticated crops. They occur in a diverse range of habitats and show a wider range of genetic diversity not found in crops. CWRs could belong to interfertile primary gene pools or to secondary or tertiary gene pools that are cross-incompatible with the cultivated modern crop varieties. Adaptation to new environments involves the ability to thrive under abiotic or biotic challenges (Ganie *et al.*, 2017). CWRs are, thus, important donors for these traits as they often occur in diverse eco-geographic conditions (Gonzalez Guzman *et al.*, 2022). CWRs have served as important donors of traits for salinity tolerance in some crop species and some of these traits are listed in Table 10.3. In addition, important salt-tolerant CWRs of selected domesticated crop species are also listed.

For the crop species and corresponding CWRs listed in Table 10.3, identification of QTLs linked to salinity tolerance involved use of traditional breeding methodologies combined with marker-assisted selection (MAS). However, utilizing CWRs in commercial breeding programmes poses a set of unique challenges such as reproductive compatibility barriers, linkage drag, poor agronomic performance when grown in ecological zones that are beyond their range of adaptability, and a wide range of phenotyping challenges (reviewed in Bohra *et al.*, 2022).

As already stated before, traits related to domestication often show simple Mendelian inheritance (loss of function or gain of function). Domestication of crops by anthropogenic intervention involved phenotypic selection of similar traits in different crop species or development of preferred crop ideotypes (Lenser and Theißen, 2013). These include features such as loss of seed shattering, increased yield or grain size, reduced plant height, variation in flowering time, colour, etc. Orthologous gene-to-trait associations across crop species showing both common and convergent evolutionary mechanisms have contributed to these parallel domestication processes (Fernie and Yan, 2019; Table 10.2). The orthology of domestication-related gene-to-trait associations across crop species has been demonstrated in certain cases using gene silencing/overexpression methodologies (e.g. awns: Hua *et al.*, 2015; Zhou *et al.*, 2021). In addition to providing proof-of-concept for domestication gene-to-trait associations, these studies also demonstrate that it is possible to make CWRs 'phenocopy' domestication traits. Expressed in another way, knowledge-based breeding strategies can be used to '*de novo*' domesticate wild species with promising traits (e.g. abiotic/biotic stress tolerance) to generate domesticated crop ideotypes resembling those raised by traditional breeding methods. For most crop species, domestication-related genes are known and desired crop ideotypes are also established (Table 10.2). Given that most abiotic stress-tolerance pathways (including salinity) are governed by multiples genes, *de novo* domestication (DND) of CWRs by introduction of 'domestication traits' has been proposed as an alternative route to

Table 10.3. QTLs identified in CWRs associated with salinity-stress tolerance. (Content created by the authors and based on references cited.)

Crop	Modern cultivar	CWR	Mapping population[a]	QTL(s)	Chromosome number	Co-localized genes/ candidate gene	Function	References
Rice	*Oryza sativa indica* cv. 93-11	*Oryza rufipogan*	ILs	*qST5-2*	5	*Acetyltransferase*, *LRP1*, *OsCML15* (calmodulin-related calcium sensor protein)	Increased salt tolerance score and relative dry weight of seedlings	Tian *et al.* (2011); Wang *et al.* (2017)
				qST10	10	*GEM* (Gram domain-containing protein)		
	O. sativa indica cv. Teqing	*O. rufipogon*	ILs	*qSTS2* *qRRW3*	2 3	– *DST* (drought and salt tolerance)	Increased salt tolerance score	Huang *et al.* (2009); Tian *et al.* (2011)
Wheat	*Triticum turgidum* L. ssp. *durum* cv. Marrocos	*Triticum monococcum* C68-101	NILs	*Nax1* *Nax2*	2AL 5AL	*TmHKT7* *TmHKT8*	Na^+ exclusion trait	Lindsay *et al.* (2004); James *et al.* (2006)
Barley	*Hordeum vulgare* ssp. *vulgare* Barque-73	*H. vulgare* ssp. *spontaneum* CPI-71284-48	DHLs	*HvNax3*	7H	*HVP10*	Na^+ exclusion trait	Shavrukov *et al.* (2010, 2013)
Cotton	*Gossypium hirsutum* CRI-12	*Gossypium tomentosum* AD3-00	F2:3	*qRL-Chr16-1*	16D	*CotAD_49680*, *CotAD_49681*, *CotAD_49685*, *CotAD_49687*, *CotAD_49688*	Increased root length and penetration	Oluoch *et al.* (2016)
Tomato	*Solanum lycopersicum* var. *cerasiforme*	*Solanum pimpinellifolium* L. and *Solanum cheesmaniae* (L. Riley) Fosberg	RILs	*lnc1.1*; *lkn1.1*	1	*LeNHX3*	Na^+ inclusion trait	Villalta *et al.* (2008)

CWRs conferring salinity tolerance

Crop	Wild relative	Gene pool	References
Maize	*Tripsacum dactyloides*	Tertiary	Mammadov *et al*. (2018); Iqbal *et al*. (2019)
Wheat	*Triticum dicoccoides*	Primary	Feuillet *et al*. (2008); Nevo and Chen (2010); Kumar *et al*. (2022)
Barley	*Hordeum spontaneum*	Primary	
Tomato	*Solanum pimpinellifolium*	Primary	Dehan and Tal (1978); Tudge (2011); Rao *et al*. (2015); Saeed and Fatima (2021); Subhasmita and Pradeepkumara (2022)
	Solanum cheesmaniae	Primary	
	Solanum pennellii	Secondary	
Cotton	*Gossypium tomentosum*	Primary	Khadi *et al*. (2010); Mammadov *et al*. (2018); Azhar and Wani (2021)
	Gossypium davidsonii	Secondary	
	Gossypium aridum	Secondary	
	Gossypium klotzschianum	Secondary	
	Gossypium trifurcatum	Secondary	
Rice	*Poteresia coarctata*	Tertiary	Bellon *et al*. (1998), Sen *et al*. (2010), Azhar and Wani (2021)
	Oryza rufipogon	Primary	
	Oryza nivara	Primary	
	Oryza ridleyi	Tertiary	
Soybean	*Glycine soja*	Primary	Chung and Singh (2008); Azhar and Wani (2021)
	Glycine canescens	Tertiary	
	Glycine latifolia	Quaternary	
Lentil	*Lens orientalis*	Primary	Wong *et al*. (2015); Azhar and Wani (2021)
	Lens ervoides	Tertiary	
	Lens lamottei	Secondary	
	Lens nigricans	Quaternary	
Chickpea	*Cicer reticulatum*	Primary	Ladizinsky and Adler (1976); Azhar and Wani (2021)
Potato	*Solanum curtilobum*	Secondary	Machida-Hirano (2015); Azhar and Wani (2021)
	Solanum juzepczuckii	Secondary	
	Solanum kurtzianum	Secondary	

[a]IL, introgression line; NIL, near isogenic line; DHL, doubled haploid line; F2:3, F2-derived F3 population; RIL, recombinant inbred line.

obtain abiotic stress-tolerant crop species (Razzaq *et al.*, 2021). DND of wild crops species was first demonstrated for tomato and subsequently for many other crop species (Zsögön *et al.*, 2018; Gasparini *et al.*, 2021). In principle, DND can be achieved using conventional breeding methods as well as state-of-the-art gene editing tools, in particular Clustered Regularly Interspaced Short Palindromic Repeats (CRISPR)-based genome editing. While conventional breeding technologies can be used to cross species within the primary gene pool, for DND, it may be more pertinent to use gene editing-based approaches when examining CWRs that occur in the secondary and tertiary gene pools. The subsequent sections examine recent biotechnological tools that can enable CRISPR-mediated DND of naturally saline-tolerant CWRs.

10.6 Pangenomes, Focused Identification of Germplasm Strategy, and Speed Breeding for Mainstreaming Crop Wild Relatives

Breakthrough technologies like next-generation sequencing (NGS) have made it possible to obtain highly accurate chromosome-level assemblies for most crop species and attention has now turned to generating *de novo* genomes of CWRs as well as resequencing of CWR accessions. NGS-based shotgun sequencing is the method of choice that can generate large volumes of high-quality sequence data (determined also by sequencing depth) and is complemented by third-generation long-read sequencing technologies (single-molecule real-time (SMRT) and nanopore sequencing) (Brozynska *et al.*, 2017). Assembly of sequence contigs is made possible by sequencing mate pair libraries to generate scaffold data that allows for assembly of most contigs. An additional step in genome assembly is to locate these assembled fragment sequences to chromosomes using genetic or physical mapping-based methods. Within genomes, DNA sequence polymorphisms include SNPs, short insertions/deletions (indels), and larger (>50 bp) structural variations (SVs) such as presence/absence variants (PAVs), copy number variants (CNVs), inversions, and inter/intra-chromosomal translocations (Jayakodi *et al.*, 2021). Accurate *de novo* genome assembly requires high computational and infrastructural investments to obtain quality reads, long read lengths, and high genome coverage (Khan *et al.*, 2020). On the other hand, reference-based genome assembly methods using genome sequences from model crop species can bias genome assembly and often misses data on large/small-scale genome rearrangements and SVs (Brozynska *et al.*, 2017; Coletta *et al.*, 2021). Polyploidy in crop species and CWRs can compound problems related to accurate genome assembly due to heterozygosity and high repetitive DNA content (Kyriakidou *et al.*, 2020; Coletta *et al.*, 2021).

Reference accessions of model plant species such as *Arabidopsis* and rice have been chosen for historical reasons for generating 'gold standard' genome sequences. However, the chosen accessions for a given species have a defined annotated set of genes (coding and non-coding) that do not reflect the entire gamut of allelic variation seen within the species, the so-called 'pangenome'. The pangenome thus defines a core 'indispensable' component of the genome of an organism shared between all individuals of the species overlaid with a 'variable' component that confers selective advantage under specific environmental conditions (e.g. abiotic/biotic stress). Pangenomics in hexaploid wheat, *Brassica*, and rice demonstrates that the variable component of pangenomes is enriched for environmental stress- and defence-response genes (Montenegro *et al.*, 2017). Comparative genomics and transcriptomics show an expanded RWP-RK transcription factor family repertoire in pearl millet and involvement of endoplasmic reticulum (ER)-related genes in heat tolerance (Yan *et al.*, 2023). The variable component of pangenomes can arise due to whole-genome duplications i.e. tandem, transposable element (TE)-mediated, or segmental duplications, introgression from related species, horizontal gene transfer, *de novo* gene birth, pseudogenization, or intra-chromosomal deletion (Bayer *et al.*, 2020). Thus, No-*OsHKT2;2/1*, a new *HKT* isoform in the salt-tolerant rice landrace 'Nona Bokra', is a functional chimera of *OsHKT2;2* and *OsHKT2;1*, arising due to a genomic deletion in chromosome 6 of rice (Oomen *et al.*, 2012). Translocation of a 12.58 kb fragment in the third intron of maize *ZmHAK4* reduces transcript levels and salinity tolerance in the maize inbred line

'HuangC' (Zhang *et al*., 2019). The role of TEs in the generation of genic diversity within a species is well documented and is also associated in many cases with loss or gain of salinity tolerance. A 390 bp LTR/Gypsy retrotransposon insertion into the second exon of *ZmHKT1* in the maize inbred line 'Chang 7'-2 and a 3.78 kb *copia* insertion in the third exon of *GmSalt3* of the soybean salt-sensitive line '85-140' results in loss of salinity tolerance (Guan *et al*., 2014; Zhang *et al*., 2018). Finally, diversity of non-coding RNAs (ncRNAs) – including microRNAs (miRNAs), long non-coding RNAs (lncRNAs), circular RNAs (circRNAs), phased secondary small interfering/*trans*-acting small interfering RNAs (phasi/tasiRNAs), and heterochromatin-associated chromosomal RNAs (hcRNAs) – in pangenomes may also be a crucial regulatory component in abiotic stress tolerance. MiRNAs have been hypothesized to originate in plants from TEs, spontaneous generation from genic inverted repeats, spliced-out introns, and by *de novo* emergence via inverted duplication of target gene fragments, as well as expansion and neofunctionalization of existing miRNA family repertoires (Baldrich *et al*., 2018). While the presence of ncRNAs in crop species has been documented, pangenomic discovery and annotation pipelines can be used to identify and differentiate between core and variable components (Tahir ul Qamar *et al*., 2020). Some of these components may target genes crucial for salinity tolerance and may show species-specific gain/loss or neofunctionalization. For example, the *miR390*-TAS3-ARF module is conserved in seed plants and upregulation of *miR390* under salinity in poplar promotes lateral root development and increases salinity tolerance (Xia *et al*., 2017; He *et al*., 2018). Thus, any SNP, indel, or SV in *miR390*, *TAS3* locus, or ARF transcript (miRNA binding site) has the potential to alter lateral root development and hence salinity tolerance.

While pangenomes are available for many domesticated crop species such as maize (*Zea mays*), rice (*Oryza sativa*), brassicas (*Brassica* spp.), and soybean (*Glycine max*) (Tay Fernandez *et al*., 2022), super pangenomes incorporating genomic information from CWRs have only recently been reported (Tao *et al*., 2019). Pangenomics also demonstrates that domestication and breeding are associated with significant gene loss in cultivars relative to CWRs (Golicz *et al*., 2016; Bayer *et al*., 2021). Differential selection pressures on specific classes of small ncRNAs (phasiRNAs, hcRNAs) during tomato domestication from wild relatives correlates with alteration of mRNA expression levels especially for genes associated with biotic and abiotic stress tolerance (Qing *et al*., 2022). PAVs in the rice pangenome (66 cultivated and wild *O. rufipogon* accessions) are enriched in abiotic/biotic stress-related genes, suggesting cultivated rice and the CWR also vary in their underlying stress responses. This is to be expected as the provenances of many CWRs are often geographically constrained and niche specific, with many CWRs showing unique ecophysiological features and associated adaptive traits conferred by SV or expression-level differences (adaptive genomic imprints) in abiotic stress-tolerance genes.

While core collections (showing wide genetic and phenotypic diversity) for many CWRs are available in gene banks worldwide, a Focused Identification of Germplasm Strategy (FIGS) can be used to map plant (crop/CWR) adaptive traits more efficiently by applying eco-geographic filtering, i.e. utilizing available geographic and agroclimatic information to efficiently identify crop/CWR accessions as donors for trait-based phenotyping (Thormann *et al*., 2014). Using FIGS, probability mapping information on saline soils was overlaid on geographical information on wheat collection sites to shortlist a subset of bread wheat accessions for phenotypic screening. The FIGS approach was more effective in shortlisting saline-tolerant accessions (21%) relative to the core collection method (3%) for screening (Street *et al*., 2016). A similar approach with CWRs would provide more rapid screening methods to identify salt-tolerant accessions.

Another rate-limiting step in plant breeding is the crop growing cycle, governed by photoperiod and/or temperature, leading to only one or two generations per year. This can be circumvented by using speed breeding (SB) technology that shortens the breeding cycle and accelerates crop research through rapid generation advancement (Ghosh *et al*., 2018). SB involves extending the plant photoperiod using supplementary lighting. Coupled with controlled temperature regimes, SB has been shown to reduce generation time in spring wheat (*Triticum aestivum*), durum wheat (*Triticum durum*), barley, chickpea, and pea and accelerate breeding in canola (Watson *et al.*, 2018). Ghosh *et al.* (2018) suggest that in addition to long photoperiods, additional physiological stress conditions (high planting density, paucity of nutrient/water, intense light) contribute to early generation times using SB methodologies. SB can be flexibly incorporated with other modern crop breeding technologies, such as high-throughput genotyping, genome editing, and genomic selection (Watson *et al.*, 2018; Hickey *et al.*, 2019). Using SB techniques for phenotyping, Riaz *et al.* (2016) identified landraces showing resistance to wheat rust at the adult stage from a collection sourced from the N.I. Vavilov Institute of Plant Genetic Resources, Russia. Similar approaches for screening CWRs can be applied for salinity tolerance but must be supplemented with field-based phenotyping data for accurate validation.

10.7 Advances in CRISPR/Cas-Based Genome Editing Technology for Crop Species

Major advances in NGS- and third-generation sequencing (TGS)-based technologies and parallel development of genome assembly tools pinpoint relationships between structural genomic variants and plant agronomic traits to be exploited by standard breeding technologies. This is also possible via genome editing using 'engineered' single-strand nucleases (SSNs). Genome editing depends on two important premises: (i) the ability to 'program' SSNs to 'home in' or target specific regions of the genome to introduce nick or double-strand breaks (DSBs); and (ii) the ability to 'edit' specific sequences in a genome. Second-generation programmable RNA-guided DNA targeting technologies, like the CRISPR and CRISPR-associated system (Cas nucleases) systems, have made DNA targeting more simpler and easier to achieve in practice, with less off-target effects as compared to first-generation SSNs (Nasti and Voytas, 2021).

Derived from accidental discoveries of bacterial immunity-conferring systems, two key components of the CRISPR technology: are (i) the Cas nuclease; and (ii) the sequence specific 'guide' RNA (gRNA) that allows the Cas protein to 'home in' on the target site within a given genome (Ishino *et al.*, 2018). Thus, in principle, by manipulating the gRNA sequence, the engineered Cas system can be programmed to target any DNA sequence for cleavage (Jinek *et al.*, 2012). The most used CRISPR/Cas9 system is derived from *Streptococcus pyogenes*. In its native form, Cas9 has both target recognition and nuclease activities (conferred by HNH and RuvC endonuclease domains), the latter generating DSBs. DSBs trigger cellular repair systems: more commonly, imprecise repair via non-homologous end-joining (NHEJ) and, less commonly, precise repair also called homology-directed repair (HDR). NHEJ-based imprecise repair mechanisms are the basis for introducing 'genome editing' changes such as SNPs and insertions/deletions (indels).

Although CRISPR/Cas9 has paved the way for precise genome editing, the limitations associated with this technology – such as sequence specificity, large length of the editing sequence, and compatibility issues of the protospacer adjacent motif (PAM) sequence – have necessitated the need for development of third-generation genome editing technologies. These methodologies: (i) rely on template-free genome editing methodologies that do not require DSBs; and (ii) utilize either a catalytically inactive version of Cas9 (deadCas9 or dCAs9) or partially active Cas9 nickase (nCas9). dCas9 or a specific variant of nCas9, when fused with either adenosine or cytosine deaminases result in chimeric proteins that are called base editors. Base editors introduce single base changes by converting A-T to G-C and C-G to T-A, respectively, avoiding DSB generation, thereby limiting occurrence of indels at target and off-target sites and have the potential to revolutionize development of agronomically important superior crop varieties (Molla and Yang, 2019; Mishra *et al.*, 2020).

In a parallel advancement, 'prime editing' utilizes a 'search and replace' strategy, altering the genomic sequence, again, without the induction of DSBs. The prime editor is a chimeric protein in which an engineered reverse transcriptase (RT) is fused to a specific variant of nCas9 (Hua *et al.*, 2020; Marzec *et al.*, 2020). The prime editing guide RNA (pegRNA) consists of a gRNA, an RT template with the desired edited sequence, and a primer binding site (PBS) for activation of RT. nCAS9 is directed to the

DNA by the pegRNA, to make a nick in the non-target DNA strand containing the PAM site. Guided by the RT template within the pegRNA, the RT engineers the desired edits from the pegRNA to the DNA (Anzalone *et al.*, 2019). Of all the possible 12 base-to-base conversions, base editing can perform only four possible transitional mutations but not the eight transversional mutations, while prime editing can be used to effect the 12 base conversions (Anzalone *et al.*, 2019). Different versions of prime editors have been developed (PE1–PE3) and prime editing has been demonstrated both in mammalian cells and in some plant species (Lu, S. *et al.*, 2022). However, the use of prime editing in plants has low efficiency and more efficient prime editors are being developed for use currently (Li *et al.*, 2022).

10.8 Delivering Where it Matters: Advances in Plant Transformation and CRISPR Construct Delivery *In Planta*

Agrobacterium- and/or particle bombardment-mediated methods have been traditionally used over the last 30 years to carry out plant transformation to deliver DNA/RNA into cells. However, these plant transformation approaches are often inefficient, involve tedious optimization of species- or genotype-specific tissue culture protocols, and are laborious and time-consuming to implement. In addition, frequencies of stable transformation are usually low (Altpeter *et al.*, 2016). Further, most transformation protocols in crop species utilize model genotypes that are often agronomically undesirable and tissue culture can lead to genomic and epigenomic changes that can affect agronomic performance. These limitations would also affect using CRISPR technology on a large scale in crop species when using tissue culture-based methods for transformation.

However, utilizing plant developmental regulators (DRs) like *Wuschel* and *Babyboom* during transgene delivery (also called morphogene-assisted transformation) circumvents the need for tissue culture to generate transgenes in recalcitrant monocot crop species or genotypes (Lowe *et al.*, 2016; Nelson-Vasilchik *et al.*, 2018). DRs in different combinations can be used for meristem induction (Maher *et al.*, 2020; Nasti *et al.*, 2021; Fast-TrACC method) or increasing the efficiency of transgene events (Chen *et al.*, 2022). Further, when co-delivered with CRISPR reagents to shoots of plants growing in soil it can lead to 'edited' plants, circumventing the need for tissue culture (Maher *et al.*, 2020; Aregawi *et al.*, 2022; Chen *et al.*, 2022). More recently Pan *et al.* (2022) have developed simultaneous genome editing and morphogen delivery-based CRISPR-Combo methods to accelerate transformation efficiencies. Replacing DRs with florigen (FT), a known flowering activator, can accelerate the breeding cycle. This is akin to physiological 'speed breeding', circumventing the need for a physical SB set-up.

10.9 CRISPR and *De Novo* Domestication of Polyploids: Present Status

As mentioned previously, while primary gene pools of CWRs can be sources of desired traits that can be transferred through traditional crosses, cross-incompatible secondary and tertiary gene pools of CWRs can be *de novo* domesticated using a CRISPR-based strategy. DND using CRISPR has been demonstrated for wild tomato (*Solanum pimpenellifolium*), with Cas9-free domesticated phenotypes retaining parental disease resistance and salt-tolerance traits (Li *et al.* 2018). By establishing an efficient tissue culture system, six agronomic domestication-related traits known in cultivated rice (seed shattering, awn length, plant height, grain size, stem thickness, and heading date) were altered in tetraploid *Oryza alta* (CCDD; a high-quality genome assembly reported prior to CRISPR modification) (Yu *et al.*, 2021). In this study, both homeologs of a gene (corresponding to C and D sub-genomes) controlling a given agronomic trait in both sub-genomes were targeted for CRISPR-based genomic editing. This landmark research effectively demonstrates two crucial aspects:

1. The possibility of genome editing in polyploid species by targeting all sub-genome homeologs. This is crucial, given that most economically important crops are polyploid (e.g. wheat, potato, cotton, soybean) and also that polyploidy is distinctly associated with salinity tolerance in many species (Shabala, 2019; Van de Peer *et al.*, 2021). Polyploid species show high K^+ retention in leaf tissues under salinity, a crucial component of tissue tolerance, driven by processes within the root (Chao *et al.*, 2013; Liu *et al.*, 2019; Fischer *et al.*, 2022). Polyploidization also causes significant alterations in methylation of stress-responsive gene-proximal TEs in tetraploid rice root tissues, causing rapid and strong expression under salinity (Wang, L. *et al.*, 2021). CRISPR editing in *O. alta* suggests that it may be possible to target domestication-related traits in other salt-tolerant polyploid crop species using genome editing. Further, this could be substituted or complemented by CRISPR/Cas9 genome editing of conserved *cis*-elements in the promoter(s) of domestication-related genes, to engineer quantitative trait variation in CWRs to obtain 'gradual' phenotypes rather than drastic 'on/off' phenotypes. *Cis*-element genome editing to obtain a continuum of quantitative traits has been demonstrated in tomato (Rodríguez-Leal *et al.*, 2017; Li *et al.*, 2018) and maize (Liu *et al.*, 2021). 'Gradual' phenotypes may be more beneficial in polyploid species as feedback mechanisms may lead to sudden plant ideotype changes when using 'on/off' genome editing.

2. Expanding the repertoire of genome editing tools to include base editing and selective editing in polyploid species. By altering the *miR156*-DD genome binding site in the *Ideal Plant Architecture* (*IPA*)-DD sub-genome component of *O. alta* using a base editor, an *miR156*-DD resistant version of *IPA*-DD was created. The presence of a naturally occurring *miR156*-CC resistant *IPA*-CC and a genome-edited *IPA*-DD (resistant version created) conferred significantly higher stem diameter relative to the wild type. This suggests that use of base editing and/or selective targeting of genomic sequences in polyploid species can lead to domestication phenotypes.

While genome editing in *O. alta* necessitated the development of robust tissue culture protocols, this can be circumvented by using DRs to rapidly generate genome-edited progeny in polyploid species without using tissue culture. More recently, Lin *et al.* (2022) have reported further refinements in CRISPR/Cas9 genome editing of wild tomato using protoplasts and cell-free CRISPR reagents (CRISPR ribonucleoprotein complex) to raise 'edited' tomato plants (targeted mutagenesis efficiency of 60%) without use of a selection marker. Kuang *et al.* (2022) have reported the genome of highly salt-tolerant sea barley grass and development of a genome editing protocol for the same species, potentially paving the way for DND.

10.10 A Possible Way Forward: Combining Technologies to Achieve *De Novo* Domestication of Saline-Tolerant Crop Wild Relatives

Based on information above, a proposed pipeline of action is suggested to *de novo* domesticate CWRs for salinity tolerance (Fig. 10.1):

1. FIGS of CWR germplasm can be used to eco-geographically 'filter' and reduce CWRs used for trait-based phenotyping. This will enable selection of accessions with optimal salt-tolerance traits for DND.

2. Whole-genome sequencing of the salt-tolerant CWR accession(s) is important as a preliminary step to identify structural genomic variants to facilitate development of optimal CRISPR

Fig. 10.1. CRISPR-based strategy to *de novo* domesticate saline-tolerant crop wild relatives (CWRs): Focused Identification of Germplasm Strategy (FIGS) to eco-geographically 'filter' accessions of CWRs followed by trait phenotyping (for salt tolerance). Whole-genome sequencing (next-generation sequencing (NGS), third-generation sequencing (TGS), or combinatorial methods) of salt-tolerant CWR accession(s), to identify structural genomic variants, is a crucial step to facilitate optimal CRISPR strategies for introduction of domestication traits. CRISPR/Cas9 methods for introducing domestication traits in CWRs (insertion/deletion/knock-in; base editing using cytosine base editors (CBEs) and adenine base editors (ABEs); prime editing) and efficient transgene delivery systems utilizing developmental regulators (DRs) can accelerate the *de novo* domestication process. Subsequently rapid generation advancement using speed breeding can be used to obtain CRISPR-modified CWRs for phenotyping in the field for salinity tolerance. (Authors' own figure.)

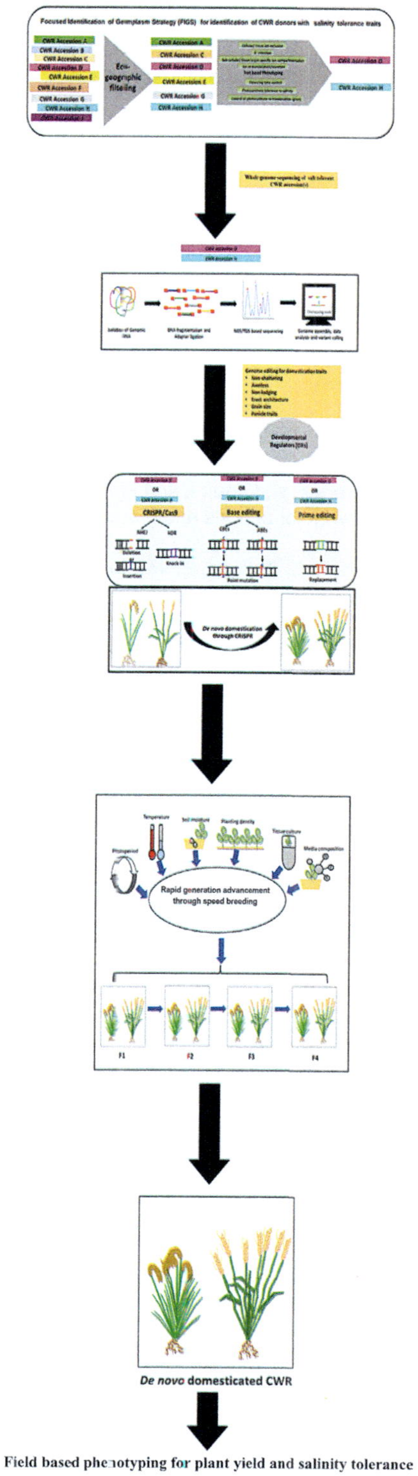
CRISPR/Cas9
Base editing
Prime editing
Rapid generation advancement through speed breeding
F1
F2
F3
F4
De novo domesticated CWR
Field based phenotyping for plant yield and salinity tolerance

strategies for introduction of domestication traits.

3. CRISPR/Cas9, base editing, or prime editing can be used to introduce domestication traits in CWRs, facilitated by use of DRs to accelerate the editing/tissue culture process.

4. *De novo* domesticated crop lines can be advanced to subsequent generations in a relatively shorter time span, again by use of rapid generation advancement techniques (SB). Alternatively, CRISPR-Combo with the use of FT can circumvent the need for SB.

5. The final *de novo* domesticated crop lines can be phenotyped for salinity tolerance and yield-related traits.

Crop domestication is a form of artificial selection, driven by anthropogenic interventions and needs. It is now possible, more than ever before, to understand the basis of crop domestication using the technologies mentioned above. The same technologies can also be harnessed to *de novo* domesticate CWRs that are naturally saline tolerant, to ensure food security in this era of climate change.

References

Abbo, S., van-Oss, R.P., Gopher, A., Saranga, Y., Ofner, I. and Peleg, Z. (2014) Plant domestication versus crop evolution: a conceptual framework for cereals and grain legumes. *Trends in Plant Science* 19(6), 351–360.

Alqahtani, M., Lightfoot, D.J., Lemtiri-Chlieh, F., Bukhari, E., Pardo, J.M. *et al.* (2021) The role of PQL genes in response to salinity tolerance in *Arabidopsis* and barley. *Plant Direct* 5(2), e00301.

Altpeter, F., Springer, N.M., Bartley, L.E., Blechl, A.E., Brutnell, T.P. *et al.* (2016) Advancing crop transformation in the era of genome editing. *The Plant Cell* 28(7), 1510–1520.

Anzalone, A.V., Randolph, P.B., Davis, J.R., Sousa, A.A., Koblan, L.W. *et al.* (2019) Search-and-replace genome editing without double-strand breaks or donor DNA. *Nature* 576(7785), 149–157.

Apse, M.P. and Blumwald, E. (2007) Na^+ transport in plants. *FEBS Letters* 581(12), 2247–2254.

Apse, M.P., Aharon, G.S., Snedden, W.A. and Blumwald, E. (1999) Salt tolerance conferred by overexpression of a vacuolar Na^+/H^+ antiport in *Arabidopsis*. *Science* 285(5431), 1256–1258.

Aregawi, K., Shen, J., Pierroz, G., Sharma, M.K., Dahlberg, J. *et al.* (2022) Morphogene-assisted transformation of *Sorghum bicolor* allows more efficient genome editing. *Plant Biotechnology Journal* 20(4), 748–760.

Assaha, D.V., Ueda, A., Saneoka, H., Al-Yahyai, R. and Yaish, M.W. (2017) The role of Na^+ and K^+ transporters in salt stress adaptation in glycophytes. *Frontiers in Physiology* 8, 509.

Azhar, M.T. and Wani, S.H. (2021) *Wild Germplasm for Genetic Improvement in Crop Plants*. Academic Press, London.

Bai, Y. and Lindhout, P. (2007) Domestication and breeding of tomatoes: what have we gained and what can we gain in the future? *Annals of Botany* 100(5), 1085–1094.

Baldrich, P., Beric, A. and Meyers, B.C. (2018) Despacito: the slow evolutionary changes in plant microRNAs. *Current Opinion in Plant Biology* 42, 16–22.

Barberon, M. and Geldner, N. (2014) Radial transport of nutrients: the plant root as a polarized epithelium. *Plant Physiology* 166(2), 528–537.

Barmukh, R., Roorkiwal, M., Garg, V., Khan, A.W., German, L. *et al.* (2022) Genetic variation in CaTIFY4b contributes to drought adaptation in chickpea. *Plant Biotechnology Journal* 20(9), 1701–1715.

Bassil, E., Tajima, H., Liang, Y.C., Ohto, M.A., Ushijima, K. *et al.* (2011) The *Arabidopsis* Na^+/H^+ antiporters *NHX1* and *NHX2* control vacuolar pH and K^+ homeostasis to regulate growth, flower development, and reproduction. *The Plant Cell* 23(9), 3482–3497.

Bayer, P.E., Golicz, A.A., Scheben, A., Batley, J. and Edwards, D. (2020) Plant pan-genomes are the new reference. *Nature Plants* 6(8), 914–920.

Bayer, P.E., Scheben, A., Golicz, A.A., Yuan, Y., Faure, S. *et al.* (2021) Modelling of gene loss propensity in the pangenomes of three *Brassica* species suggests different mechanisms between polyploids and diploids. *Plant Biotechnology Journal* 19(12), 2488–2500.

Bellon, M.R., Brar, D.S., Lu, B.R. and Pham, J.L. (1998) Rice genetic resources. In: Dowling, N.G., Greenfield, S.M. and Fischer, K.S. (eds) *Sustainability of Rice in the Global Food System*. Pacific Basin Study Center/International Rice Research Institute, Davis, California/Manila, pp. 251–283.

Benito, B., Haro, R., Amtmann, A., Cuin, T.A. and Dreyer, I. (2014) The twins K^+ and Na^+ in plants. *Journal of Plant Physiology* 171(9), 723–731.

Berthomieu, P., Conéjéro, G., Nublat, A., Brackenbury, W.J., Lambert, C. *et al.* (2003) Functional analysis of *AtHKT1* in *Arabidopsis* shows that Na^+ recirculation by the phloem is crucial for salt tolerance. *The EMBO Journal* 22(9), 2004–2014.

Blumwald, E., Aharon, G.S. and Apse, M.P. (2000) Sodium transport in plant cells. *Biochimica et Biophysica Acta (BBA) – Biomembranes* 1465(1–2), 140–151.

Bohra, A., Tiwari, A., Kaur, P., Ganie, S.A., Raza, A. *et al.* (2022) The key to the future lies in the past: insights from grain legume domestication and improvement should inform future breeding strategies. *Plant & Cell Physiology* 63(11), 1554–1572.

Brozynska, M., Copetti, D., Furtado, A., Wing, R.A., Crayn, D. *et al.* (2017) Sequencing of Australian wild rice genomes reveals ancestral relationships with domesticated rice. *Plant Biotechnology Journal* 15(6), 765–774.

Byrne, P., Richards, C. and Volk, G.M. (2020) From wild species to landraces and cultivars. In: Volk, G.M. and Byrne, P. (eds) *Crop Wild Relatives and their Use in Plant Breeding*. Colorado State University, Fort Collins, Colorado. Available at: https://colostate.pressbooks.pub/cropwildrelatives/chapter/from-wild-species-to-landraces-and-cultivars/ (accessed 26 February 2024).

Byrt, C.S., Platten, J.D., Spielmeyer, W., James, R.A., Lagudah, E.S. *et al.* (2007) *HKT1;5*-like cation transporters linked to Na^+ exclusion loci in wheat, Nax2 and Kna1. *Plant Physiology* 143(4), 1918–1928.

Cao, Y., Liang, X., Yin, P., Zhang, M. and Jiang, C. (2019) A domestication-associated reduction in K^+-preferring HKT transporter activity underlies maize shoot K^+ accumulation and salt tolerance. *New Phytologist* 222(1), 301–317.

Chao, D.Y., Dilkes, B., Luo, H., Douglas, A., Yakubova, E. *et al.* (2013) Polyploids exhibit higher potassium uptake and salinity tolerance in *Arabidopsis*. *Science* 341(6146), 658–659.

Cheeseman, J.M. (2015) The evolution of halophytes, glycophytes and crops, and its implications for food security under saline conditions. *New Phytologist* 206(2), 557–570.

Chen, J., Li, S., He, Y., Li, J. and Xia, L. (2022) An update on precision genome editing by homology-directed repair in plants. *Plant Physiology* 188(4), 1780–1794.

Chen, Q., Li, W., Tan, L. and Tian, F. (2021) Harnessing knowledge from maize and rice domestication for new crop breeding. *Molecular Plant* 14(1), 9–26.

Chen, T., Shabala, S., Niu, Y., Chen, Z.H., Shabala, L. *et al.* (2021) Molecular mechanisms of salinity tolerance in rice. *The Crop Journal* 9(3), 506–520.

Choudhary, O.P. and Kharche, V.K. (2018) Soil salinity and sodicity. In: *Soil Science: An Introduction*. Indian Society of Soil Science, New Delhi, pp. 353–384.

Chung, G. and Singh, R.J. (2008) Broadening the genetic base of soybean: a multidisciplinary approach. *Critical Reviews in Plant Sciences* 27(5), 295–341.

Colmenero-Flores, J.M., Martínez, G., Gamba, G., Vázquez, N., Iglesias, D.J. *et al.* (2007) Identification and functional characterization of cation–chloride cotransporters in plants. *The Plant Journal* 50(2), 278–292.

Colmer, T.D. and Flowers, T.J. (2008) Flooding tolerance in halophytes. *New Phytologist* 179(4), 964–974.

DeHaan, L., Larson, S., López-Marqués, R.L., Wenkel, S., Gao, C. and Palmgren, M. (2020) Roadmap for accelerated domestication of an emerging perennial grain crop. *Trends in Plant Science* 25(6), 525–537.

Dehan, K. and Tal, M. (1978) Salt tolerance in the wild relatives of the cultivated tomato: Responses of *Solanum pennellii* to high salinity. *Irrigation Science* 1(1), 71–76.

Della Coletta, R., Qiu, Y., Ou, S., Hufford, M.B. and Hirsch, C.N. (2021) How the pan-genome is changing crop genomics and improvement. *Genome Biology* 22(1), 3.

Demidchik, V. (2006) Physiological roles of nonselective cation channels in the plasma membrane of higher plants. In: Baluška, F., Mancuso, S. and Volkmann, D. (eds) *Communication in Plants*. Springer, Berlin, Heidelberg, Germany, pp. 235–248.

Demidchik, V. and Maathuis, F.J. (2007) Physiological roles of nonselective cation channels in plants: from salt stress to signalling and development. *New Phytologist* 175(3), 387–404.

Demidchik, V. and Tester, M. (2002) Sodium fluxes through nonselective cation channels in the plasma membrane of protoplasts from *Arabidopsis* roots. *Plant Physiology* 128(2), 379–387.

Denham, T., Barton, H., Castillo, C., Crowther, A., Dotte-Sarout, E. *et al.* (2020) The domestication syndrome in vegetatively propagated field crops. *Annals of Botany* 125(4), 581–597.

DeWitt, N., Guedira, M., Lauer, E., Sarinelli, M., Tyagi, P. *et al.* (2019) Sequence based mapping identifies *AWNS1*, a candidate transcription repressor underlying awn suppression at the *B1* locus in wheat. *bioRxiv*, 519264. https://doi.org/10.1101/519264.

Dorweiler, J., Stec, A., Kermicle, J. and Doebley, J. (1993) *Teosinte glume architecture 1*: a genetic locus controlling a key step in maize evolution. *Science* 262(5131), 233–235.

El Mahi, H., Pérez-Hormaeche, J., De Luca, A., Villalta, I., Espartero, J. *et al.* (2019) A critical role of sodium flux via the plasma membrane Na^+/H^+ exchanger *SOS1* in the salt tolerance of rice. *Plant Physiology* 180(2), 1046–1065.

FAO (2016) *The State of Food and Agriculture 2016: Climate Change, Agriculture and Food Security.* Food and Agriculture Organization of the United Nations, Rome. Available at: http://www.fao.org/3/i6030e/i6030e.pdf (accessed 26 February 2024).

Fernie, A.R. and Yan, J. (2019) *De novo* domestication: an alternative route toward new crops for the future. *Molecular Plant* 12(5), 615–631.

Feuillet, C., Langridge, P. and Waugh, R. (2008) Cereal breeding takes a walk on the wild side. *Trends in Genetics* 24(1), 24–32.

Fischer, S., Flis, P., Zhao, F.J. and Salt, D.E. (2022) Transcriptional network underpinning ploidy-related elevated leaf potassium in neo-tetraploids. *Plant Physiology* 190(3), 1715–1730.

Flowers, T.J. and Colmer, T.D. (2015) Plant salt tolerance: adaptations in halophytes. *Annals of Botany* 115(3), 327–331.

Fuller, D.Q., Harvey, E. and Qin, L. (2007) Presumed domestication? Evidence for wild rice cultivation and domestication in the fifth millennium BC of the Lower Yangtze region. *Antiquity* 81(312), 316–331.

Ganie, S.A. (2020) RNA chaperones: potential candidates for engineering salt tolerance in rice. *Crop Science* 60(2), 530–540.

Ganie, S.A. (2021) Amino acids other than proline and their participation in abiotic stress tolerance. In: Wani, S.H., Gangola, M.P. and Ramadoss, B.R. (eds) *Compatible Solutes Engineering for Crop Plants Facing Climate Change*. Springer, Cham, Switzerland, pp. 47–96.

Ganie, S.A., Debnath, A.B., Gumi, A.M. and Mondal, T.K. (2017) Comprehensive survey and evolutionary analysis of genome-wide miRNA genes from ten diploid *Oryza* species. *BMC Genomics* 18(1), 711.

Gasparini, K., dos Reis Moreira, J., Peres, L.E.P. and Zsögön, A. (2021) *De novo* domestication of wild species to create crops with increased resilience and nutritional value. *Current Opinion in Plant Biology* 60, 102006.

Gepts, P. (2010) Crop domestication as a long-term selection experiment. *Plant Breeding Reviews* 24(2), 1–44.

Ghosh, S., Watson, A., Gonzalez-Navarro, O.E., Ramirez-Gonzalez, R.H., Yanes, L. *et al.* (2018) Speed breeding in growth chambers and glasshouses for crop breeding and model plant research. *Nature Protocols* 13(12), 2944–2963.

Golicz, A.A., Bayer, P.E., Barker, G.C., Edger, P.P., Kim, H. *et al.* (2016) The pangenome of an agronomically important crop plant *Brassica oleracea*. *Nature Communications* 7(1), 13390.

Gong, J.M., Waner, D.A., Horie, T., Li, S.L., Horie, R. *et al.* (2004) Microarray-based rapid cloning of an ion accumulation deletion mutant in *Arabidopsis thaliana*. *Proceedings of the National Academy of Sciences USA* 101(43), 15404–15409.

Gonzalez Guzman, M., Cellini, F., Fotopoulos, V., Balestrini, R. and Arbona, V. (2022) New approaches to improve crop tolerance to biotic and abiotic stresses. *Physiologia Plantarum* 174(1), e13547.

Gross, B.L. and Olsen, K.M. (2010) Genetic perspectives on crop domestication. *Trends in Plant Science* 15(9), 529–537.

Guan, R., Qu, Y., Guo, Y., Yu, L., Liu, Y. *et al.* (2014) Salinity tolerance in soybean is modulated by natural variation in *GmSALT3*. *The Plant Journal* 80(6), 937–950.

Hammer, K. (1984) The domestication syndrome (in German). *Kulturpflanze* 32, 11–34.

Hao, H., Li, Z., Leng, C., Lu, C., Luo, H. *et al.* (2021) Sorghum breeding in the genomic era: opportunities and challenges. *Theoretical and Applied Genetics* 134(7), 1899–1924.

Harlan, J.R., De Wet, J.M.J. and Price, E.G. (1973) Comparative evolution of cereals. *Evolution* 27(2), 311–325.

Hassani, A., Azapagic, A. and Shokri, N. (2020) Predicting long-term dynamics of soil salinity and sodicity on a global scale. *Proceedings of the National Academy of Sciences USA* 117(52), 33017–33027.

Hazell, P.B. (2002) *Green Revolution: Curse or Blessing?* International Food Policy Research Institute, Washington, DC. Available at: https://www.ifpri.org/publication/green-revolution (accessed 26 February 2024).

He, F., Xu, C., Fu, X., Shen, Y., Guo, L. *et al.* (2018) The MicroRNA390/TRANS-ACTING SHORT INTERFERING RNA3 module mediates lateral root growth under salt stress via the auxin pathway. *Plant Physiology* 177(2), 775–791.

Henderson, S.W., Wege, S., Qiu, J., Blackmore, D.H., Walker, A.R. *et al.* (2015) Grapevine and *Arabidopsis* cation–chloride cotransporters localize to the Golgi and trans-Golgi network and indirectly influence long-distance ion transport and plant salt tolerance. *Plant Physiology* 169(3), 2215–2229.

Henderson, S.W., Wege, S. and Gilliham, M. (2018) Plant cation–chloride cotransporters (CCC): evolutionary origins and functional insights. *International Journal of Molecular Sciences* 19(2), 492.

Hickey, L.T., Hafeez, A.N., Robinson, H., Jackson, S.A., Leal-Bertioli, S. *et al.* (2019) Breeding crops to feed 10 billion. *Nature Biotechnology* 37(7), 744–754.

Hillman, G.C. and Davies, M.S. (1990) Measured domestication rates in wild wheats and barley under primitive cultivation, and their archaeological implications. *Journal of World Prehistory* 4(2), 157–222.

Hong, Y., Guan, X., Wang, X., Kong, D., Yu, S. *et al.* (2023) Natural variation in *SlSOS2* promoter hinders salt resistance during tomato domestication. *Horticulture Research* 10(1), uhac244.

Houston, K., Qiu, J., Wege, S., Hrmova, M., Oakey, H. *et al.* (2020) Barley sodium content is regulated by natural variants of the Na^+ transporter *HvHKT1;5*. *Communications Biology* 3(1), 258.

Hua, K., Jiang, Y., Tao, X. and Zhu, J.K. (2020) Precision genome engineering in rice using prime editing system. *Plant Biotechnology Journal* 18(11), 2167.

Hua, L., Wang, D.R., Tan, L., Fu, Y., Liu, F. *et al.* (2015) *LABA1*, a domestication gene associated with long, barbed awns in wild rice. *The Plant Cell* 27(7), 1875–1888.

Huang, S., Spielmeyer, W., Lagudah, E.S., James, R.A., Platten, J.D. *et al.* (2006) A sodium transporter (HKT7) is a candidate for *Nax1*, a gene for salt tolerance in durum wheat. *Plant Physiology* 142(4), 1718–1727.

Huang, X.Y., Chao, D.Y. Gao, J.P., Zhu, M.Z., Shi, M. and Lin, H.X. (2009) A previously unknown zinc finger protein, DST, regulates drought and salt tolerance in rice via stomatal aperture control. *Genes & Development* 23(15), 1805–1817.

Hübner, C.A. and Rust, M.B. (2007) Physiology of cation–chloride cotransporters. In: *Chloride Movements Across Cellular Membranes*. Advances in Molecular and Cell Biology, Vol. 38. Elsevier, Amsterdam, pp. 241–277.

Iqbal, M.Z., Cheng, M., Su, Y., Li, Y., Jiang, W. *et al.* (2019) Allopolyploidization facilitates gene flow and speciation among corn, *Zea perennis and Tripsacum dactyloides. Planta* 249(6), 1949–1962.

Ishikawa, T., Shabala, L., Zhou, M., Venkataraman, G., Yu, M. *et al.* (2022) Comparative analysis of root Na^+ relation under salinity between *Oryza sativa* and *Oryza coarctata*. *Plants (Basel)* 11(5), 656.

Ishino, Y., Krupovic, M. and Forterre, P. (2018) History of CRISPR–Cas from encounter with a mysterious repeated sequence to genome editing technology. *Journal of Bacteriology* 200(7), e00580–17.

James, R.A., Davenport, R.J. and Munns, R. (2006) Physiological characterization of two genes for Na^+ exclusion in durum wheat, *Nax1* and *Nax2*. *Plant Physiology* 142(4), 1537–1547.

Jayakodi, M., Schreiber, M., Stein, N. and Mascher, M. (2021) Building pan-genome infrastructures for crop plants and their use in association genetics. *DNA Research* 28(1), dsaa030.

Jegadeeson, V., Kumari, K., Pulipati, S., Parida, A. and Venkataraman, G. (2019) Expression of wild rice *Porteresia coarctata PcNHX1* antiporter gene (*PcNHX1*) in tobacco controlled by *PcNHX1* promoter (*PcNHX1p*) confers Na^+-specific hypocotyl elongation and stem-specific Na^+ accumulation in transgenic tobacco. *Plant Physiology and Biochemistry* 139, 161–170.

Jinek, M., Chylinski, K., Fonfara, I., Hauer, M., Doudna, J.A. and Charpentier, E. (2012) A programmable dual-RNA–guided DNA endonuclease in adaptive bacterial immunity. *Science* 337(6096), 816–821.

Khadi, B.M., Santhy, V. and Yadav, M.S. (2010) Cotton: an introduction. In: Zehr, U.B. (ed.) *Cotton*. Springer, Berlin, Heidelberg, Germany, pp. 1–14.

Khan, A.W., Garg, V., Roorkiwal, M., Golicz, A.A., Edwards, D. and Varshney, R.K. (2020) Super-pangenome by integrating the wild side of a species for accelerated crop improvement. *Trends in Plant Science* 25(2), 148–158.

Kronzucker, H.J. and Britto, D.T. (2011) Sodium transport in plants: a critical review. *New Phytologist* 189(1), 54–81.

Kuang, L., Shen, Q., Chen, L., Ye, L., Yan, T. *et al.* (2022) The genome and gene editing system of sea barleygrass provide a novel platform for cereal domestication and stress tolerance studies. *Plant Communications* 3(5), 100333.

Kumar, A., Choudhary, A., Kaur, H. and Mehta, S. (2022) A walk towards wild grasses to unlock the clandestine of gene pools for wheat improvement: a review. *Plant Stress* 3, 100048.

Kyriakidou, M., Anglin, N.L., Ellis, D., Tai, H.H. and Strömvik, M.V. (2020) Genome assembly of six polyploid potato genomes. *Scientific Data* 7(1), 88.

Ladizinsky, G. and Adler, A. (1976) Genetic relationships among the annual species of *Cicer* L. *Theoretical and Applied Genetics* 48(4), 197–203.

Lenser, T. and Theißen, G. (2013) Molecular mechanisms involved in convergent crop domestication. *Trends in Plant Science* 18(12), 704–714.

Li, T., Yang, X., Yu, Y., Si, X., Zhai, X. *et al.* (2018) Domestication of wild tomato is accelerated by genome editing. *Nature Biotechnology* 36(12), 1160–1163.

Li, X., Zhou, L., Gao, B.Q., Li, G., Wang, X. *et al.* (2022) Highly efficient prime editing by introducing same-sense mutations in pegRNA or stabilizing its structure. *Nature Communications* 13(1), 1669.

Lin, C.S., Hsu, C.T., Yuan, Y.H., Zheng, P.X., Wu, F.H. *et al.* (2022) DNA-free CRISPR–Cas9 gene editing of wild tetraploid tomato *Solanum peruvianum* using protoplast regeneration. *Plant Physiology* 188(4), 1917–1930.

Lindsay, M.P., Lagudah, E.S., Hare, R.A. and Munns, R. (2004) A locus for sodium exclusion (*Nax1*), a trait for salt tolerance, mapped in durum wheat. *Functional Plant Biology* 31(11), 1105–1114.

Liu, L., Gallagher, J., Arevalo, E.D., Chen, R., Skopelitis, T. *et al.* (2021) Enhancing grain-yield-related traits by CRISPR–Cas9 promoter editing of maize *CLE* genes. *Nature Plants* 7(3), 287–294.

Liu, Y., Yu, Y., Sun, J., Cao, Q., Tang, Z. *et al.* (2019) Root-zone-specific sensitivity of K^+- and Ca^{2+}-permeable channels to H_2O_2 determines ion homeostasis in salinized diploid and hexaploid *Ipomoea trifida*. *Journal of Experimental Botany* 70(4), 1389–1405.

Lowe, K., Wu, E., Wang, N., Hoerster, G., Hastings, C. *et al.* (2016) Morphogenic regulators *Baby boom* and *Wuschel* improve monocot transformation. *The Plant Cell* 28(9), 1998–2015.

Lu, C., Kuang, J., Shao, T., Xie, S., Li, M. *et al.* (2022) Prime editing: an all-rounder for genome editing. *International Journal of Molecular Sciences* 23(17), 9862.

Lu, S., Fang, C., Abe, J., Kong, F. and Liu, B. (2022) Current overview on the genetic basis of key genes involved in soybean domestication. *aBIOTECH* 3(2), 126–139.

Maathuis, F.J. (2014) Sodium in plants: perception, signalling, and regulation of sodium fluxes. *Journal of Experimental Botany* 65(3), 849–858.

Machida-Hirano, R. (2015) Diversity of potato genetic resources. *Breeding Science* 65, 26–40.

Maher, M.F., Nasti, R.A., Vollbrecht, M., Starker, C.G., Clark, M.D. and Voytas, D.F. (2020) Plant gene editing through *de novo* induction of meristems. *Nature Biotechnology* 38(1), 84–89.

Mammadov, J., Buyyarapu, R., Guttikonda, S.K., Parliament, K., Abdurakhmonov, I.Y. and Kumpatla, S.P. (2018) Wild relatives of maize, rice, cotton, and soybean: treasure troves for tolerance to biotic and abiotic stresses. *Frontiers in Plant Science* 9, 886.

Martínez-Atienza, J., Jiang, X., Garciadeblas, B., Mendoza, I., Zhu, J.K. *et al.* (2007) Conservation of the salt overly sensitive pathway in rice. *Plant Physiology* 143(2), 1001–1012.

Marzec, M., Brąszewska-Zalewska, A. and Hensel, G. (2020) Prime editing: a new way for genome editing. *Trends in Cell Biology* 30(4), 257–259.

Mäser, P., Eckelman, B., Vaidyanathan, R., Horie, T., Fairbairn, D.J. *et al.* (2002) Altered shoot/root Na^+ distribution and bifurcating salt sensitivity in *Arabidopsis* by genetic disruption of the Na^+ transporter AtHKT1. *FEBS Letters* 531(2), 157–161.

Meyer, R.S., DuVal, A.E. and Jensen, H.R. (2012) Patterns and processes in crop domestication: an historical review and quantitative analysis of 203 global food crops. *New Phytologist* 196(1), 29–48.

Mishra, R., Joshi, R.K. and Zhao, K. (2020) Base editing in crops: current advances, limitations and future implications. *Plant Biotechnology Journal* 18(1), 20–31.

Molla, K.A. and Yang, Y. (2019) CRISPR/Cas-mediated base editing: technical considerations and practical applications. *Trends in Biotechnology* 37(10), 1121–1142.

Montenegro, J.D., Golicz, A.A., Bayer, P.E., Hurgobin, B., Lee, H. *et al.* (2017) The pangenome of hexaploid bread wheat. *The Plant Journal* 90(5), 1007–1013.

Munns, R. and Tester, M. (2008) Mechanisms of salinity tolerance. *Annual Review of Plant Biology* 59, 651–681.

Munns, R., James, R.A., Xu, B., Athman, A., Conn, S.J. *et al.* (2012) Wheat grain yield on saline soils is improved by an ancestral Na^+ transporter gene. *Nature Biotechnology* 30(4), 360–364.

Nasti, R.A. and Voytas, D.F. (2021) Attaining the promise of plant gene editing at scale. *Proceedings of the National Academy of Sciences USA* 118(22), e2004846117.

Nasti, R.A., Zinselmeier, M.H., Vollbrecht, M., Maher, M.F. and Voytas, D.F. (2021) Fast-TrACC: a rapid method for delivering and testing gene editing reagents in somatic plant cells. *Frontiers in Genome Editing* 2, 621710.

Nayyeripasand, L., Garoosi, G.A. and Ahmadikhah, A. (2021) Genome-wide association study (GWAS) to identify salt-tolerance QTLs carrying novel candidate genes in rice during early vegetative stage. *Rice (New York, N.Y.)* 14(1), 9.

Nelson-Vasilchik, K., Hague, J., Mookkan, M., Zhang, Z.J. and Kausch, A. (2018) Transformation of recalcitrant sorghum varieties facilitated by *Baby Boom* and *Wuschel2*. *Current Protocols in Plant Biology* 3(4), e20076.

Nevo, E. and Chen, G. (2010) Drought and salt tolerances in wild relatives for wheat and barley improvement. *Plant, Cell & Environment* 33(4), 670–685.

Oh, D.H., Zahir, A., Yun, D.J., Bressan, R.A. and Bohnert, H.J. (2009) *SOS1* and halophytism. *Plant Signaling & Behavior* 4(11), 1081–1083.

Oka, H.I. and Morishima, H. (1971) The dynamics of plant domestication: cultivation experiments with *Oryza perennis* and its hybrid with *O. sativa*. *Evolution* 25(2), 356–364.

Olías, R., Eljakaoui, Z., Li, J.U.N., De Morales, P.A., Marin-Manzano, M.C. *et al.* (2009) The plasma membrane Na^+/H^+ antiporter *SOS1* is essential for salt tolerance in tomato and affects the partitioning of Na^+ between plant organs. *Plant, Cell & Environment* 32(7), 904–916.

Oluoch, G., Zheng, J., Wang, X., Khan, M.K.R., Zhou, Z. *et al.* (2016) QTL mapping for salt tolerance at seedling stage in the interspecific cross of *Gossypium tomentosum* with *Gossypium hirsutum*. *Euphytica* 209(1), 223–235.

Oomen, R.J., Benito, B., Sentenac, H., Rodríguez-Navarro, A., Talón, M. *et al.* (2012) HKT2;2/1, a K^+-permeable transporter identified in a salt-tolerant rice cultivar through surveys of natural genetic polymorphism. *The Plant Journal* 71(5), 750–762.

Pan, C., Li, G., Malzahn, A.A., Cheng, Y., Leyson, B. *et al.* (2022) Boosting plant genome editing with a versatile CRISPR-Combo system. *Nature Plants* 8(5), 513–525.

Paran, I. and Van Der Knaap, E. (2007) Genetic and molecular regulation of fruit and plant domestication traits in tomato and pepper. *Journal of Experimental Botany* 58(14), 3841–3852.

Pardo, J.M., Cubero, B., Leidi, E.O. and Quintero, F.J. (2006) Alkali cation exchangers: roles in cellular homeostasis and stress tolerance. *Journal of Experimental Botany* 57(5), 1181–1199.

Pickersgill, B. (2018) Parallel vs. convergent evolution in domestication and diversification of crops in the Americas. *Frontiers in Ecology and Evolution* 6, 56.

Poncet, V., Martel, E., Allouis, S. Devos, K., Lamy, F. *et al.* (2002) Comparative analysis of QTLs affecting domestication traits between two domesticated × wild pearl millet (*Pennisetum glaucum L.*, Poaceae) crosses. *Theoretical and Applied Genetics* 104(6), 965–975.

Pourkheirandish, M. and Komatsuda, T. (2007) The importance of barley genetics and domestication in a global perspective. *Annals of Botany* 100(5), 999–1008.

Purugganan, M.D. (2019) Evolutionary insights into the nature of plant domestication. *Current Biology* 29(14), R705–R714.

Qing, Y., Zheng, Y., Mlotshwa, S., Smith, H.N., Wang, X. *et al.* (2022) Dynamically expressed small RNAs, substantially driven by genomic structural variants, contribute to transcriptomic changes during tomato domestication. *The Plant Journal* 110(6), 1536–1550.

Qiu, Q.S., Guo, Y., Dietrich, M.A., Schumaker, K.S. and Zhu, J.K. (2002) Regulation of *SOS1*, a plasma membrane Na^+/H^+ exchanger in *Arabidopsis thaliana*, by *SOS2* and *SOS3*. *Proceedings of the National Academy of Sciences USA* 99(12), 8436–8441.

Qu, M., Essemine, J., Xu, J., Ablat, G., Perveen, S. *et al.* (2020) Alterations in stomatal response to fluctuating light increase biomass and yield of rice under drought conditions. *The Plant Journal* 104(5), 1334–1347.

Rahman, S., Islam, S., Yu, Z., She, M., Nevo, E. and Ma, W. (2020) Current progress in understanding and recovering the wheat genes lost in evolution and domestication. *International Journal of Molecular Sciences* 21(16), 5836.

Rao, E.S., Kadirvel, P., Symonds, R.C., Geethanjali, S., Thontadarya, R.N. and Ebert, A.W. (2015) Variations in *DREB1A* and *VP1.1* genes show association with salt tolerance traits in wild tomato (*Solanum pimpinellifolium*). *PLoS ONE* 10(7), e0132535.

Razzaq, A., Saleem, F., Wani, S.H., Abdelmohsen, S.A., Alyousef, H.A. *et al.* (2021) *De-novo* domestication for improving salt tolerance in crops. *Frontiers in Plant Science* 12, 681367.

Remigereau, M.S., Lakis, G., Rekima, S., Leveugle, M., Fontaine, M.C. *et al.* (2011) Cereal domestication and evolution of branching: evidence for soft selection in the *Tb1* orthologue of pearl millet (*Pennisetum glaucum* [L.] R. Br.). *PLoS ONE* 6(7), e22404.

Ren, Z.H., Gao, J.P., Li, L.G., Cai, X.L., Huang, W. *et al.* (2005) A rice quantitative trait locus for salt tolerance encodes a sodium transporter. *Nature Genetics* 37(10), 1141–1146.

Riaz, A., Periyannan, S., Aitken, E. and Hickey, L. (2016) A rapid phenotyping method for adult plant resistance to leaf rust in wheat. *Plant Methods* 12(1), 17.

Rodríguez-Leal, D., Lemmon, Z.H., Man, J., Bartlett, M.E. and Lippman, Z.B. (2017) Engineering quantitative trait variation for crop improvement by genome editing. *Cell* 171(2), 470–480.

Rodríguez-Navarro, A. (2000) Potassium transport in fungi and plants. *Biochimica et Biophysica Acta, Reviews on Biomembranes* 1469(1), 1–30.

Rodríguez-Rosales, M.P., Jiang, X., Gálvez, F.J., Aranda, M.N., Cubero, B. and Venema, K. (2008) Overexpression of the tomato K^+/H^+ antiporter *LeNHX2* confers salt tolerance by improving potassium compartmentalization. *New Phytologist* 179(2), 366–377.

Rubio, F., Nieves-Cordones, M., Horie, T. and Shabala, S. (2020) Doing 'business as usual' comes with a cost: evaluating energy cost of maintaining plant intracellular K^+ homeostasis under saline conditions. *New Phytologist* 225(3), 1097–1104.

Saeed, A. and Fatima, N. (2021) Wild germplasm: shaping future tomato breeding. In: Azhar, M.T. and Wani, S.H. (eds) *Wild Germplasm for Genetic Improvement in Crop Plants*. Academic Press, London, pp. 201–214.

Sakuma, S., Golan, G., Guo, Z., Ogawa, T., Tagiri, A. *et al.* (2018) Unleashing floret fertility by a mutated homeobox gene improved grain yield during wheat evolution under domestication. *bioRxiv*, 434985. https://doi.org/10.1101/434985.

Sen, P., Kar, M.K., Panda, B., Behura, N., Bose, L.K. and Misra, R.N. (2010) Exploitation of secondary and tertiary gene pool in the genus *Oryza*. *Oryza* 47(1), 1–12.

Shabala, S. (2013) Learning from halophytes: physiological basis and strategies to improve abiotic stress tolerance in crops. *Annals of Botany* 112(7), 1209–1221.

Shabala, S. (2019) Linking ploidy level with salinity tolerance: NADPH-dependent 'ROS–$Ca2^+$ hub' in the spotlight. *Journal of Experimental Botany* 70(4), 1063–1067.

Shavrukov, Y., Gupta, N.K., Miyazaki, J., Baho, M.N., Chalmers, K.J. *et al.* (2010) *HvNax3* – a locus controlling shoot sodium exclusion derived from wild barley (*Hordeum vulgare* ssp. *spontaneum*). *Functional & Integrative Genomics* 10(2), 277–291.

Shavrukov, Y., Bovill, J., Afzal, I., Hayes, J.E., Roy, S.J. *et al.* (2013) *HVP10* encoding V-PPase is a prime candidate for the barley *HvNax3* sodium exclusion gene: evidence from fine mapping and expression analysis. *Planta* 237(4), 1111–1122.

Shi, H., Ishitani, M., Kim, C. and Zhu, J.K. (2000) The *Arabidopsis thaliana* salt tolerance gene *SOS1* encodes a putative Na^+/H^+ antiporter. *Proceedings of the National Academy of Sciences USA* 97(12), 6896–6901.

Shi, H., Quintero, F.J., Pardo, J.M. and Zhu, J.K. (2002) The putative plasma membrane Na^+/H^+ antiporter *SOS1* controls long-distance Na^+ transport in plants. *The Plant Cell* 14(2), 465–477.

Singh, J. and van der Knaap, E. (2022) Unintended consequences of plant domestication. *Plant & Cell Physiology* 63(11), 1573–1583.

Stetter, M.G., Vidal-Villarejo, M. and Schmid, K.J. (2020) Parallel seed color adaptation during multiple domestication attempts of an ancient new world grain. *Molecular Biology and Evolution* 37(5), 1407–1419.

Stitzer, M.C. and Ross-Ibarra, J. (2018) Maize domestication and gene interaction. *New Phytologist* 220(2), 395–408.

Street, K., Bari, A., Mackay, M. and Amri, A. (2016) How the Focused Identification of Germplasm Strategy (FIGS) is used to mine plant genetic resources collections for adaptive traits. In: Maxted, N., Dulloo, M.E. and Ford-Lloyd, B.V. (eds) *Enhancing Crop Genepool Use: Capturing Wild Relative and Landrace Diversity for Crop Improvement*. CAB International, Wallingford, UK, pp. 54–63.

Subhasmita, S. and Pradeepkumara, N. (2022) Concept of gene pool and distant hybridization in vegetable crops. *International Journal of Current Microbiology and Applied Sciences* 11(5), 148–159.

Tahir ul Qamar, M., Zhu, X., Khan, M.S., Xing, F. and Chen, L.L. (2020) Pan-genome: a promising resource for noncoding RNA discovery in plants. *The Plant Genome* 13(3), e20046.

Tao, Y., Zhao, X., Mace, E., Henry, R. and Jordan, D. (2019) Exploring and exploiting pan-genomics for crop improvement. *Molecular Plant* 12(2), 156–169.

Tay Fernandez, C.G., Nestor, B.J., Danilevicz, M.F., Gill, M., Petereit, J. *et al.* (2022) Pangenomes as a resource to accelerate breeding of under-utilised crop species. *International Journal of Molecular Sciences* 23(5), 2671.

Thormann, I., Endresen, D.T.F., Rubio-Teso, M.L., Iriondo, M.J., Maxted, N. and Parra-Quijano, M. (2014) *Predictive Characterization of Crop Wild Relatives and Landraces: Technical Guidelines Version 1*. Bioversity International, Rome.

Tian, L., Tan, L., Liu, F., Cai, H. and Sun, C. (2011) Identification of quantitative trait loci associated with salt tolerance at seedling stage from *Oryza rufipogon*. *Journal of Genetics and Genomics* 38(12), 593–601.

Tudge, C. (2011) *The Engineer in the Garden*. Random House, London.

UN Report (2013) World population projected to reach 9.6 billion by 2050. Available at: https://news.un.org/en/story/2013/06/442212 (accessed 27 February 2024).

United States Salinity Laboratory Staff (1954) *Diagnosis and Improvement of Saline and Alkali Soils*. United States Department of Agriculture, Agriculture Handbook No. 60. U.S. Government Printing Office, Washington, DC.

Uozumi, N., Kim, E.J., Rubio, F., Yamaguchi, T., Muto, S. *et al.* (2000) The *Arabidopsis HKT1* gene homolog mediates inward Na^+ currents in *Xenopus laevis* oocytes and Na^+ uptake in *Saccharomyces cerevisiae*. *Plant Physiology* 122(4), 1249–1260.

Van de Peer, Y., Ashman, T.L., Soltis, P.S. and Soltis, D.E. (2021) Polyploidy: an evolutionary and ecological force in stressful times. *The Plant Cell* 33(1), 11–26.

Venkataraman, G., Shabala, S., Véry, A.A., Hariharan, G.N., Somasundaram, S. *et al.* (2021) To exclude or to accumulate? Revealing the role of the sodium HKT1;5 transporter in plant adaptive responses to varying soil salinity. *Plant Physiology and Biochemistry* 169, 333–342.

Villalta, I., Reina-Sánchez, A., Bolarín, M.C., Cuartero, J., Belver, A. *et al.* (2008) Genetic analysis of Na^+ and K^+ concentrations in leaf and stem as physiological components of salt tolerance in tomato. *Theoretical and Applied Genetics* 116(6), 869–880.

Wang, L., Cao, S., Wang, P., Lu, K., Song, Q. *et al.* (2021) DNA hypomethylation in tetraploid rice potentiates stress-responsive gene expression for salt tolerance. *Proceedings of the National Academy of Sciences USA* 118(13), e2023981118.

Wang, S., Cao, M., Ma, X., Chen, W., Zhao, J. *et al.* (2017) Integrated RNA sequencing and QTL mapping to identify candidate genes from *Oryza rufipogon* associated with salt tolerance at the seedling stage. *Frontiers in Plant Science* 8, 1427.

Wang, Z., Hong, Y., Zhu, G., Li, Y., Niu, Q. *et al.* (2020) Loss of salt tolerance during tomato domestication conferred by variation in a Na^+/K^+ transporter. *The EMBO Journal* 39(10), e103256.

Wang, Z., Hong, Y., Li, Y., Shi, H., Yao, J. *et al.* (2021) Natural variations in *SlSOS1* contribute to the loss of salt tolerance during tomato domestication. *Plant Biotechnology Journal* 19(1), 20.

Watson, A., Ghosh, S., Williams, M.J., Cuddy, W.S., Simmonds, J. *et al.* (2018) Speed breeding is a powerful tool to accelerate crop research and breeding. *Nature Plants* 4(1), 23–29.

Wedger, M.J. and Olsen, K.M. (2018) Evolving insights on weedy rice. *Ecological Genetics and Genomics* 7, 23–26.

White, P.J. and Karley, A.J. (2010) Potassium. In: Hell, R. and Mendel, R.R. (eds) *Cell Biology of Metals and Nutrients*. Plant Cell Monographs. Springer, Berlin, Heidelberg, Germany, pp. 199–224.

Wong, M.M., Gujaria-Verma, N., Ramsay, L., Yuan, H.Y., Caron, C. *et al.* (2015) Classification and characterization of species within the genus *Lens* using genotyping-by-sequencing (GBS). *PLoS ONE* 10(3), e0122025.

Wu, D., Sun, Y., Wang, H., Shi, H., Su, M. *et al.* (2018) The *SlNAC8* gene of the halophyte *Suaeda liaotungensis* enhances drought and salt stress tolerance in transgenic *Arabidopsis thaliana*. *Gene* 662, 10–20.

Wu, P., Yang, L., Guo, G., Hu, J., Qiu, D. *et al.* (2021) Molecular mapping and identification of a candidate gene for new locus *Hg2* conferring hairy glume in wheat. *Plant Science* 307, 110879.

Würschum, T., Liu, G., Boeven, P.H., Longin, C.F.H., Mirdita, V. *et al.* (2018) Exploiting the *Rht* portfolio for hybrid wheat breeding. *Theoretical and Applied Genetics* 131(7), 1433–1442.

Xia, R., Xu, J. and Meyers, B.C. (2017) The emergence, evolution, and diversification of the miR390-TAS3-ARF pathway in land plants. *The Plant Cell* 29(6), 1232–1247.

Xu, R. and Sun, C. (2021) What happened during domestication of wild to cultivated rice. *The Crop Journal* 9(3), 564–576.

Yan, H., Sun, M., Zhang, Z., Jin, Y., Zhang, A. *et al.* (2023) Pangenomic analysis identifies structural variation associated with heat tolerance in pearl millet. *Nature Genetics* 55(3), 507–518.

Yu, H., Lin, T., Meng, X., Du, H., Zhang, J. *et al.* (2021) A route to *de novo* domestication of wild allotetraploid rice. *Cell* 184(5), 1156–1170.

Zhang, G.H., Su, Q., An, L.J. and Wu, S. (2008) Characterization and expression of a vacuolar Na^+/H^+ antiporter gene from the monocot halophyte *Aeluropus littoralis*. *Plant Physiology and Biochemistry* 46(2), 117–126.

Zhang, M., Cao, Y., Wang, Z., Wang, Z.Q., Shi, J. *et al.* (2018) A retrotransposon in an HKT1 family sodium transporter causes variation of leaf Na^+ exclusion and salt tolerance in maize. *New Phytologist* 217(3), 1161–1176.

Zhang, M., Liang, X., Wang, L., Cao, Y., Song, W. *et al.* (2019) A HAK family Na^+ transporter confers natural variation of salt tolerance in maize. *Nature Plants* 5(12), 1297–1308.

Zhang, M., Li, Y., Liang, X., Lu, M., Lai, J., Song, W. and Jiang, C. (2023) A teosinte-derived allele of an HKT1 family sodium transporter improves salt tolerance in maize. *Plant Biotechnology Journal* 21(1), 97–108.

Zhang, X., Zhu, T., Li, Z., Jia, Z., Wang, Y. *et al.* (2022) Natural variation and domestication selection of *ZmSULTR3;4* is associated with maize lateral root length in response to salt stress. *Frontiers in Plant Science* 13, 992799.

Zhou, L., Zhu, C., Fang, X., Liu, H., Zhong, S. *et al.* (2021) Gene duplication drove the loss of awn in sorghum. *Molecular Plant* 14(11), 1831–1845.

Zsögön, A., Čermák, T., Naves, E.R., Notini, M.M., Edel, K.H. *et al.* (2018) *De novo* domestication of wild tomato using genome editing. *Nature Biotechnology* 36(12), 1211–1216.

Index

Note: Page numbers in **bold** type refer to **figures** and page numbers in *italic* type refer to *tables*.